INFRARED ASTRONOMY

IV Canary Islands Winter School of Astrophysics

W0051212

Edited by

A. Mampaso, M. Prieto and F. Sánchez

Instituto de Astrofísica de Canarias. Tenerife. Spain

PUBLISHED BY THE PRESS SYNDICATE OF THE UNIVERSITY OF CAMBRIDGE
The Pitt Building, Trumpington Street, Cambridge, United Kingdom

CAMBRIDGE UNIVERSITY PRESS
The Edinburgh Building, Cambridge CB2 2RU, UK
40 West 20th Street, New York NY 10011–4211, USA
477 Williamstown Road, Port Melbourne, VIC 3207, Australia
Ruiz de Alarcón 13, 28014 Madrid, Spain
Dock House, The Waterfront, Cape Town 8001, South Africa

http://www.cambridge.org

First published 1993
First paperback edition 2003

A catalogue record for this book is available from the British Library

ISBN 0 521 46462 5 hardback
ISBN 0 521 54810 1 paperback

What do we understand of the birth and death of stars? What is the nature of the tiny dust grains that permeate our Galaxy and other galaxies? And how likely is the existence of brown dwarfs, extrasolar planets or other sub-stellar mass objects? These are just a few of the exciting questions that can now be addressed in a new era of infrared observations.

IR astronomy has been revolutionised over the past few years by the widespread availability of large, sensitive IR arrays and the success of IR satellites. Several IR space missions due for launch over the next few years also promise an exciting future. For these reasons the IV Canary Islands Winter School of Astrophysics was dedicated to the burgeoning field of IR astronomy. Its primary goal was to provide an introduction to the important new observations and physical ideas that are emerging in this wide-ranging field of research.

Lectures from leading researchers, renowned for their teaching abilities, are gathered in this volume. These nine chapters provide an excellent introduction as well as a thorough and up-to-date review of developments – essential reading for graduate students and professionals.

INFRARED ASTRONOMY

IV Canary Islands Winter School of Astrophysics

CONTENTS

LIST OF PARTICIPANTS

Ackerman, Erich	*Max-Planck-Institüt. Germany*
Alonso, Almudena	*Univ. Complutense de Madrid. Spain*
Alonso, Oscar	*Univ. Complutense de Madrid. Spain*
Ballesteros, Ezequiel	*Inst. de Astrofísica de Canarias. Spain*
Barrado, David	*Univ. Complutense de Madrid. Spain*
Barrera, José E.	*Inst. de Astrofísica de Canarias. Spain*
Brandl, Bernard	*Max-Planck-Institüt. Germany*
Beaulieu, Jean Philippe	*Inst. d´Astrophysique de Paris. France*
Becklin, Eric E.	*Univ. of California at Los Angeles. USA*
Berdugina, Svetlana	*Crimean Astrophysical Obs. Ukrania*
Beskrovnaya, Nina	*Central Astronomical Obs. Russia*
Briceño, Cesar	*CIDA. Venezuela*
Burgos, Jesús	*Inst. de Astrofísica de Canarias. Spain*
Cabrera, Fernando	*Inst. de Astrofísica de Canarias. Spain*
Carrillo, René	*Inst. de Astronomía, UNAM. Mexico*
Cerviño, Miguel	*LAEFF (INTA). Spain*
Comerón, Fernando	*Univ. de Barcelona. Spain*
Claes, Peter	*Inst. d´Astrophysique. Belgium*
Dallier, Richard	*Observatoire de Meudon. France*
De Graauw, Thijs	*Laboratory for Space Research. Netherlands*
Degirmenci, Omer L.	*Ege Universitesi Observatory. Turkey*
Domínguez, Francisco B.	*Univ. Complutense de Madrid. Spain*
Duc, Pierre-Alain	*Service d´Astrophysique, CE Saclay. France*
Eff-Darwich, Antonio M.	*Inst. de Astrofísica de Canarias. Spain*
Erdem, Ahmet	*Ege Universitesi Observatory. Turkey*
Everall, Chris	*Southampton Univ. U.K.*
Faziliddin, T. Shamshiev	*Inst. of Theoretical Astronomy. Russia*
Friedli, Daniel	*Geneve Observatory. Switzerland*
Galli, Danielle	*Univ. of California at Berkeley. USA*
Gilmore, Gerard	*Inst. of Astronomy. Cambridge. U.K.*
Golombek, Daniel	*Space Telesc. Sci. Inst. USA*
González, Rosa A.	*Univ. of California at Berkeley. USA*
Grison, Philippe	*Inst. d´Astrophysique de Paris. France*
Guglielmo, Francois	*Obs. de Meudon. DESPA. France*
Guichard, José S.	*Inst. de Astronomía, UNAM. Mexico*
Ikhsanov, Nazar R.	*Central Astronomical Obs. Russia*
Jansen, Rolf A.	*Kapteyn Laboratorium. Netherlands*
Jiménez, Raul	*Univ. Autónoma de Madrid. Spain*
Joseph, Robert D.	*IRTF. Inst. for Astronomy. USA*
Kaas, Anlaug A.	*Univ. of Oslo. Norwey*
Krenz, Thomas	*Max-Planck-Institüt. Germany*
Laine, Seppo J.	*Univ. of Florida. USA*
Lançon, Ariane	*Inst. d´Astrophysique de Paris. France*
Likkel, Lauren	*Univ. of Illinois. USA*
López-Cruz, Omar	*Univ. of Toronto. Canada*

Lorente, Rosario — *Univ. Complutense de Madrid. Spain*
Mampaso, Antonio — *Inst. de Astrofísica de Canarias. Spain*
Marconi, Alessandro — *Univ. of Florence. Italy*
McLean, Ian S. — *Univ. of California at Los Angeles. USA*
Mitrou, C.K. — *Univ. of Athens. Greece*
Miroshnichenko, Anatoly — *Central Astronomical Obs. Russia*
Najarro, Francisco — *Univ. Sternwarte München. Germany*
Palla, Francesco — *Osservatorio Astrofisico di Arcetri. Italy*
Papadopoulos, Padeli P. — *Univ. Toronto. Canada*
Pérez-Olea, Diego E. — *Univ. Autónoma de Madrid. Spain*
Plets, Hans — *Inst. voor Sterrenkunde. Belgium*
Popescu, Cristina C. — *Astron. Inst. Romanian Acad. Romania*
Pottasch, Stuart R. — *Kapteyn Astronomical Inst. Netherlands*
Prada, Francisco — *Queen's University of Belfast. Ireland*
Prieto, Mercedes — *Inst. de Astrofísica de Canarias. Spain*
Rabello, Mª Cristina — *Inst. de Astrofísica de Canarias. Spain*
Rodríguez, Jesús — *Inst. de Astrofísica de Canarias. Spain*
Rodríguez, Mónica — *Inst. de Astrofísica de Canarias. Spain*
Romero, Gustavo — *Inst. Argentino de Radioastronomía. Argentina*
Rosa, Daniel — *Univ. de La Laguna. Spain*
Sahu, Kailash C. — *Kapteyn Astronomical Inst. Netherlands.*
Schneider, Nicola — *Univ. Köln. Germany*
Singh, Prabhojot — *Indian Inst. of Astrophysics. India*
Solano, Enrique — *Vilspa-Inta IUE Obs. Spain*
Steele, Iain A. — *Univ. of Leicester. U.K.*
Telesco, Charles M. — *Space Science Laboratory, NASA-MSFC. USA*
Testi, Leonardo — *Osservatorio Astrofisico di Arcetri. Italia*
Testori, Juan C. — *Inst. Argentino de Radioastronomía. Argentina*
Torgovkina, Elizaveta — *Sternberg Astron. Inst. Russia*
Ulla, Ana — *LAEFF (INTA). Spain*
Ustjugov, Sergey — *Sternberg Astronomical Inst. Russia*
Vavilova, Irina B. — *Astron. Obs. of Kief Univ. Ukrania*
Vazdekis, Alexandre — *Univ. de La Laguna. Spain*
Vela, Eduardo — *Inst. de Astrofísica de Canarias. Spain*
Wickramasinghe, N. Chandra — *Univ. of Wales. U.K.*
Wickramasinghe, Thulsi — *Univ. of Pennsylvania. USA*
Yudin, Boris — *Sternberg Astronomical Instit. Russia*
Yudin, Ruslan — *Central Astronomical Obs. Russia*
Yun, Joao L. — *Univ. of Boston. USA*
Zubko, Victor — *Ukranian Acad. of Sciences. Ukrania*

PREFACE

The fourth "Canary Islands Winter School of Astrophysics" was held in Playa de las Américas (Adeje, Tenerife) from the 7th to 18th December 1992, organized by the Instituto de Astrofísica de Canarias and the Universidad Internacional Menéndez-Pelayo. A total of 85 participants from 21 countries world-wide attended the meeting.

This volume contains a series of nine courses which were delivered during the School. The aim of the lectures was to portray a thorough, up-dated view on the field of Infrared Astronomy.

The last School dedicated to Infrared Astronomy before this one (the legendary ISA Course held in Erice, Italy) took place in 1977, fifteen years ago. Since then, dramatic changes in our understanding of the Infrared Universe -pushed forward by the corresponding advances in telescopes, instruments and detector capabilities- have strongly influenced all branches of Astrophysics. One of the primary goals at the present School was to put the new generation of astrophysicists into contact with the extremely important new findings which have surfaced from recent infrared research, and to present and discuss the fundamental physical ideas emerging from those results.

We wish to express our gratitude to the Commission of the European Community (Human Capital and Mobility Programme -Euroconferences), the Dirección General de Investigación Científica y Técnica (Spanish Ministry of Education and Science) and the Government of the Canary Islands for helping us to fund the School. They, together with HOTESA and their "Hotel Gran Tinerfe", made it possible to allocate grants to over 75% of attendees.

Special thanks are due to the nine lecturers -E.E. Becklin, G. Gilmore, T. de Graauw, R.D. Joseph, I.S. McLean, F. Palla, S.R. Pottasch, C.M. Telesco and N.C. Wickramasinghe- and the participants, for their contribution to a memorable and enjoyable School, and to the former for having provided us with their manuscripts in such a short time. The School Secretary, Ms L. González, and Ms M. Murphy, together with other members of the IAC's staff, ensured that the School ran smoothly, and their unstinting collaboration is sincerely acknowledged.

La Laguna, June, 1993

A. Mampaso
M. Prieto
F. Sánchez

Instituto de Astrofísica de Canarias
38200. La Laguna. Tenerife. SPAIN.

STAR FORMATION

Francesco Palla

Osservatorio Astrofisico di Arcetri
L.go E. Fermi, 5
Firenze, Italy

...per aspera ad astra

1 INFRARED ASTRONOMY AND STAR FORMATION

In a report of the current status of the observational studies of star formation published in 1982, Wynn-Williams introduced the subject by boldly stating *"Protostars are the Holy Grail of infrared astronomy"*, one of the most (ab)used astronomical quotations ever. At the time of the review, searches had been made at IR wavelengths towards a restricted sample of objects, guided by the theoretical expectation that protostars would undergo a phase of high IR luminosity during the main accretion phase (Larson 1969). The aim of these studies, and of those that followed, was to get an unambiguous example of Spitzer's definition (1948) of a protostar as *"an isolated interstellar cloud undergoing inexorable gravitational contraction to form a single star"*.

This definition appears now rather restrictive, since observations have shown that the structure of star forming regions is highly complex, rarely revealing well isolated, noninteracting interstellar clouds. In addition, observations in the near infrared at high spatial resolution have shown that many young stars are not single, but do have companions (Zinnecker & Wilking 1992). However, despite great effort, resulting from improvements in instrumentation and progress in the theoretical models, the same discouraging conclusion reached by Winn-Williams in 1982 still holds true: no conclusive identification of a genuine protostar has yet been made. Nevertheless, the motivation for continuing the search using the IR band remains still unquestionable.

In these lectures I will present an overview of the main properties of the star formation process with a special emphasis on stars of low- and intermediate-mass, for which observations are the most detailed and a consistent theoretical framework has been developed and tested. The first Section describes the infrared properties of star forming regions, and the modes of star formation within giant molecular clouds. The second lecture is devoted to the analysis of the gravitational collapse of dense molecular cores and the protostellar accretion phase. The third lecture reviews the theoretical foundation of pre–main-sequence evolution, and describes the circumstellar environment of young stars. The last lecture focuses on the properties of young embedded stellar clusters, and the derivation of infrared and bolometric luminosity functions.

1.1 The Quest for Protostars

By definition, a protostar is an object that derives its energy from the conversion of infall kinetic energy to radiation, and that has not yet reached hydrostatic equilibrium. The photons produced at the accretion shock, either bounding the protostellar core or a circumstellar disk, are effectively absorbed by the large amount of residual circumstellar matter and reemitted at far-infrared wavelengths, typically longward of 30μm. In 1982, the list of candidates consisted of about 30 infrared sources, all of them with bolometric luminosities in excess of 10^3 $L_\odot$. Such a high luminosity implies the presence of embedded sources with masses greater than ~ 3 $M_\odot$, presumably fully formed stars of the early spectral types. (Yorke & Shustov 1981). Since the distribution of stellar masses, the initial mass function (IMF), peaks at values less than 1 $M_\odot$ (Salpeter 1955; Scalo 1986), it is clear that all the searches carried out until then were unable to probe the most common byproduct of star formation, solar type stars.

The fact that none of the candidate protostars satisfied Spitzer's definition derived not from a shortage of candidates (albeit the number was indeed small), but from the lack of knowledge of their evolutionary status. In order to recognize a protostar, two requirements must be fulfilled: first, the infrared continuum spectrum should have the expected characteristics; second, the infalling gas should show evidence of dynamical collapse in the line profile, as indicated by an inverse P-Cygni profile with the redshifted absorption appearing against either a continuum source or the emission line. A decade ago these constraints posed unsurmountable limits to succeed in the quest: it is the progress both in the observational capabilities, in the infrared as well as in the submillimeter and millimeter wavebands, and in the theoretical understanding of the basic physical processes that have made the problem of star formation and the quest for protostars much less frustrating ten years later. The main events that have marked this period of time are schematically listed below. Of course, this list is by no means complete, but it is merely intended as a guideline for the following discussion.

<u>Dense Molecular Cores</u>: it has been recognized that high density molecular cloud cores, as traced by NH_3, CS and $C^{18}O$ molecules, represent the initial conditions for the star formation process, mainly of low- and intermediate-mass stars (Myers & Benson 1983).

<u>IRAS</u>: the satellite has provided the first complete map of most of the sky ($\sim 95\%$). The Point Source Catalogue (PSC), with its approximately 250000 entries, has offered the possibility of approaching the statistical aspects of star formation in a systematic and unbiased way (Beichman *et al.* 1984).

<u>Winds & Outflows</u>: it has been demonstrated that, opposite to the theoretical expectation, the main kinematic signature of star forming clouds is *outflow* rather than *infall*. The earliest evolutionary phases of *all* stars are marked by the occurrence of ejection of matter in the form of powerful winds, as traced by P-Cygni profiles, H-H objects, maser emission, and optical jets (Lada C. 1985).

<u>IR Classification of YSOs</u>: an empirical scheme based on the shape of the broad-band (1-100μm) spectral energy distribution has been devised to classify young stellar objects (Lada C. 1987), and to infer their evolutionary state (Shu, Adams & Lizano 1987).

<u>NIR Array Cameras</u>: a veritable revolution in instrumentation that allows to image large areas of star forming regions, with both high spatial resolution and sensitivity. The IR cameras have revealed the existence of many visually obscured stellar clusters in giant molecular clouds; probed the low-mass stellar population associated with HII regions, thus overcoming the basic limitation of previous surveys, and shed light on the stellar Luminosity Function at birth (Gatley *et al.* 1991; see also chapter by I. McLean).

1.2 IRAS and Star Formation

The IRAS survey, based on four wavebands centred at 12, 25, 60 and 100 μm, has produced a catalogue of point sources essentially complete, at the 10σ level, to 0.7, 0.65, 0.8, and 3.0 Jy, respectively. This enormous database has been used to select, extract and analyse sample of sources solely on infrared criteria. For studies of star formation, the classification of IRAS sources using their spectral energy distributions (SED) has proved extremely useful to separate young stellar objects in different evolutionary states. Also, the use of color-color plots, based on the measured flux densities, has allowed the definition of selection criteria to distinguish different classes of objects. The results are summarized in Table I. The location of young stellar objects in selected color-color diagrams is illustrated in Figure 1. An analysis of the problems encountered in selecting highly complete and not contaminated samples of young stellar objects in the IRAS Point Source Catalogue can be found in Prusti *et al.* (1991).

Dense Cores & Protostars: The dense cores observed in ammonia by Myers and Benson (1983) have been used as prime targets to search for embedded sources of low luminosity. In fact, IRAS had the adequate sensitivity at long wavelengths to detect all low luminosity objects in nearby star forming regions (Beichman *et al.* 1986). For example, the IRAS survey samples the entire stellar population for $L_* \gtrsim 0.5\ L_\odot$ in the Taurus-Auriga complex (Kenyon *et al.*

Class of Object	Log (F_{25}/F_{12})	Log (F_{60}/F_{25})	Log (F_{100}/F_{60})	Log (F_{60}/F_{12})	Ref.
Dense Cores	$0.4 - 1.0$	$0.4 - 1.3$	$0.1 - 0.7$	—	1
Maser H_2O	$0.5 - 1.1$	$0.4 - 1.7$	$-0.1 - 1.5$	—	2
UC HII regions	> 0.6	—	—	> 1.3	3
T Tauri stars	$0.03 - 0.58$	$-0.26 - 0.51$	$0.0 - 0.4$	—	4
Herbig Ae/Be	$-0.1 - 0.6$	$-0.2 - 1.0$	$-0.2 - 0.5$	—	5

1 Emerson (1987); 2 Wouterloot & Walmsley (1986); 3 Wood & Churchwell (1989); 4 Harris *et al.* (1988); 5 Berrilli *et al.* (1990)

Table I: *IRAS selection criteria for young stellar objects*

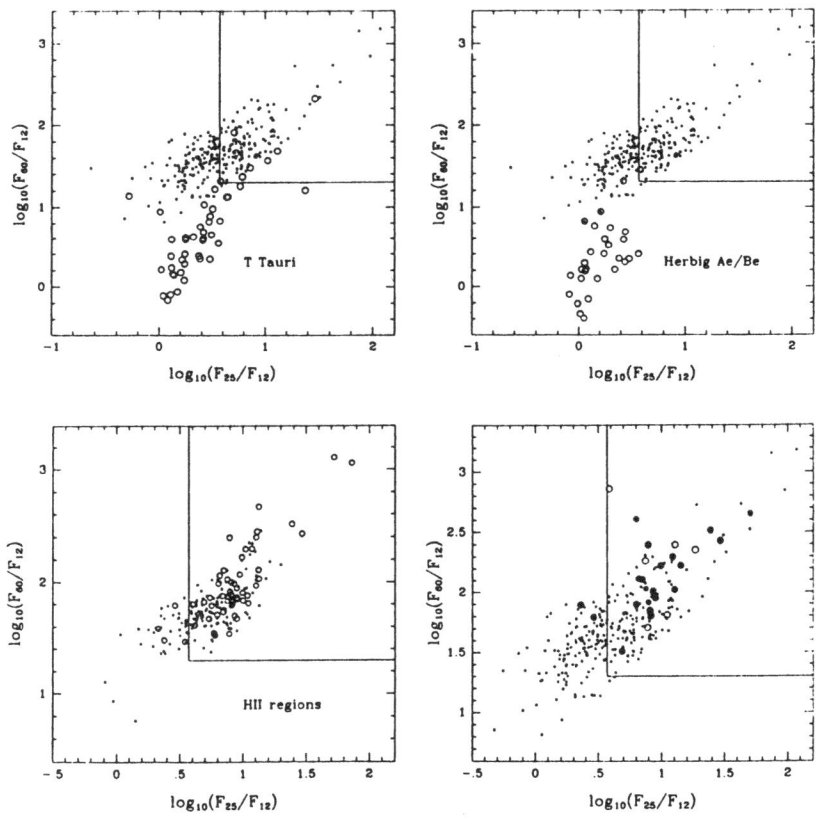

Figure 1: *Color-color diagrams of various classes of objects. Empty circles refer to T Tauri stars, Herbig Ae/Be stars, and HII regions. The small filled circles show the location of IRAS sources associated with dense molecular clouds and ultracompct HII regions. The solid lines delimit the colors for embedded massive stars proposed by Wood & Churchwell (1989).*

1990), and in Ophiucus (Wilking 1989). At the distance of Orion ($\sim$450 pc), the nearest giant molecular complex, the census is limited to sources with $L_* \gtrsim 6$ $L_\odot$ (Strom *et al.* 1989a). Thus, the stellar populations hosted in these regions cannot be directly compared on the basis of the IRAS data alone, but candidate protostars can be searched systematically in the closer clouds.

The association of the IRAS point sources with the dense cores has been secured in approximately half of the known cores, and typically one to few stars are found per core (Benson & Myers 1989). The remaining half of the cores are thought to be in the process of forming stars. The nature of the embedded sources has been determined by follow-up studies at optical and NIR wavelengths. In most cases the objects are young stars (T Tauri-like), still surrounded by large amounts of circumstellar matter. The rest of the sources remain optically invisible, with an estimated extinction in excess of 30 mag: they represent the best candidates of protostars (Myers *et al.* 1987; Ohashi *et al.* 1991). The knowledge of the relative proportion of the two classes of objects in several star forming regions has given an indirect estimate of the duration of the embedded phase, $\tau_{emb} \sim 10^5$ yr. If the typical source has a solar mass, τ_{emb} would imply an average mass accretion rate of $\dot{M}_{acc} \sim 10^{-5} M_\odot yr^{-1}$, in agreement with theoretical predictions (Stahler, Shu & Taam 1980). However, the accretion rate derived from the measured bolometric luminosity can be substantially smaller, $\dot{M}_{acc} \sim 10^{-7} M_\odot yr^{-1}$ (Myers *et al.* 1987; Kenyon *et al.* 1990), suggesting that very few sources can be catched during the main accretion phase.

Massive Stars & HII regions: The formation of massive stars takes also place in dense cores within molecular cloud complexes. A powerful two color selection criterion has been developed by Wood & Churchwell (1989) to identify massive stars still embedded inside molecular clouds. These sources are characterized by the largest flux densities at 100 μm, implying that the emitting dust is quite cool ($T_d \sim$30 K); their spectra peak at $\sim 100 \mu$m and the shape does not show an appreciable variation from source to source. As shown in the color-color plots of Fig. 1, the population of embedded massive stars, and their associated ultracompact HII regions, occupies a limited and well defined portion of the diagram, away from other entries in the PSC. Their distinctive colors have been used to estimate the total number and distribution of embedded massive stars in the Galaxy, the timescale of the embedded phase (typically $\sim$10-20% of the main-sequence lifetime), and the current rate of massive star formation ($\sim 3 \times 10^{-3}$ O stars yr^{-1}). Follow-up studies in molecular lines, NH_3, CS, ^{13}CO and H_2O maser transition (Churchwell *et al.* 1990; Palla *et al.* 1991; Cesaroni *et al.* 1992) have largely confirmed the association between the IRAS sources, dense molecular gas and UC HII regions.

T Tauri and Herbig Ae/Be stars: A large fraction of the catalogued optically visible, pre–main-sequence (PMS) stars have an IRAS association. The studies by Harris *et al.* (1988) and Weintraub (1990) of a sample of T Tauri stars in Taurus have shown that their colors are tightly confined by the bounds given in Table 1. The higher mass counterparts, the Herbig Ae/Be stars, have very similar color, although the scatter is larger (Berrilli *et al.* 1990). This is somewhat surprising, given the different luminosities of the two classes of stars. In both cases, the color corresponds to uncorrected temperatures of few hundred Kelvin. The analysis of the spectral energy distribution has provided the first indirect evidence of the presence of circumstellar disks around young stars (Rucinski 1985). We will return in Section 3 to the IR properties of the circumstellar matter associated with PMS stars.

1.3   The IR classification scheme

Studies of the *shape* of the spectral energy distribution (SED) of young stellar objects (YSOs) have proved very useful to determine their nature and evolutionary state. Since the emitted spectrum from a YSO depends on the distribution and physical properties (density, temperature and composition) of the surrounding dust and gas, it is natural to expect a dependence of its shape on the evolutionary state of the source. A protostellar object deeply embedded in the parent cloud should have an infrared signature markedly different than that of a mature PMS star, where most of the material has been accumulated onto the central object. An empirical classification scheme has been developed, based on the slope of the SED longward of 2.2 μm (Lada C. 1987; Wilking 1989). For each SED, a spectral index $\alpha \equiv -dlog(\nu F_\nu)/dlog\nu$, with F_ν the flux density at wavelength λ, is computed between 2.2 and 25 μm, and the resulting morphological classes are shown in Figure 2.

Class I sources have $\alpha > 0$, indicating a rise in the SED that often continues even at longer wavelengths. The IR excess is very conspicuous and the SED is much broader than that of a single temperature blackbody function. *Class II* sources have $-2 < \alpha < 0$. Their SEDs fall towards longer wavelenghts, but are still broad, due to a significant amount of circumstellar dust. Finally, *Class III* sources have $\alpha < -2$, and the SED resembles that of a normal, reddened stellar photosphere. As it always is the case in astronomy, the classification represents a powerful tool, but the scheme is not so rigid. A variety of YSOs have been found with intermediate properties, and additional subclasses (labeled "D") have been introduced to account for double-peaked SEDs.

The interpretation in terms of an evolutionary sequence is sketched in the

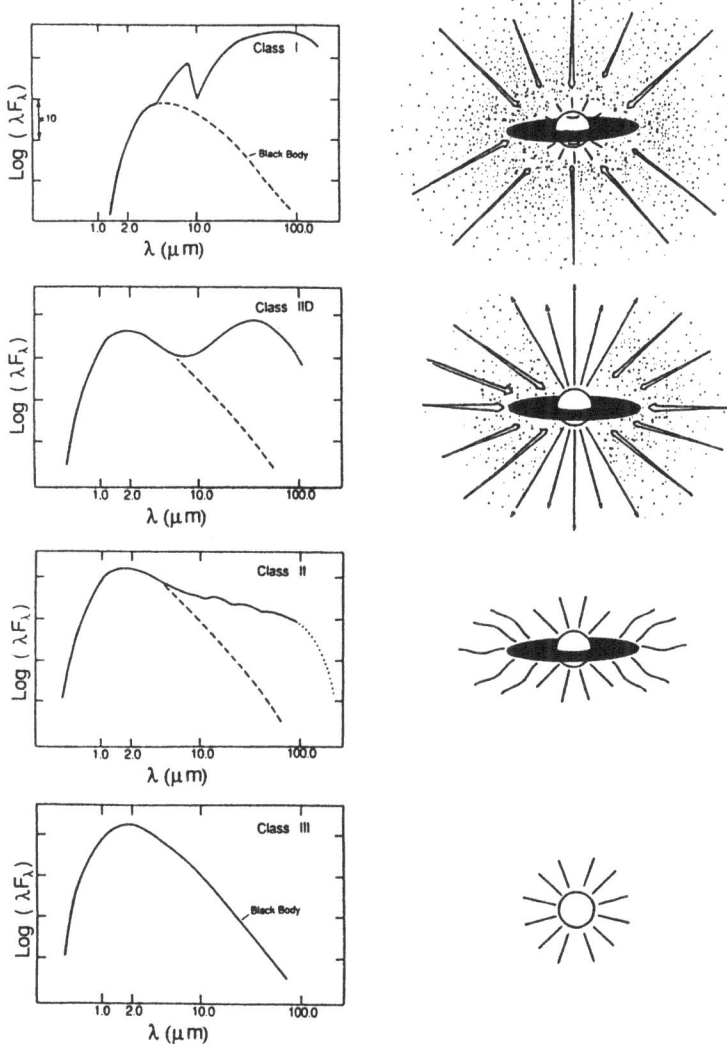

Figure 2: *Classification scheme (left) and evolutionary sequence (right) for young stellar objects (adapted from Lada C. 1991 and Shu et al. 1987).*

right panel of Fig. 2 (Shu *et al.* 1987). *Class I* sources are thought to represent accreting protostars, surrounded by luminous disks, with extent of $\sim 10-100$AU, and by infalling, spherically symmetric envelopes with typical sizes of $\sim 10^4$AU. The stage of partial clearing of the obscuring matter where the central star begins to be optically visible, due to the action of a powerful stellar wind, corresponds to Class IID. A *Class II* source results from the final disruption of the envelope, and corresponds to a classical T Tauri or Herbig Ae/Be star surrounded by spatially thin, optically thick circumstellar disk of radius ~ 100AU. Finally, the absence of the IR excess implies the disappearance of the circumstellar structures, disks and envelopes, and the approach to the conditions of a normal main sequence star (*Class III*). The evolutionary scenario in four stages (the initial stage of core formation has been omitted in this discussion; cf. Shu *et al.* 1993) has been widely used to interpret the observations towards individual star forming regions, and to elucidate the nature of Class I sources, the elusive protostars.

The reason for the lack of success in the identification of a genuine example of a protostar is now becoming more clear. It has been suggested that Class I sources are in fact more evolved objects than true protostars, having already assembled most of the final stellar mass (André *et al.* 1993). Support to this view comes from studies at millimeter and sub-millimeter wavelengths of a sample of deeply embedded sources. Continuum observations at these wavelengths probe the density and temperature distribution of the circumstellar material, and allow a direct estimate of the mass in the form of dust and of the total luminosity of the source (Mezger *et al.* 1991; Zinnecker et al. 1992). By definition, Class I sources should be surrounded by large amounts of dust, but the observations have surprisingly shown that this is not the case: the mass in the envelopes does not differ significantly from that measured around more evolved PMS stars (Beckwith *et al.* 1990). Once again, the mistery remains. Perhaps, the final answer will come from the detailed study of a class of sources so deeply hidden within the parent cloud that they cannot be detected *even* at infrared wavelengths. An example of the SED emitted from such a source, VLA 1623, is shown in Figure 3. Unlike Class I objects, the distinctive feature here is that the SED is well fitted by a single temperature blackbody function. André *et al.* (1993) have argued that VLA 1623 is the prototype of an entirely new class of (proto)stellar objects, called *Class 0*, and proposed a new scheme of age ordering of YSOs based on their sub-mm luminosities. The interesting aspect of this source is that, despite its extreme youth, it drives a well developed bipolar CO outflow. The final proof that this, or similar sources are indeed the long sought for protostars must however await the appropriate kinematic evidence of gravitational infall.

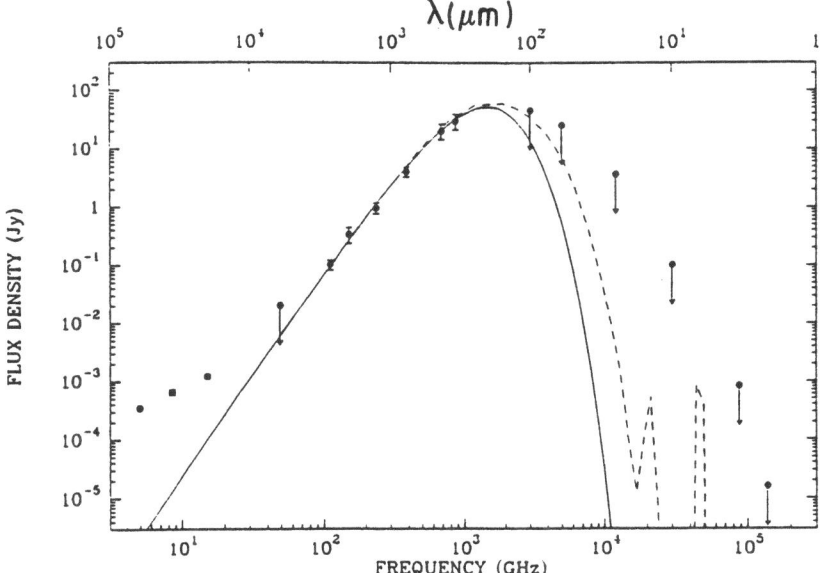

Figure 3: *The spectral energy distribution of VLA 1623, the prototype of a class of YSOs that only show at wavelengths longer than 10μm. The solid curve is a fit with a single temperature blackbody function. The dashed curve represents radiative transfer modelling (from André et al. 1993).*

1.4   Modes and distribution of star formation

The knowledge of the distribution of star forming regions within giant molecular clouds (GMCs) provides a clue to understanding the star formation mechanisms. Schematically, two extreme situations have been envisaged as being the dominant modes of star formation: a *spontaneous* mode, in which the formation process occurs without external influence, and an externally *stimulated* mode. Which of the two is prevalent in molecular clouds, however, is not well understood. The advent of NIR array cameras has offered the appropriate tool to properly address this question. Let us consider each of these schemes in some detail.

1.4.1   Stimulated Star Formation

The variety of physical agents that can induce star formation under external influences has been thoroughly presented by Elmegreen (1992). The basic idea is that external forces, represented by the action of stellar winds, ionization fronts, supernova explosions, result in a compression of pre-exisiting clouds, altering their stable configuration and making them suitable for gravitational collapse. The propagation of shock fronts rises the gas density and enhances the local rate of turbulent, magnetic, or rotational energy dissipation, leaving the gas in a state close to virialization. Clouds can then collapse, and, because of the higher initial density, they do so in a faster time scale. On the other hand, the presence of strong shocks also influences the gas bulk motions, enhancing the kinetic energy of the clouds. Their overall stability against collapse is thus favored, and in some cases, shocks and ionization fronts can even induce the evaporation of a large fraction of the original interstellar material. Numerical simulations of the interaction of ionized radiation and expanding HII regions with globules have shown the twofold effects on the interstellar condensations: an increase both in the *mass loss* rate, by evaporation of the neutral matter from the surfaces, and in the *mass accretion* rate, by compression of the gas to higher densities (Klein *et al.* 1985; Bertoldi & McKee 1990).

From an observational viewpoint, the outcome of the triggered mode is that star formation should take place predominantly near sites of previous generations of massive stars, and not throughout a GMC. The best example of such a situation is provided by the linear sequences of stellar subgroups in OB associations (Blaauw 1964; 1991). But stimulated star formation can also affect the large scale distribution of the interstellar gas, creating holes, shells and bubbles (Tenorio-Tagle & Bodenheimer 1988). On an even larger scale, strong dynamical effects can induce the formation of galactic fountains and shape the structure of a galaxy as a whole (Franco 1992). More convincing examples of the *small scale* range of induced star formation are the bright rimmed globules observed by Sugitani *et al.* (1991) and Tauber *et al.* (1993). These globules have the same physical characteristics of the quiescent dense cores, but there is evidence that the embedded IRAS point sources could represent stars of intermediate mass. Thus, more massive stars are produced by this process than would otherwise if the clouds were in isolation.

1.4.2   Spontaneous Star Formation

According to this idea, the evolution of an individual star forming cloud is determined solely by the balance of the destabilizing effects of self-gravity and the restoring forces due to thermal pressure, plus turbulence, rotation and magnetic fields. Each of these energy reservoirs has to be dissipated *before* gravitational collapse can occur. Thus, the typical timescale for star formation is not governed by the gravitational free-fall time, that at the typical densities of molecular clouds is of order 10^{5-6} yr, but by the dissipation time, that can amount to $\sim 10^7$ yr (e.g. Elmegreen 1991). The physical processes that govern the spontaneous mode are conducive of a situation where a dense core, of the type discussed by

Myers and collaborators, gradually builds up inside the larger molecular cloud, presumably via the action of ambipolar diffusion (Lizano & Shu 1989). This stage then corresponds to the first phase of the evolutionary scenario outlined above, and represents the initial conditions for the actual phase of gravitational collapse.

Observationally, the spontaneous mode predicts a more or less uniform spatial distribution of stars throughout a GMC complex. Examples of the location of young stars in several regions were discussed by Larson (1982), who noted that while in Taurus most emission line stars are scattered through the cloud (and similarly in the streamers of Ophiucus), the opposite is true in Orion: the majority of the stars is in large clusters, with high degree of central condensations. However, the data used by Larson were representative only of limited portions of the molecular clouds in Orion, and definite conclusions about the nature of star formation could not be drawn. At present, the situation has greatly improved, thanks to the developments of IR array cameras and millimeter wave telescopes, and a more definite picture is emerging.

Clustered Mode: In order to obtain a complete census of both the stellar population (embedded and optically revealed stars) *and* the dense molecular gas, large-scale mapping over extended areas must be carried out. The results of recent unbiased, systematic surveys in individual molecular clouds in the Orion complex have revealed that the clustered mode is indeed prevalent (Lada E. 1992; Lada E. *et al.* 1993; Tatematsu *et al.* 1993; Umemoto *et al.* 1993). Figure 4 shows the distribution of stellar clusters and dense gas in L1630. The impressive findings of the survey in this cloud indicate that almost all the sources detected at 2.2μm and associated with the molecular cloud are contained in four clusters only; that these clusters occupy a very small fraction ($\sim$18%) of the surveyed area and are coincident with the most massive cores revealed in CS. The star formation efficiency can be quite high within these cores ($\sim$30-40%), although overall is rather modest ($\sim$3-4%), in accordance with previous estimates.

Thus, it appears that the clustered mode of star formation prevails over the distributed mode even in the case of low-mass stars. A conclusion supported also by the discovery through infrared camera work of hundreds of such stars associated with luminous OB stars in, e.g., S106, M17, W3OH (see last Section). If Orion represents an average GMC, then Lada E. *et al.* (1993) suggest that the clustered mode should account for the formation of the bulk of the stars in our Galaxy, independent of their mass. The problem of star formation is thus shifted to the question of how massive cores form in the first place, and produce rich clusters of stars.

Isolated mode: This subsection cannot be complete without a mention to the

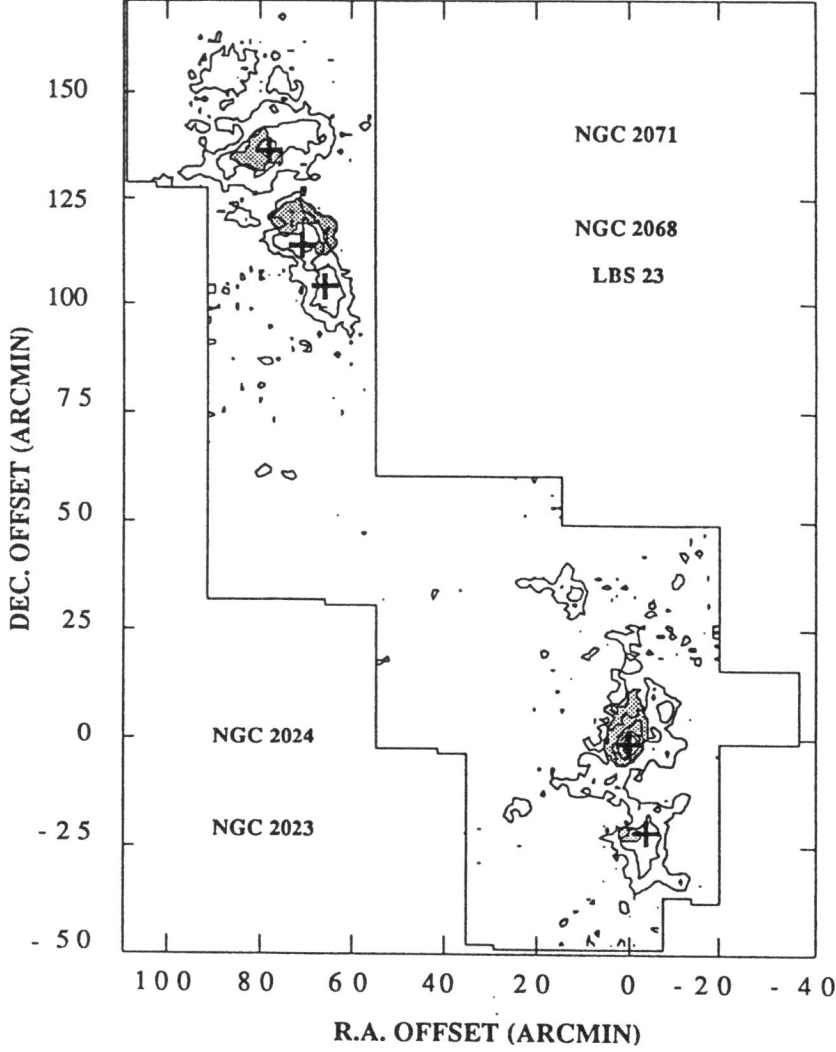

Figure 4: *Maps of the embedded stellar cluster (shaded regions) and dense gas traced by CS emission (intensity contours) in L1630 in Orion. Crosses mark the positions of the most massive CS cores. The survey consists of 3000 1'x1' fields imaged at 2.2μm, down to a completeness limit of 13 mag, and 13000 points with 1' spacing observed in CS (from Lada E. 1992).*

best examples of isolated cases of star formation, the so-called Bok globules (Bok & Reilly 1947). These small ($\sim$0.1 pc), cold ($\sim$10 K), isolated dark clouds have long been suspected to be sites of low-mass star formation, but until recently direct evidence has been missing. In fact, detailed analysis have emphasized their overall dynamical stability against gravitational collapse, with possibly very few exceptions (e.g. B5, B335). Infrared studies have now confirmed Bok's original suggestion. It has been shown that a large fraction of them hosts far-infrared point sources (Yun & Clemens 1990; indeed, they argue that *all* Bok globules harbour a newly formed star), and molecular outflows (Parker *et al.* 1991; Yun & Clemens 1992). The high frequency of the outflows, their estimated duration ($\sim 10^5$yr) and energetics (total kinetic energy$\sim$gravitational binding energy) strongly indicate their major role in driving the early evolution of low-mass young stars. The next Section describes in detail the physical processes that govern the formation of stars under the conditions of this *isolated mode*.

2   ISOLATED STAR FORMATION

The formation of a protostar from the quiescent conditions typical of dense cores occurs through the gradual accumulation of the interstellar gas onto an accreting core. This process requires a large decrease in gravitational potential energy. A large fraction of this energy is radiated away in radial or disk accretion shocks that form as a result of the abrupt change in the velocity of the freely-falling gas. If no net energy is absorbed by the circumstellar material, the resulting luminosity is given to a good approximation by

$$L_{\text{proto}} = \frac{GM_*\dot{M}_{\text{acc}}}{R_*} , \tag{1}$$

where M_* and R_* are the instantaneous mass and radius of the protostellar core, and $\dot{M}_{\text{acc}} = dM_*/dt$ the mass accretion rate. Thus, estimates of the luminosity emitted during this phase rely on the knowledge of two fundamental quantities: *the mass accretion rate*, and the *mass-radius relation*. The former is determined by the dynamics of the gravitational collapse, while the relation between the mass and the radius are established by processes occuring in the protostellar interior. The radiation produced at the shock is absorbed, reradiated and thermalized in the optically thick dusty infalling envelope. Most of the observable radiation is emitted at mid- and far-infrared wavelengths. The exact shape of the emergent spectrum depends on the density and temperature distribution of the dust component, which are set by the dynamics of collapse. This section describes how current models of protostellar formation and evolution have constrained both the value of the mass accretion rate, and the structural properties of the central core. Some relevant observational tests will be presented at the end.

2.1   Protostellar collapse: the mass accretion rate

An estimate of $\dot{M}_{\text{acc}}$ requires the solution of the dynamical collapse problem. In the idealized case of the collapse of a marginally unstable cloud, such a solution has been found semi-analitically by Shu (1977). In this theory, the rate at which a protostar is built up is

$$\dot{M}_{\text{acc}} = \alpha \, \frac{a_{\text{eff}}^3}{G} . \tag{2}$$

In eq. (2), a_{eff} is the effective isothermal sound speed, G the gravitational constant, and α a constant of order unity, If the cloud is supported only by thermal pressure, the expression for the sound speed is simply $a_{\text{eff}} = \sqrt{(KT/m)}$, where

T is the gas kinetic temperature. Otherwise, a_{eff} should also include the contribution of magnetic and turbulent pressures. In the conditions found in dense cores with very little nonthermal support, the gas kinetic temperature varies between 10 and 30 K, corresponding to $a_{eff} = 0.1 - 0.3$ km s^{-1}. This yields an average accretion rate of order $\dot{M}_{acc} \sim$ several$\times 10^{-6} - 10^{-5}$ $M_{\odot}$yr^{-1}. Then, for (sub)solar protostars, the duration of the embedded phase is expected to last typically $t_{acc} = M_*/\dot{M} \gtrsim 10^5$yr. The remarkable property of the form of $\dot{M}_{acc}$ in eq. (2) is that it depends only on one parameter, the sound speed, which is controlled by the conditions in the parent cloud, and not by the properties of the central protostar. Paradoxically, eq. (2) also implies that the higher the temperature of the cloud, the faster the accretion, a result that goes against physical intuition. This situation is in fact radically different from the more familiar case of the Bondi-type solution of the accretion problem, in which the mass accretion rate does depend on the mass of the star, and is inversely proportional to the velocity (and hence the temperature) of the gas. In Shu's solution, a faster accretion rate merely implies higher initial density for the onset of gravitational collapse. The predictions of the *inside-out* collapse model, and the estimate of $\dot{M}_{acc}$, have been extensively used as a diagnostic tool in the interpretation of molecular cloud observations (e.g Evans 1991; Myers and Fuller 1992). In the following, I will give a brief account of its derivation, in the light of the results of numerical simulations of the collapse phase.

2.1.1 Classical Collapse Models

The onset of the dynamical collapse of a self-gravitating cloud requires the specification of the initial conditions. In the first numerical models (e.g. Larson 1969), these were assumed to correspond to the physical conditions in interstellar clouds at the end of the fragmentation phase, a process by which a massive cloud breaks up into smaller and smaller subunits, clumps and fragments. Considering thermal and gravitational effects alone, the requirement that the clumps be self-gravitating leads to the determination of a critical mass, the well-known Jeans criterion. For an isothermal, non-magnetic, non-rotating sphere composed primarily of molecular hydrogen, neutral helium and heavy metals in solar proportions, the minimum mass that a cloud of given (ρ, T) must have to be unstable is given by

$$M_J = 8.5 \cdot 10^{22} \left(\frac{T}{\mu}\right)^{3/2} \rho^{-1/2} \; g \simeq 30 \; M_{\odot} \; T^{3/2} \; n^{-1/2} \; , \qquad (3)$$

where μ is the mean molecular weight, T the gas temperature, ρ and n the mass density per unit volume and the total number of particles per cubic cm. Rewriting eq. (3) in terms of a critical density, ρ_{cr}, a blob of gas of mass M_{cl} will be unstable to contraction or collapse if the density is higher than the critical value. For a cloud with $M_{cl} = 1\ M_\odot$ and temperature $T = 10K$, the critical density is about $\rho_{cr} \sim 10^{-19}$ g cm^{-3}, in reasonable agreement with the high densities measured in dense cores. Thus, Larson suggested that protostellar clouds should have initially an almost uniform density, and that the initial density would be close to the critical one. Starting with these initial conditions, the numerical simulations of protostellar collapse have identified three main stages: an initial *isothermal* contraction, during which the gravitational potential energy is effectively radiated away by the dust and the collapse proceeds at constant temperature; an *adiabatic* phase, corresponding to the transition from low to high optical depth, as a result of the increased density; and, finally, the formation of an *hydrostatic core*, that grows in mass as the collapse proceeds. The basic physical processes that govern each of these phases have been extensively reviewed and need not be repeated here (for a detailed account see Bodenheimer 1992). Suffice to mention only the most general ones, and underline the important fact that they do not depend on the restrictive assumptions under which the numerical calculations were carried out originally (mainly, spherical symmetry). More realistic, and complex, 2- and 3-dimensional collapse models with less stringent geometry constraints have been developed in the meantime (as reviewed by, e.g., Tscharnuter 1991), but the main results of the simple case can still be used as useful guidelines.

Gravitational collapse occurs in a highly *non-homologous* way. The central regions of the cloud collapse faster than the outer parts, pressure gradients develop fast, and the nonhomology sets in already during the initial isothermal phase. Independent of the initial conditions, whether uniform or centrally condensed, the matter develops a density distribution of the type $\rho \propto r^{-2}$, where r is the radial distance from the cloud center. This behavior is illustrated in Figure 5. In a time slightly longer than the free-fall time, the structure of the cloud resembles that of an asymptotically singular isothermal sphere, modified at the center for the presence of a stable hydrostatic core, which contains only few percent of the total cloud mass. The core accretes matter from a freely-falling inner envelope, characterized by a density and velocity profile of the type $\rho \propto r^{-3/2}$ and $v \propto r^{-1/2}$. The outer parts of the clouds are in nearly static equilibrium.

On the basis of these results, Shu (1977) demonstrated that the singular isothermal sphere,

$$\rho = \frac{a_{\text{eff}}^2}{2\pi G r^2}\,, \tag{4}$$

has a self-similar analytic solution that reproduces well the dynamical collapse model. Since a power-law of this type does not have a typical scale associated with it, the solution does not define a characteristic mass scale to be associated

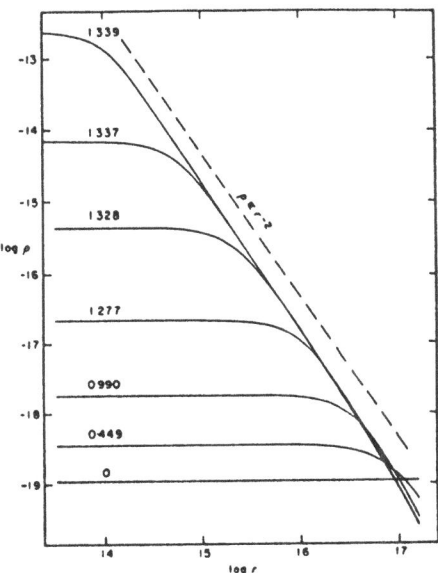

Figure 5: *The density distribution at various times during the isothermal phase of gravitational collapse (from Larson 1969).*

with the formation of an ordinary star. What the self-similar solution defines is the *rate* at which the central star is built up by accretion. It is exactly this rate that appears in eq. (2). The collapse is said to proceed from *inside-out*, since the inner parts collapse first, and the outer parts later on after being reached by an expansion wave that propagates outward at the sound speed, a_{eff}.

2.1.2 More realistic models

The assumptions that enter in the derivation of $\dot{M}_{\text{acc}}$ represent an oversimplification of the realistic case where forces other than thermal compete in maintaining the stability of molecular clouds. It is clear that molecular clouds cannot be collapsing as a whole on a dynamical time scale, as the Jeans criterion would predict, or the resulting star formation rate in the Galaxy would be far too high (Zuckerman & Evans 1974). Thus, the concept of the Jeans mass appears of little use in the definition of the appropriate initial conditions, and eq. (3) should be replaced by a more realistic one. Among the various mechanisms of cloud support, magnetic fields are the most likely to play the dominant role. A magnetic field of strength B can support a molecular cloud of radius R, provided its mass is less than a magnetic critical mass

$$M_{\text{cr}}^{B} = 0.12 \frac{\Phi}{\sqrt{G}} \simeq 10^3 \left(\frac{B}{30\mu G} \right) \left(\frac{R}{2pc} \right)^2 M_{\odot} , \qquad (5)$$

19

where Φ is the magnetic flux through the cloud (Mouschovias 1976). Depending on the cloud mass, M_{cl}, two situations can be envisaged. Supercritical clouds ($M_{cl} \gg M_{cr}$) cannot be supported by magnetic fields alone even if they were perfectly frozen in the gas (no magnetic dissipation), and they would collapse on a magnetic diluted free-fall time scale. Subcritical clouds ($M_{cl} \ll M_{cr}$) are supported by the magnetic field against collapse (even if the external pressure increases) and evolve on the timescale that characterizes the diffusion of the magnetic field. Observational evidence that magnetic fields of the magnitude required by eq. (5) are present in dense cores and in molecular cloud complexes has been obtained by Zeeman splitting measurements (see the review by Heiles *et al.* 1993).

At the scale of the dense cores, the value of M_{cr}^{B} is of the same order as M_J, namely a few solar masses. The main difference between the two cases is in the time scale of evolution. In dense cores, the gas is lightly ionized (the fractional ionization is about 10^{-7}), and the relevant mechanism of magnetic diffusion is due to *ambipolar diffusion* (Mestel & Spitzer 1956), a process in which the fluid of charged particles can slowly drift with respect to the fluid of neutral particles, the two fluid being coupled by collisions of atomic and molecular species. Studies of the quasi-static evolution of molecular cores have shown that the configuration tends to acquire the density profile of the singular isothermal sphere (eq. (4)) at the time the central regions become gravitationally unstable and undergo dynamical collapse (Lizano & Shu 1989). Recent observations towards several dense cores have revealed a high degree of central condensation (Myers & Fuller 1992). Collapse calculations starting from these new initial conditions have been presented by Galli & Shu (1993a,b), who considered in detail the effects of magnetic field and plasma drift. The main result is that the collapse still takes place in an inside-out fashion, but now propagating outward as a fast magnetosonic wave, travelling faster in the direction parallel to the initial magnetic field. Surprisingly, the value of $\dot{M}_{acc}$ obtained in the nonmagnetic case is *not* significantly altered by the presence of a magnetic field. The reason is that the gas is slowed down, because of the Lorentz force opposing gravity, but the information travels outward faster, because of the increased characteristic speed of the system. The two effects cancel out, and the numerical value of $\dot{M}_{acc}$ remains unchanged.

However, the dynamics of the gas is significantly affected by the magnetic field. The strong pinching Lorentz force deflects the infalling gas toward the equatorial plane, with the formation of an inner disequilibrium structure. The disk instantaneous radius is given by

$$r_B = 0.12 \left(\frac{G^2 B^4}{a_{\text{eff}}}\right)^{1/3} t^{7/3} \simeq 600 \left(\frac{B_0}{30\mu\text{G}}\right)^{4/3} \left(\frac{a_{\text{eff}}}{0.35\text{kms}^{-1}}\right)^{-1/3} \left(\frac{t}{10^5\text{yr}}\right)^{7/3} AU.$$

$$(6)$$

Disk-like structures over scales of several hundred AUs have been mapped recently in molecular line emission around a few YSOs (Beckwith & Sargent 1993). The scale of the magnetic pseudo-disk is much larger than that of the centrifugal disk, that forms as a result of the effects of the initial small, but not negligible rotation of the molecular cloud. According to the estimates of Terebey, Shu & Cassen (1984), the value of the centrifugal radius is

$$r_C = 0.06 \left(a_{eff}\Omega^2\right) t^3 \simeq 7 \left(\frac{a_{\text{eff}}}{0.35\text{kms}^{-1}}\right) \left(\frac{\Omega}{4 \times 10^{-14}\text{s}^{-1}}\right)^2 \left(\frac{t}{10^5\text{yr}}\right)^3 AU , \quad (7)$$

where Ω is the initial rotation rate of the core. Inside r_C, infalling matter encounters a centrifugal barrier in the equatorial plane, and accumulates in a disk. The various elements that characterize the collapse of magnetized, rotating molecular cores over a range of different scales are schematically shown in Figure 6.

2.2 Protostellar evolution

Having established the magnitude of the mass accretion rate, we now discuss the properties of protostars, as they grow from the small hydrostatic core formed at the end of the adiabatic phase to a stellar sized object during the main accretion phase (cf. 2.1.2). The purpose here is to show how a *mass-radius* is established, that can then be used in eq. (1) to estimate the protostellar luminosity. In the inside-out collapse scenario, the structure and evolution of the protostar are solely determined by the specification of $\dot{M}_{\text{acc}}$. The evolution of the infalling envelope and the hydrostatic core are effectively decoupled, owing to the existence of a region of very low opacity (*the opacity gap*) in the outer parts of the envelope. The photons produced at the accretion shock can escape freely, instead of being pushed back into the core by the infalling matter. This results in a great simplification of the problem, that otherwise would require the joint solution of the full hydrodynamical equations for the dynamical collapse and a detailed radiation transfer. Despite great efforts, fully self-consistent models do not yet exist (see Boss 1993). The quasi-hydrostatic evolution of the protostar can be treated as a problem in stellar structure, with the modifications introduced by a time varying mass and nonstandard surface boundary conditions provided by an accretion shock (Stahler *et al.* 1980). These models have shown the key role of the nuclear burning of interstellar deuterium in determining the properties of

The inside—out collapse scenario with magnetic field

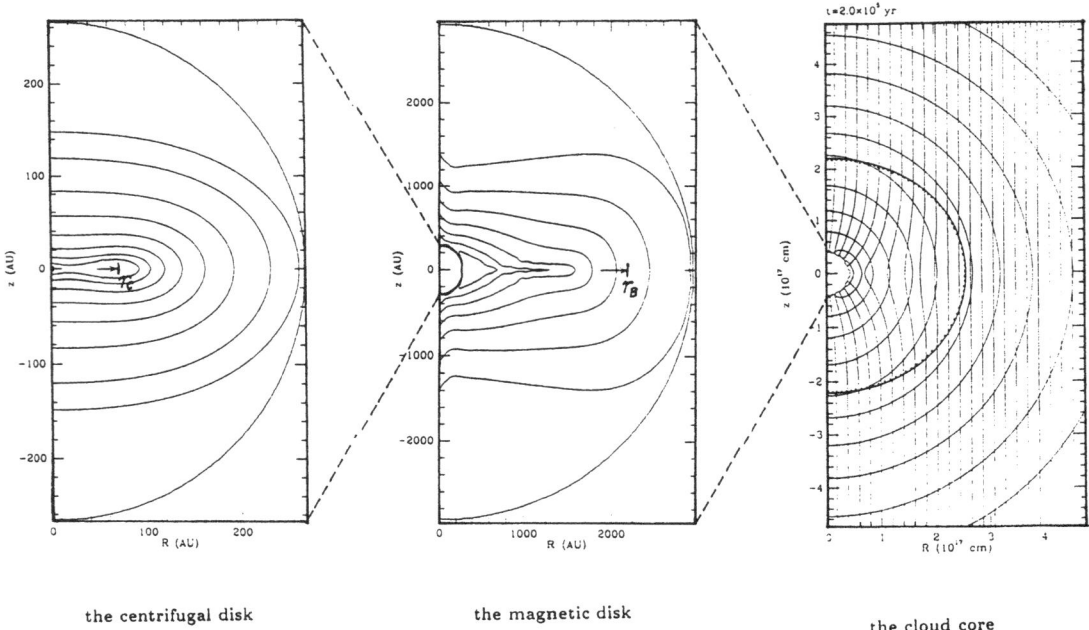

the centrifugal disk   the magnetic disk

the cloud core

Figure 6: *Global view of the collapse of a magnetized, rotating dense core. The right panel shows the isodensity contours and magnetic field lines over a scale of about 0.1 pc. The dotted line represents the infall magnetosonic wave that propagates faster in the direction perpendicular to the field. In the middle panel, on scales of* $\sim 10^3$ *AU, the dynamics is dominated by magnetic and gravitational forces that shape the density distribution in the equatorial plane. The left panel shows the isodensity contours in the innermost* $\sim 10^2$ *AU. The positions of* r_B *and* r_C *(see text) are indicated in the middle and left panels (adapted from Galli & Shu 1993).*

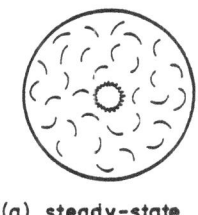

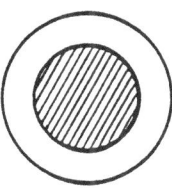

(a) steady-state   (b) radiative   (c) depleted   (d) shell
    burning   barrier   interior   burning

Figure 7: *A schematic view of the main phases of deuterium burning in proto-stars (from Palla & Stahler 1990).*

the accreting core. The protostellar evolution in a mass interval that covers low- and intermediate-mass stars, $0.01 \lesssim M_*/M_\odot \lesssim 10$, is marked by the occurrence of 4 distinct phases of deuterium burning, independently of the assumptions of the accretion process (Stahler 1988a; Palla & Stahler 1991). These phase are sketched in the cartoon of Figure 7 and will be described below. The resulting mass-radius relation for a protostar accreting at a rate $\dot{M}_{\mathrm{acc}} = 10^{-5} \ M_\odot \ \mathrm{yr}^{-1}$ is shown in Figure 8.

Central D-burning: for low-mass protostars ($M_* \lesssim 1 \ M_\odot$), deuterium burning occurs near the center for temperatures exceeding 10^6 K. The energy input is sufficient to turn and maintain the star convectively unstable. As more mass is gathered, the newly accreted deuterium is instantly transported by convective eddies to the center, and a situation of steady-state burning is created. In this condition, the deuterium generated luminosity, L_D is given by

$$L_D \ = \ \dot{M}_{\mathrm{acc}}\delta \ = \ 15 L_\odot \left(\frac{\dot{M}_{\mathrm{acc}}}{10^{-5} \ M_\odot \mathrm{yr}^{-1}} \right) , \qquad (8)$$

where δ is the nuclear energy per unit mass available in interstellar matter (assuming a [D/H] value of 2.5×10^{-5}). The extreme temperature sensitivity of the energy generation rate is such that deuterium acts as an effective thermostat, preventing the central temperature from rising above $T_c \sim 10^6$ K as the star gains mass. Since the amount of energy available in deuterium, δ, is comparable to the gravitational binding energy of the star, GM_*/R_*, the thermostatic effect results in a core radius that increases almost linearly with mass during the phase of active burning, as shown in Fig. 8 for $0.3 \lesssim M_*/M_\odot \lesssim 1$.

Radiative barrier: The evolution of protostars more massive than solar is marked by the return to radiative stability and the onset of gravitational contraction. These events occur at about the same time, as can be seen by the following argument. Convection ceases to be effective when the equality

$$L_D = L_{\text{rad}}(M_*, R_*) \,, \tag{9}$$

where L_{rad} is the luminosity that a star of given mass and radius can carry radiatively, is met. For low-mass stars, L_{rad} is very small, but it quickly rises, due to the strong dependence on stellar mass, $L_{\text{rad}} \sim M_*^{5.5} R_*^{-1/2}$. The requirement on the equality of the two luminosities provides a condition on the star's gravitational contraction time, t_{KH}, at the time of radiative equilibrium, i.e.

$$t_{\text{KH}} \equiv \frac{GM_*^2}{R_* L_{\text{rad}}} \approx \frac{GM_*^2}{R_* L_D} \approx \frac{GM_*^2}{R_* \dot{M}\delta} \approx \frac{M_*}{\dot{M}} \equiv t_{\text{acc}}. \tag{10}$$

Eq. (10) establishes that the star is losing heat by radiation at the same time as it accretes new matter, and therefore it undergoes a phase of rapid gravitational contraction. Note that for low-mass stars, the accretion time is always much shorter than t_{KH}. Due to the sensitivity of L_{rad} on M_*, the equality of the two time scales is confined to a rather narrow range of masses, between 2 and 3 $M_\odot$. If nothing else happened in the interior, the rapid contraction would lead to the conditions appropriate for hydrogen burning in the center, and the star would then join the main-sequence while still accreting. As a consequence, there should be no stars of mass greater than $2-3$ $M_\odot$ in the pre–main-sequence phase (e.g. Larson 1972). This result is in clear contradiction with the observational evidence of intermediate-mass stars, the so-called Herbig Ae/Be stars, whose location in the H-R diagram is well above the ZAMS. The reason why this expected behavior does not occur is subtle, and is again related to the effects of deuterium burning. Physically, the transition to radiative equilibrium does not occur throughout the star all at once, but begins in a localized region (the radiative barrier). Interior to the barrier, matter quickly burns its nuclear fuel that can no longer be replenished by newly accreted deuterium, and an inert core develops (cf. the second and third panel of Fig. 7).

The onset of D-shell burning and gravitational contraction: Matter exterior to the barrier still contains substantial deuterium. When the thermodynamical conditions are appropriate, deuterium once more ignites, this time in a convective shell located at the barrier (as indicated in the last panel in Fig. 7). This shell slowly shrinks toward the surface, due to the overall tendency of the star to become more and more radiatively stable. The release of nuclear energy in the lower density subsurface regions results in a dramatic swelling of the star, that doubles its radius. This is shown in the steep, almost vertical rise of the curve in Fig. 8. The expansion of the radius counteracts the gravitational pull, and retards the beginning of the phase of gravitational contraction. As a consequence, the fusion of ordinary hydrogen, and thus the arrival on the Main sequence is

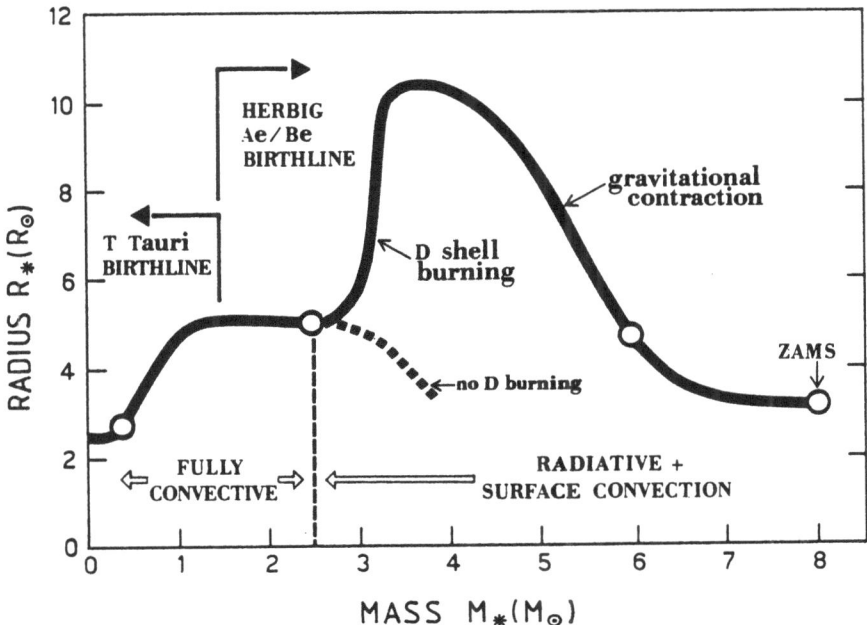

Figure 8: *The evolution of the radius vs. mass in accreting protostars. The open circles represent, from left to right, the ignition of central deuterium, the start of D-shell burning, the ignition of central hydrogen via the CN cycle, and the final arrival on the ZAMS at a mass $M_* \sim 8\ M_\odot$. To gauge the importance of D-shell burning, the dashed curve shows the path of the radius obtained ignoring it (adapted from Palla & Stahler 1990).*

postponed to higher masses, $\simeq 8\ M_\odot$ in the specific case considered here, thus solving the inconsistency with the observations.

Under the assumption that protostars end their accretion phase in a time much shorter than the Kelvin-Helmoltz time, the mass-radius relation shown in Fig. 8 can be used to construct a theoretical birthline for stars of low- and intermediate-mass in the H-R diagram. The birthline is the locus along which young stars first appear as optically visible stars (Stahler 1983). Accordingly, there should be no observed PMS stars whose position in the H-R diagram is *above* the birthline, since the largest radius a PMS star of a given mass can achieve is the protostellar radius at that mass. The resulting birthline is shown in Figure 9, together with the location of T Tauri stars and Herbig Ae/Be stars. It is evident that the upper envelope of the stellar distribution is generally well matched by the theoretical birthline. Note also that only the stars located near the birthline are associated with molecular outflows, indicating their extreme youth (Levreault 1988). The agreement is also suggestive of the fact that the specialized assumptions of the calculations (spherical symmetry, constant mass accretion rate) are not essential to the outcome of mass-radius relation. We

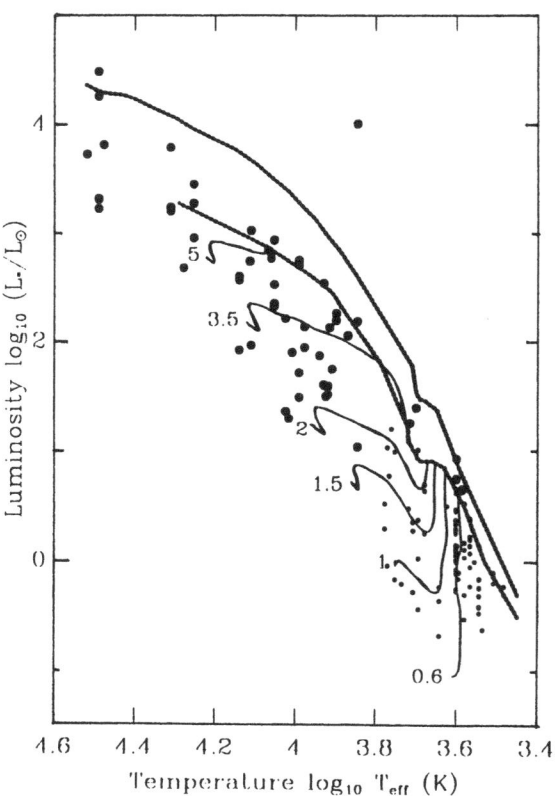

Figure 9: *Location of the stellar birthline in the H-R diagram for accretion rates of* 10^{-5} *(lower dotted curve) and* 10^{-4} *(upper dotted curve)* $M_{\odot}yr^{-1}$. *Symbols give the positions of young PMS stars. Labeled curves are PMS evolutionary tracks for the indicated stellar mass (from Palla & Stahler 1993).*

discuss this point in the following subsection.

2.2.1   Effects of the accretion flow

Variations in the geometry of the accretion flow mainly reflect the influence of rotation and magnetic fields in the parent cloud. Angular momentum conservation imposes that the infalling material does not strike the stellar surface directly, but first forms a circumstellar disk from which it spirals into the star (e.g. Mercer-Smith *et al.* 1984). The largest uncertainty in accounting for this effect in numerical models comes from the lack of detailed knowledge of the physical mechanism responsible for the accretion through the disk, and an *ad hoc* prescription must be used (see Adams & Lin 1993 for a summary of accretion disk models). Qualitatively, the global effect of accretion through a disk is to yield matter landing onto the star with a lower specific entropy than that which hits the surface directly, because of the heat loss through radiation from the disk faces. In actual calculations, where this effect has been mimecked by altering the boundary conditions for the core, the reuslting protostellar radius results smaller than in the shock case. However, the sequence of events described above remains unaltered with the protostar undergoing the same dramatic swelling at the time of deuterium shell-burning (Palla & Stahler 1992).

As for a variation of the mass accretion rate, there are observational indications that stars of increasing mass may be formed from molecular cloud cores with large nonthermal internal motions (e.g Myers, Ladd & Fuller 1991). These motions would provide support to the cloud against collapse, and would tend to increase $\dot{M}_{acc}$ well above the thermal value given by eq. (2). On the other hand, if the fiducial accretion rate was much *lower* than the thermal one, the resulting intermediate-mass stars would run through the entire PMS evolution during protostellar accretion, a result in clear conflict with the observations. The effect of increasing $\dot{M}$ is opposite to that of assuming disk accretion; namely, the entropy level of the protostar becomes higher, since each gas element landing on the star has less time to radiate away its heat. The various evolutionary phases are once again recovered: the shape of the mass-radius remains unaltered, even though each event is shifted to higher masses. The resulting birthline shown in Fig. 9 by the upper dotted curve does not improve the agreement with the observational data. These considerations lead to the conclusion that an accretion rate $\dot{M} \sim 10^{-5} M_\odot \mathrm{yr}^{-1}$ can represent the typical value for the formation of stars in a mass range from few tenths of a solar mass to about 10 $M_\odot$.

2.2.2 Conditions for star formation

The last statement implies that the vast majority of stars, apart from the most massive stars, form with nearly the same accretion rate. For it to be true, the temperature and density structure of the local regions forming such stars cannot show wide variations. Do molecular cloud observations support this statement? This question brings us back to the problem of specifying the initial conditions for the star formation process. While for low-mass star forming clouds the physical state of the dense cores is rather well defined, the case for more massive stars is much more complicated by the damage these short lived stars exert on their surroundings. As discussed in Myers & Fuller (1992), the understanding of the conditions for the formation of more massive stars has several additional complications, such as the lack of a clear link between the properties of the parent cores and their end products; the dominance of poorly understood nonthermal motions; and the tendency of massive stars to appear in clusters, suggesting some degree of interaction with their cluster neighbors (see also last section).

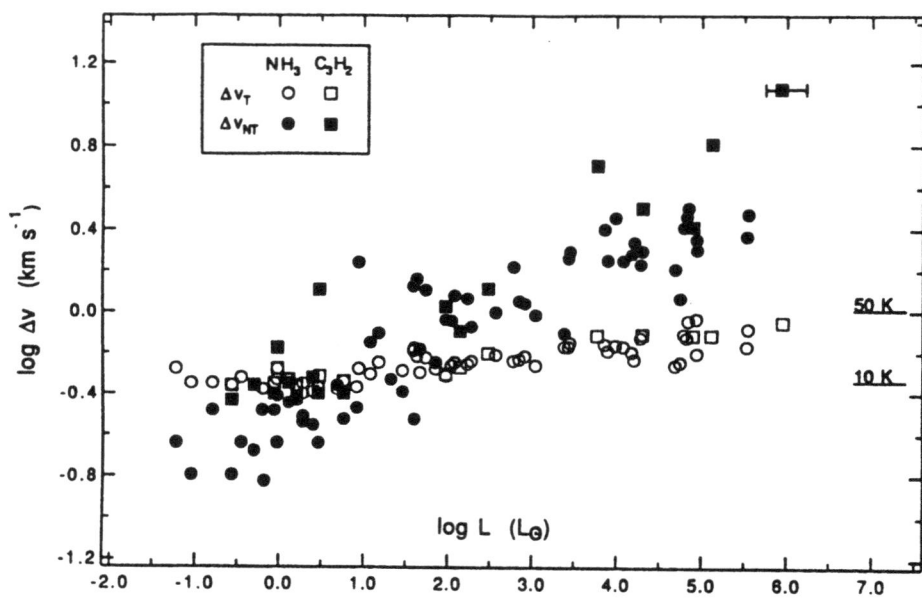

Figure 10: *The correlation between thermal* Δv_T *(open symbols) and nonthermal* Δv_{NT} *(filled symbols) line widths in about 80 dense cores associated with IRAS sources of luminosity L (circles and squares) (from Myers, Ladd & Fuller 1991).*

Finally, brighter stars are generally farther away than low-mass stars, and the molecular line observations of the associated gas are necessarily of lower spatial resolution. However, one trend is clear: the existence of a positive correlation between the luminosity of embedded infrared sources and the *nonthermal* line widths of nearby cloud material in a rather large, although inhomogeneous, sample of cloud cores (Myers *et al.* 1991). The *thermal* line widths, on the other hand, are nearly the same for all sources, indicating a rather remarkable uniformity of cloud temperature. The observed trend is shown in Figure 10.

Myers and collaborators contend that the increased line widths reflect higher turbulent velocities which were present in the dense cores prior to their collapse to form the presently observed stars. It would follow, therefore, that more massive stars form with higher accretion rates. More recently, Myers & Fuller (1992) have presented a relation between the stellar mass and the observed line width in NH_3 in two regions associated with massive star formation (L 1688, L 1204) that supports this view. They have also suggested a model of the equilibrium of dense cores accounting for the presence of both thermal and nothermal motions, that complements the isothermal model of Shu (1977) and extends it to higher masses. This way they can explain the formation of stars in the range $0.3 - 30 M_\odot$. Interestingly, from this model, the average mass accretion rate for a $10 M_\odot$ star is of the order of $10^{-5} M_\odot \mathrm{yr}^{-1}$.

Finally, it is important to underline the role played by IR observations in this context. The notion that a wide range of stellar masses forms under similar conditions is corroborated by the NIR array camera observations of embedded clusters. They have clearly shown that regions once thought to form exclusively high-mass stars, do indeed host a large low-mass population, and in a proportion that does not seem to require radical departures from the standard IMF (e.g. Zinnecker *et al.* 1993). It seems difficult to imagine that the same units in which collective star formation took place almost simultaneously, were characterized by extremely different physical conditions. We will return to the discussion of NIR observations of embedded clusters in the last Section.

2.2.3   Some observational tests

So far, direct spectroscopic evidence for protostellar collapse has been rare and inconclusive. The main difficulty is that, even in the case of well isolated regions, other effects, such as turbulence, rotation and outflows, compete to mask the genuine signature of dynamical collapse in the molecular line profiles. Nonetheless, a number of claims have been made at different times, most notably in the case of the source IRAS 16293-2422 (Walker *et al.* 1986). The asymmetry in the line profile of CS transitions was interpreted as a unique evidence for collapse, but later analysis showed that another explanation involving rotation of the dense core and foreground absorption produces a better agreement with the data (Menten *et al.* 1987). High resolution millimeter interferometric observations have revealed the existence of a double source, possibly in an accretion phase, making it a good case for a protostellar binary system (Walker *et al.* 1993). Stronger evidence for large scale infall has been found in distant regions of massive star formation (e.g. Rudolph *et al.* 1990), but the observations do not apply to the collapse of a single object. Very recently, a strong case for inside-out collapse has been advocated for B335, an isolated Bok globule at a distance of 250 pc, based on high resolution observations of H_2CO and CS lines and detailed radiative transfer models (Zhou 1992, Zhou *et al.* 1993). Evidence comes from a consistency of several physical quantities, including the infall radius, the density structure in the static envelope following the expected r^{-2} dependence, the velocity field in the inner regions, as well as the shape of the spectral lines. Figure 11, taken from Zhou *et al.* (1993), shows the comparison between the observed H_2CO line profiles at two frequencies and the theoretical predictions of Shu's model. These authors suggest that B335 is the best candidate for the Holy Grail of star formation, a conclusion that, of course, should be tested by independent observations.

As for the infrared properties of Class I sources, spectral modeling of several

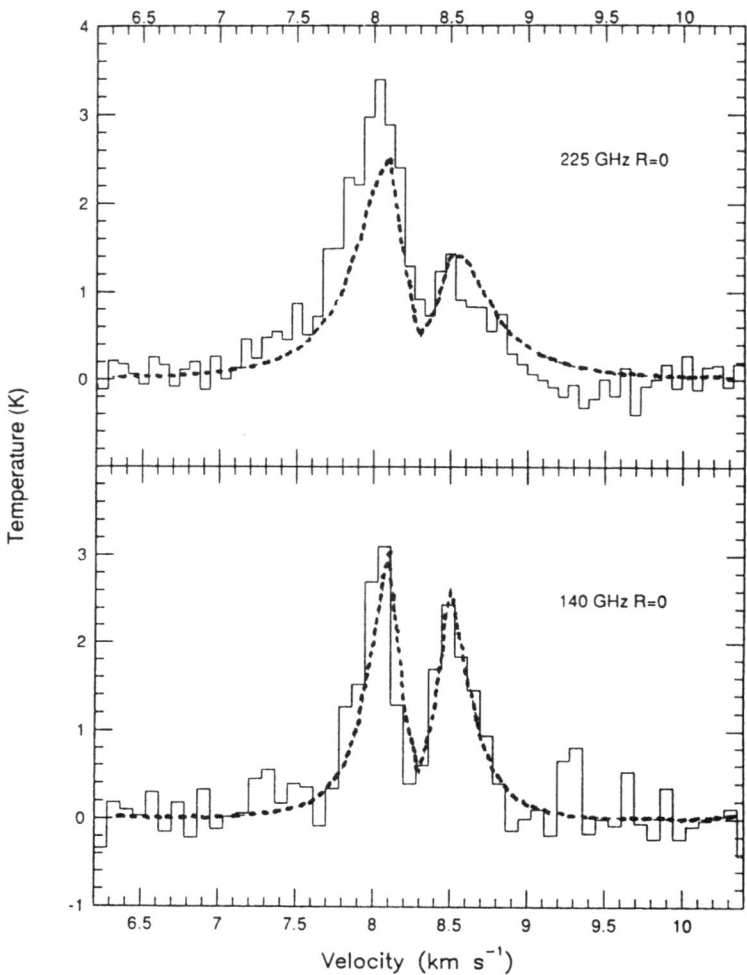

Figure 11: *Evidence for collapse in B335. The solid lines show the observed H_2CO line profiles at two frequencies toward the center of the globule. The dashed lines are the predictions of Shu's collapse model with an infall radius of 0.033 pc (from Zhou et al. 1993).*

well observed examples in the Taurus and Ophiucus dark clouds have clarified the structure of their dusty envelope (Adams, Lada & Shu 1987; Ladd *et al.* 1991). Typically, such a source consists of three distinct regions with different velocity and density fields: on the large scale (~ 0.1 pc), the original r^{-2} density distribution; on small scale (down to ~ 350 AU), an infalling envelope, approximately spherical, and, on even smaller scales, a thin disk around the central star. The presence of a geometrically thin, but optically thick disk is necessary to produce enough emission at mid-infrared wavelengths (5-20 μm). In fact, if the density distribution in the envelope extended all the way to the central source, the extinction would be so high that no near- or mid-infrared photons would escape. Quite the contrary, Class I sources do show strong emission at these wavelengths, and the inferred size of the emitting region corresponds to about few hundreds AU. The presence of a disk ensures both large mid-infrared fluxes and lack of excessive extinction. This interpretation has become a paradigm in most studies of YSOs.

However, two caveats are in order. The first one is that the match to the SED depends on the dust opacity properties at these critical wavelengths. As it has been shown by Butner *et al.* (1991) in the case of a prototypical source, L1551 IRS 5, using the dust mixture of Mathis, Mezger & Panagia (1983) instead of Draine & Lee (1984) produces a fit in good agreement with the observations *without* including the contribution from a disk. Secondly, for sufficiently bright sources, it is possible to resolve the far-infrared emission at 50 and 100 μm, using high-resolution scanning techniques. Again, in the case of L1551 IRS 5, the emitting region is extended ruling out its origin in a disk. In conclusion, extreme care should be used in deriving conclusions on circumstellar disks from modeling of SEDs (see also Butner, Natta & Evans 1993). The best direct evidence for the presence of a compact disk (45$\pm$20 AU) around this source still comes from the interferometric measurements at 2.7 mm by Keene & Masson (1990).

3 EARLY STELLAR EVOLUTION

Once the main phase of accretion is completed, the stellar core emerges as an optically visible star along the birthline. The physical process by which infall stops is still not known, although stellar winds and the associated bipolar flows are most likely to play a fundamental role. The circumstellar matter surrounding these young stars, partly distributed in a disk and the rest in an extended envelope, still emits copiously at infrared wavelengths: the emergent spectral energy distribution departs substantially from that of a normal stellar photosphere, showing "excess" emission extending from the near- to the far-infrared. The origin of this excess has been commonly attributed to the presence of accretion disks. Depending on the mass of the star, young visible stars are classified as T Tauri (low-mass, typically $\sim 1 \ M_\odot$), or Herbig Ae/Be (intermediate-mass, $\lesssim 10 M_\odot$) stars. Their early stellar evolution can be followed in detail, once the initial conditions for the gravitational contraction phase towards the Main Sequence are specified. The birthline joins the ZAMS at $M_* \sim 8 \ M_\odot$. Thus, for stars with masses above this value, there is no optical pre-main-sequence phase: the rapid gravitational contraction in these objects has led to hydrogen ignition already during the accretion phase. Massive stars are born *on* the ZAMS. In this Section, the basic features of theoretical models of pre–main-sequence (PMS) evolution will be outlined. The effects of circumstellar disks, and the modifications induced by mass accretion on the resulting evolutionary tracks will then be discussed. Finally, an overview of the global properties of the circumstellar environment of observed PMS stars, as probed by infrared observations, is presented.

3.1 Pre–Main-Sequence Evolution

Unlike the previous hydrodynamical protostellar phase, the PMS evolution of a star can be followed by models in hydrostatic and thermal equilibrium. This models are much easier to construct and less subject to numerical inaccuracies. In the classic studies of Henyey *et al.* (1955) and Hayashi (1961) it was established that PMS stars of all masses derive their luminosity primarily by gravitational contraction through a sequence of quasi-static configurations. In these calculations, the initial configuration was essentially arbitrary, and it was assumed that *all* stars begin this contraction as fully convective, with radii nearly two orders of magnitude larger than their final main sequence values (Iben 1965; Ezer & Cameron 1965). Based on simple energy conservation arguments, Cameron (1962) estimated that $R_*^{\mathrm{init}} \sim 50 \ R_\odot (M_*/M_\odot)$. Thus, convection is due to, and maintained by, the large stellar radius and low effective temperatures.

For solar-type stars with large radii, it was also shown that the configuration quickly becomes convectively unstable even if it was initially in radiative equilibrium (Bodenheimer 1966; cf. Stahler 1988b for a general discussion). These two conditions, distended configurations and full convection, have become a standard assumption in all models calculations that ignored the prior accretion history (e.g. Mazzitelli 1989).

3.1.1 Protostellar initial conditions

The results of protostellar evolution described in Section 2 clearly show that the standard assumptions are no longer appropriate. First, protostellar cores never achieve large radii. The core radius remains typically a factor of ten smaller than predicted by Cameron's estimate. From the mass-radius relation of Fig. 8, we see that the maximum value is approximately 10 $R_\odot$ for a $M_* \sim 4\ M_\odot$, at the time of the swelling due to deuterium shell-burning. It is reasonable to assume that these radii represent the *maximum* values that a star of a given mass can achieve prior to gravitational contraction. Second, the internal structure departs significantly, at least for intermediate-mass objects, from the assumption of thermal convection. In low-mass stars, convection is due to central nuclear burning. Protostars more massive than about $2 M_\odot$ are radiatively stable in the inner regions, and possess a thick, subsurface mantle of deuterium. This deuterium must ignite in a shell to fuse to helium during the approach to the main sequence. The disparate state of the two regions implies that the stars are *thermally unrelaxed* at the beginning of the PMS evolution. Thus, in contrast to the situation described by the classical theory, these stars undergo *nonhomologous* quasi-static contraction (Palla & Stahler 1993).

Another important aspect is that for a PMS star of fixed mass, the amount of deuterium available is that which a protostar of the same mass did acquire in the course of its accretion history. The variation of the fractional deuterium concentration, f_D, in a protostar accreting at $\dot{M} = 10^{-5}\ M_\odot$ yr^{-1} is shown in Figure 12. Here, f_D is calculated relative to the interstellar concentration, assuming a standard [D/H] value of 2.5×10^{-5} (Geiss & Reeves 1981; McCullough 1992). Also shown is ΔM_D, the stellar mass fraction at any time which has a non-zero deuterium abundance. We see that f_D declines quite rapidly for $M_* \geq 1 M_\odot$, even though the star is fully convective, and therefore accreting deuterium as fast as it is being consumed. The fall of f_D continues until the radiative barrier appears, and thereafter deuterium exists only in a subsurface mantle. Although the concentration there rises to almost unity, the mass of unburned deuterium ΔM_D decreases very rapidly. The total deuterium content, as measured by $f_D \Delta M_D$, continues to decline monotonically for masses higher

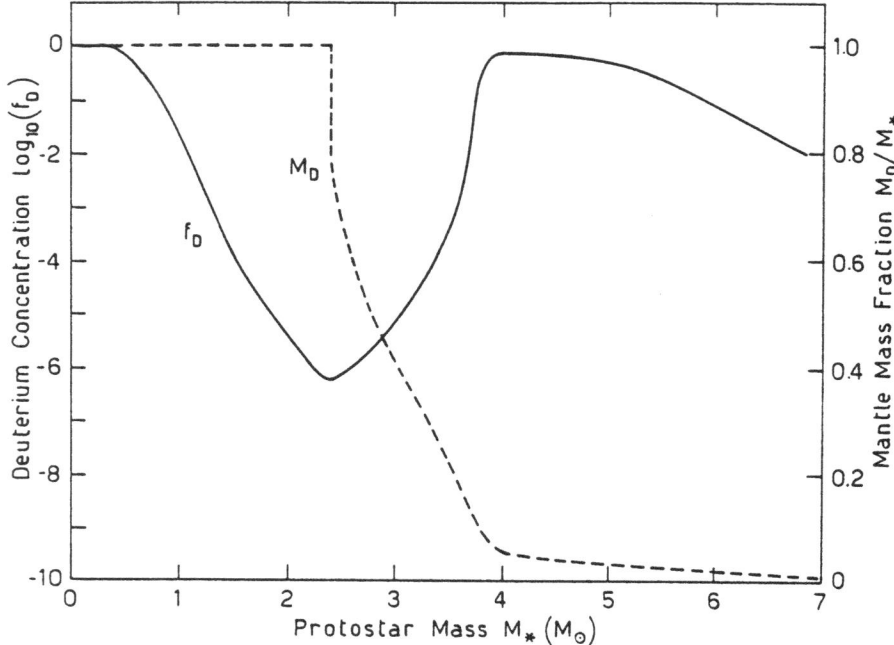

Figure 12: *Evolution of the fractional deuterium concentration, f_D, and of the unburned deuterium mantle, ΔM_D, in protostars. The abrupt change in the curves at $M_* = 2.4 M_\odot$ marks the appearance of the radiative barrier and the rapid loss of interior supply of deuterium (from Palla & Stahler 1993).*

than $\sim 4 M_\odot$. These results mark a net departure from classical PMS theory. It appears, therefore, that the consideration of more realistic initial conditions should have a strong effect on the basic properties of the standard evolution. In the following, the case for low- and intermediate-mass PMS stars will be discussed separately.

3.1.2 PMS evolution of low-mass stars

An updated summary of the PMS evolution of nonrotating, nonmagnetic stars without mass loss or accretion has been presented by D'Antona (1992). In a recent study, D'Antona & Mazzitelli (1993) have reconsidered the evolution of stars with population I composition in the mass range 0.015 $M_\odot$ to 2 $M_\odot$. The main motivation comes from the need of better constraints on the location of tracks and isochrones in the H-R diagram, to determine ages and masses of observed stars. Also, the abundance of the light elements ^{7}Li and ^{9}Be observed in several T Tauri stars poses constraints on their internal structure and on the mechanisms of angular momentum transport (e.g. Pinsonneault *et al.* 1992). Uncertainties in the absolute location of the evolutionary tracks in the H-R diagram are mostly due to limited knowledge in the input physics, particularly for low temperature opacities and the treatment of convective transport. Several new sets of opacities have become available recently, that cover both the atmospheric and subatmospheric layers (Alexander *et al.* 1989; Kurucz 1991), and the inner regions where the burning of light elements takes place (Rogers & Iglesias 1992). The models of D'Antona & Mazzitelli (1993) show that the estimates of the effective temperatures are not strongly dependent on the opacity for stars in the range 0.08 to 2 $M_\odot$; while large variations occur for smaller masses (~ 0.04 dex in $\log T_{\rm eff}$ for $M_* = 0.03 \, M_\odot$).

The largest departure from the standard theory of low-mass PMS stars concerns the history of deuterium burning. Salpeter (1954) was the first to recognize that deuterium ignition could impede the gravitational contraction of a star of fixed mass. This effect was confirmed by the detailed calculations of Bodenheimer (1966), Grossman & Graboske (1971), and Mazzitelli & Moretti (1980), that introduced the concept of a "deuterium main sequence" (DMS) to explain the slowing down of the evolution at the time the central temperature was high enough to ignite deuterium. Stahler (1988a) has shown that this concept is misleading, since nuclear burning occurs already in the accretion phase for stars with masses above a few tenths of a solar mass: the locus of the DMS is nothing else than the birthline. For lower mass stars, protostars theory still implies deuterium burning during the contraction phase, while for masses as low as 0.018 (D'Antona & Mazzitelli 1993) degeneracy effects set in and prevent nuclear burning. Note also that, according to the behavior of f_D shown in Fig. 12, most of the available deuterium has been used up during the accretion phase. Thus, solar-type PMS stars begin their contraction phase by simply descending along the vertical portion of the Hayashi track.

In addition to deuterium, other light elements, ^{6}Li, ^{7}Li, and ^{9}Be are depleted in the course of PMS contraction when the central temperature becomes sufficiently high ($T_c \gtrsim 2 - 3.5 \times 10^6$ K). However, their relativily tiny abundances render the effect energetically and structurally unimportant. On the other hand, these trace elements are of cosmological interest, and their depletion histories (especially for ^{7}Li) provide insights into problems as diverse as the age estimates of young clusters (Balachandran 1992), and galactic chemical evolution (D'Antona & Matteucci 1991). Until recently, the observations have implied larger abundances of ^{7}Li in T Tauri stars than in young clusters, but the work of Magazzú *et al.* (1992) has shown a remarkable constant abundance, with enhanced depletion only in low luminosity T Tauri stars. A full account of the problem of lithium depletion in young stars can be found in D'Antona (1991).

3.1.3 PMS evolution of intermediate-mass stars

For stars more massive than solar, the imprint of the prior accretion history persists much longer than in the low-mass, convective case. The main reason is that the stellar interior is not thermally relaxed, and the stars must undergo a phase of global readjustment. The evolution of stars in the mass range $1 \leq M_*/M_\odot \leq 8$ has been followed by Palla & Stahler (1993), and the new evolutionary tracks are displayed in the H-R diagram of Figure 13. The main results can be schematically summarized as follows.

Fully convective stars: Stars in the range $1 - 2.5\ M_\odot$ are fully convective due to surface cooling as soon as they appear as optically visible objects. The evolution is undisturbed by the fusion of residual deuterium: its concentration is so low, that it is quickly consumed (on a time scale of few years). Stars contract along the Hayashi track, but their path is much reduced (cf. the case for $M_* = 2\ M_\odot$ in Fig. 13).

Partially convective stars: Stars in the range $2.5 - 4\ M_\odot$ undergo thermal relaxation and nonhomologous contraction. When they are first optically visible, they are *underluminous*: i.e., the evolutionary track begins at low luminosity and then moves up to join the radiative portion (cf. the case for $M_* = 3.0\ M_\odot$ in Fig. 13). The Hayashi phase is skipped entirely.

Fully radiative stars: Stars more massive than $4\ M_\odot$ appear immediately on the radiative portion of the classical track, and begin to contract homologously under their own gravity.

In comparison with standard Iben's calculations (1965), the new tracks occupy a much reduced portion of the H-R diagram. In addition, the surface temperatures are *higher*, since the outermost layers are fully ionized, and the dominant opacity source in this region is no longer the H^- ion. More importantly, the new tracks imply a substantial decrease in the evolutionary time scale: indeed, Herbig Ae/Be stars are younger than previously believed, thus alleviating the problem of maintaining the observed level of activity. In conclusion, the combination of protostellar and PMS theory does seem to account for at least one basic property of T Tauri and Herbig Ae/Be stars, their distribution in the H-R diagram. As shown in Fig. 9, the upper envelope of the stellar distribution is well matched by the theoretical birthline obtained with the fiducial value of $10^{-5}\ M_\odot \mathrm{yr}^{-1}$.

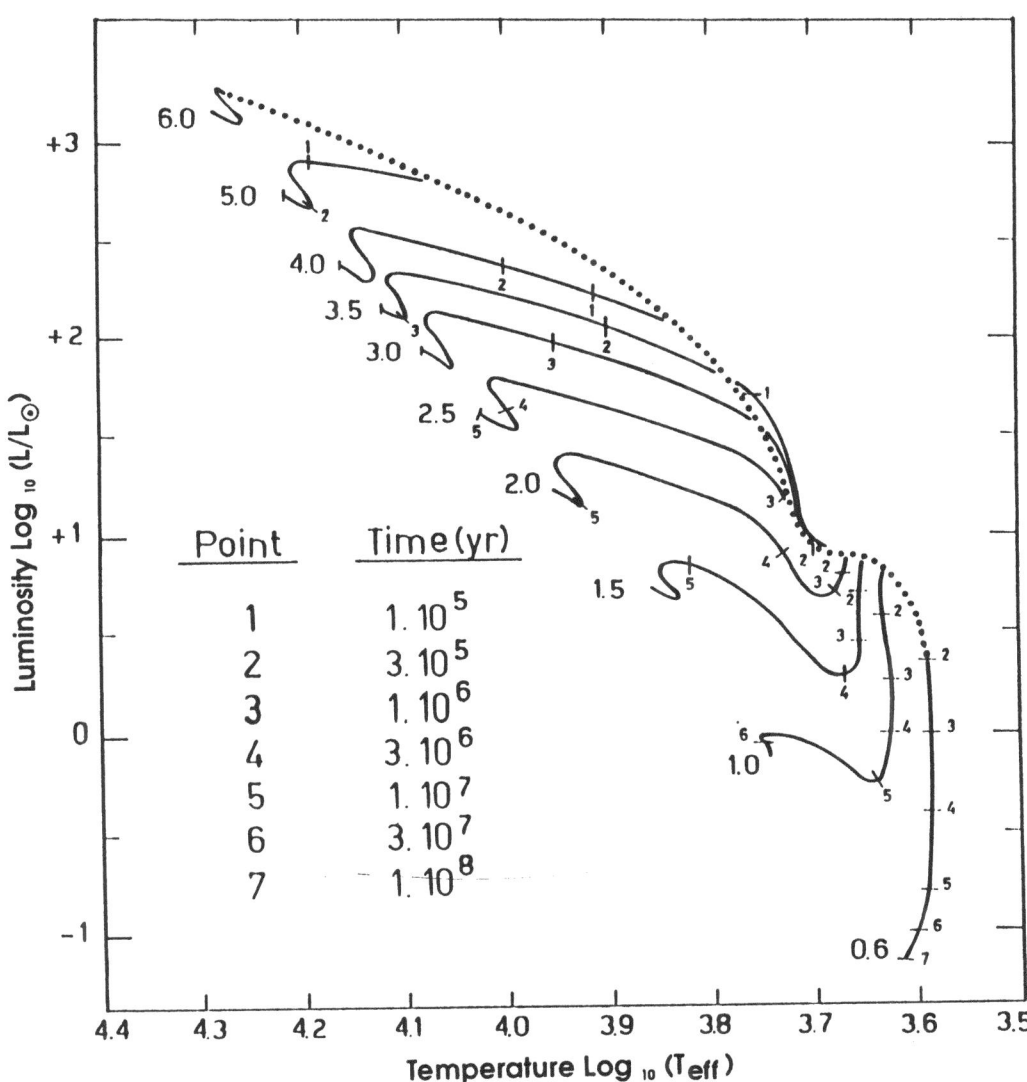

Figure 13: *Evolutionary tracks in the H-R diagram for low- and intermediate-mass stars. Each track is labeled by the corresponding mass, in solar units. Tick marks indicate points in the evolution whose time is given in the table. For each track, the evolution starts at the birthline (dotted curve), and ends at the ZAMS (not shown) (from Palla & Stahler 1993).*

Francesco Palla

3.2 Effects of mass accretion on PMS evolution

The PMS evolution described so far, although improved by the proper account of the results of protostellar theory, is still rather conventional, in the sense that it assumes that each star evolves in isolation at constant mass. However, it is now recognized that the real situation is much more complex, since both T Tauri and Herbig Ae/Be stars are surrounded by, and interacting with, circumstellar disks of appreciable mass. Accretion of matter is expected to occur within these disks, and from them onto the stellar surface. Although the nature of the physical process responsible for accretion is still under debate, extended and optically thick disks are required to explain the large infrared and ultraviolet excesses and the continuum radiation that veils the optical spectrum (Rucinski 1985; Adams, Lada & Shu 1987; Kenyon & Hartmann 1987; Basri & Bertout 1989). As discussed in Hartmann & Kenyon (1990), the disks can affect the interpretation of observed PMS stars by altering their position in the H-R diagram in two ways: first, the system luminosity is not simply that of the central star, but it must include a contribution from the disk itself; second, accretion from the disk onto the star can modify its final mass. Whether this effect is substantial, depends on the mass accretion rate, which must be inferred from the observations. This is not an easy task, since the different sources of the observed infrared excesses are difficult to separate, and also because understanding the spectral shape requires a proper treatment of the radiative transfer in the disk and envelope, including reprocessing and scattering of the stellar and disk photons.

Current estimates for accretion rates derive from the analysis of the excess optical continuum radiation in T Tauri stars that also show infrared excess. The disk luminosity comes from two components,

$$L_{\text{disk}} = \frac{GM_* \dot{M}_{\text{acc}}}{2R_*} + L_{\text{rep}} , \qquad (11)$$

where $\dot{M}_{\text{acc}}$ is the accretion rate onto a star of mass M_* and radius R_*, and L_{rep} is the luminosity due to absorption and reradiation of light from the central star. Typical values give $L_{\text{rep}} \sim 0.25 - 0.5L_*$ (Adams et al. 1987), where L_* is the stellar luminosity. Thus, unless $L_{\text{disk}} \gtrsim L_*$, the identification of the excess luminosity as *accretion* luminosity is not secure. Note that typically $L_{\text{disk}} \sim 0.5L_*$, as inferred from observations (e.g. Hartmann & Kenyon 1990; Strom et al. 1989b). If the optical excess is assumed to arise in a boundary layer (BL) between star and disk, then the standard theory of viscous accretion disks (Lynden-Bell & Pringle 1974) predicts that up to half the accretion luminosity is emitted as radiation,

$$L_{\text{BL}} \leq \frac{GM_* \dot{M}_{\text{acc}}}{2R_*} . \qquad (12)$$

38

Thus, the measurement of the BL luminosity can provide a direct measure of the amount of mass actually falling onto the star. Independent estimates of $\dot{M}_{acc}$ on a rather large sample of T Tauri stars in the Taurus-Auriga molecular cloud yield values of the order $5 \times 10^{-8} - 10^{-7} M_\odot yr^{-1}$ (Hartmann & Kenyon 1990; Bertout & Regev 1992). Note, however, that in absence of angular momentum loss, accretion at these rates onto the stellar surface would spin up the central star to half the break-up velocity in a few 10^6 yr (Hartmann & Stauffer 1989). Since T Tauri stars, as well as Herbig Ae/Be stars, rotate at typically 1/10-1/5 of the break-up velocity, it is likely that the strong winds commonly seen in these objects carry away the angular momentum excess. In support of this hypothesis, Cabrit *et al.* (1990) found a tight correlation between mass-accretion and mass-loss diagnostics in classical T Tauri stars, suggesting that the gravitational energy released during accretion is the ultimate source for driving the strong winds. Thus, it is not clear yet what fraction of the total accreted mass is incorporated into the central star.

The question, then, is whether the inferred rates are high enough to modify the evolutionary tracks. Hartmann & Kenyon (see also Hillenbrand *et al.* 1992) argue that indeed they affect the evolution significantly, and estimate from qualitative arguments that about 10% of the final mass of a T Tauri star is accreted through a disk. A quantitative evaluation of the effect on actually computed tracks including mass accretion is shown in Figure 14. Here, the evolution of a star with initial mass 0.6 and 1.0 $M_\odot$ is followed in the case of steady mass accretion at rates $10^{-8}, 10^{-7}$, and 10^{-6} $M_\odot$ yr^{-1}. As we can see, as long as $\dot{M}_{acc}$ remains smaller than $\sim 2 \times 10^{-7}$ $M_\odot yr^{-1}$, the global effect is rather inconspicuous, and hardly changes the path followed by the stars along the Hayashi track. Incidentally, it is interesting to note that a correct estimate of the effect of mass accretion on fully convective stars can be found by simple energy considerations. During the convective phase, the evolution of the stellar luminosity, L, is described by the equation,

$$\frac{d}{dt}\left(-\frac{3GM_*^2}{7R_*}\right) = -L - \frac{GM_*\dot{M}_{acc}}{R_*} + \dot{M}_{acc}\delta , \qquad (13)$$

where the term on the left hand side represents the temporal change of the gravitational potential energy; while the second and third terms on the right hand side express the gain of (negative) energy by mass accretion, and the released luminosity of freshly accreted deuterium. This equation can be recast in the form of a nondimensional ordinary differential equation, whose solution is straightforward. The same equation does not describe the evolution of more massive stars that possess radiative interiors. The effects of mass accretion on the tracks can be gauged only by a full numerical calculations.

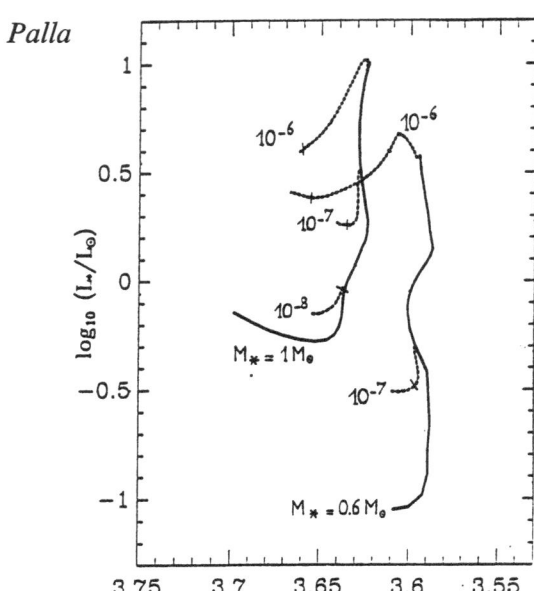

Figure 14: *Effects of mass accretion on the PMS evolutionary tracks of stars of $M_* = 0.6$ and 1.0 $M_\odot$. Three values of $\dot{M}_{acc}$ are considered, $\dot{M} = 10^{-8}, 10^{-7}$, and 10^{-6} $M_\odot yr^{-1}$ and the resulting tracks are shown by the dotted lines. The solid line is for the evolution at constant mass. Evolutionary times are indicated by tick marks (from Parigi 1992).*

By analogy with the low-mass counterparts, Herbig Ae/Be stars are also expected to undergo substantial accretion of matter. The excess infrared emission has been interpreted as arising from active circumstellar disk, and estimates of the characteristic accretion rates have been derived (e.g. Hillenbrand *et al.* 1992; Lada & Adams 1992; Natta *et al.* 1992). The conclusion of these studies is that accretion rates are about 100 times as large as those typical for T Tauri stars, i.e. in excess of $10^{-6} M_\odot yr^{-1}$. Also, it appears that there is a trend with the stellar mass of the type $\dot{M}_{acc} \sim M_*^{2.2}$, indicating that disk accretion rates are higher among disks surrounding higher mass stars. We will comment below on the plausibility of these estimates. In the present context, suffice to consider the effects that large accretion might have on the evolution of these stars. With the rates quoted above, it turns out that the accretion luminosity is in many cases comparable to, or even greater than, the stellar luminosity. Thus, we have

$$\frac{GM_*\dot{M}_{acc}}{R_*} \approx L_* \tag{14}$$

Now, M_* always exceeds the disk mass, M_{disk}. Hence, the above relation implies

$$t_{disk} \equiv \frac{M_{disk}}{\dot{M}_{acc}} \leq \frac{M_*}{\dot{M}_{acc}} \approx \frac{GM_*^2}{R_*L_*} \equiv t_{KH} . \tag{15}$$

In other words, the time for the disk to be depleted is less than the characteristic evolutionary time of the central star. Over the life of the star, the disk must be continually replenished, presumably by matter from a more extended, infalling envelope. If that is the case, one is led back to the difficulty that motivated the accretion disk hypothesis, namely that a more isotropic distribution of dust would make the central star optically invisble. The other inconsistency of very fast accretion rates in Herbig Ae/Be stars is that these stars during the PMS evolution would never depart significantly from the original birthline. Therefore, very few stars should be found near the ZAMS, in contradiction with the claim that the stars showing the largest IR excess are also those closest to the ZAMS (Hillenbrand *et al.* 1992).

As a final comment, if high accretion rates accompany the PMS evolution of all intermediate-mass stars and the mass dependence of $\dot{M}_{acc}$ is valid, then we should expect to see a segregation effect in the distribution of the stellar mass on the ZAMS in young clusters. Namely, there should be a deficit of stars of mass around 2-3 $M_{\odot}$. In fact, lower mass stars accreting at the rates typical of T Tauri stars would end the PMS phase with practically the same initial mass; while higher mass stars would rapidly accumulate a mass comparable to their initial value. Since, the equality between the accretion and the Kelvin-Helmoltz time scales is met for stars of mass $\gtrsim 2\ M_{\odot}$ (cf. eq. (10)), then relatively fewer stars should arrive on the ZAMS with this mass. Interestingly, such a dip in the observed luminosity function at magnitudes corresponding to ZAMS spectral types mid-A to mid-F has been marginally found in two young clusters, NGC 3293 and NGC 2362 (see Wilner & Lada 1991).

3.3 Infrared properties of PMS stars

The distinctive property of T Tauri and Herbig Ae/Be stars is that their SED in the infrared can be well described beyond about 5 μm by a power-law form $\nu F_{\nu} \propto \lambda^{-n}$, with the spectral index n varying between 4/3 and 0 (Rucinski 1985; Rydgren & Zak 1987; Beckwith *et al.* 1990). Examples of the observed SED for a sample of T Tauri stars are shown in Figure 15. The slow fall off of the spectrum has been explained in terms of emission from an accretion disk, in which the emitting dust is distributed with a temperature profile $T_r \propto r^{-p}$. The index n is related to p via $n = 4 - 2/p$. In order to cover the observed range of n, p must vary between 3/4 (steep spectrum sources) and 1/2 (flat spectrum sources). The fact that in most cases the observed SED is shallower than that predicted by the *active* accretion disk model, i.e. the mean value of n is appreciably smaller than 4/3, implies that the temperature law differs substantially from the canonical value $p = 3/4$ (Shakura & Sunyaev 1973; Lynden-Bell & Pringle 1974). Note that the same value of p describes the temperature profile of a disk that merely absorbs and reprocesses the radiation from the central star (so-called *passive*

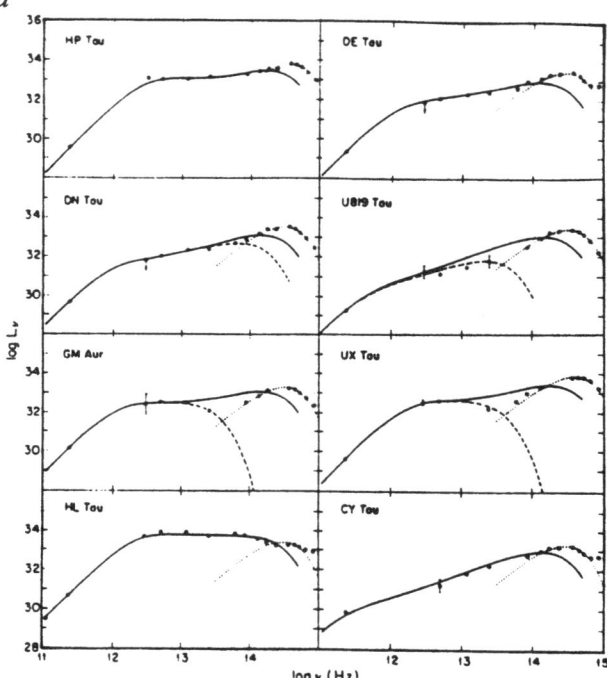

Figure 15: *Spectral energy distributions from optical/NIR to mm-wavelengths for a subsample of T Tauri stars. The observed data points extend to 1.3 mm. The solid and dashed lines represent disk model spectra calculated under various assumptions. The dotted lines are Planck functions at the stellar effective temperatures fitted to the short wavelength data. A large variation in the spectral index, n, is clearly visible (from Beckwith et al. 1990).*

disk; Adams *et al.* 1988). Thus, active disks cannot be distinguished from passive ones on the basis of the shape of the SED alone. Notwithstanding these discrepancies between theory and observations, the accretion disk hypothesis has become a paradigm in the current interpretation of the infrared excess from PMS stars (for a dissenting view, see Mitskevich 1993, who proposes a model of dusty *clumps* distributed in an extended, quasi-spherical envelope without the need of an accretion disk). Apart from modelling SEDs, independent evidence for circumstellar disks comes from a variety of indicators, such as the blue-shifted forbidden-line emission (Appenzeller *et al.* 1984; Edwards *et al.* 1987); the presence of strong millimeter and sub-millimeter continuum radiation (Beckwith *et al.* 1990); and optical and near-infrared polarization maps (Bastien & Ménard 1990).

In order to explain the extreme flat spectrum sources and the anomalous temperature distributions, various mechanisms have been suggested. These include geometric flaring at large distances in purely reprocessing disks (Kenyon & Hartmann 1987); non-viscous mechanisms of mass and angular momentum transport (in this case the temperature law index is reduced to an adjustable free parameter in spectral matching, Adams *et al.* 1988); non-local mechanisms generated by eccentric gravitational instabilities (Adams, Ruden & Shu 1989); and the presence of unresolved infrared companions (e.g. Leinert & Haas 1989). A very interesting, and quite natural solution has been presented by Natta (1993), who suggests that the presence of a tenuous ($\tau_{vis} \ll 1$), dusty envelope surrounding the star/disk system can well account for the heating of the dust at large disk radii, and thus give rise to a flat temperature profile. The effect of the envelope is to absorb and scatter part of the stellar radiation back onto the disk, increasing the disk luminosity by an amount proportional to the envelope optical depth, τ_{vis}. The emerging spectrum is so sensitive to the density distribution, α, and the optical depth, that small variations in these two parameters can explain the observed star to star variation of the spectral index n.

Additional information on the properties of the circumstellar environment of PMS stars comes from the analysis of their near-infrared JHK colors. Observations at wavelengths between 1-2 μm are useful probes of the regions in the immediate vicinity of the central source, where the temperature regime corresponds to the thermal evaporation of the emitting dust. Lada & Adams (1992) and Hillenbrand *et al.* (1992) have shown that different types of YSOs, from protostars to Herbig Ae/Be stars, tend to occupy well defined portions of color-color diagrams, and that their location is determined to a large extent by their evolutionary state. As an example, the distribution of low- and intermediate-mass stars is illustrated in Figure 16. The main conclusion of these studies is that the pattern of color displayed by classical T Tauri stars can be well reproduced by the same circumstellar disk models used to fit the overall SED (see also Calvet, Hartmann & Kenyon 1993).

The situation for the Herbig Ae/Be stars is much less clear. Indeed, the standard disk models fail to account for the JHK color, due to extreme near-infrared excesses. In the majority of these stars, the observed spectrum rises rapidly shortward of ~ 3 μm, where it shows a local peak, and then falls off with a slope $\lambda^{-4/3}$. To explain the near-infrared peak a modification of the standard disk model is required. This is accomplished by imposing the presence of an inner hole in the dust distribution at temperatures exceeding ~ 2000 K (close to the conditions of dust evaporation). In this way, optically thin disk material inside this radius emits less radiation than optically thick dust located at larger disk radii, thus explaining the decline in the SED at $\lambda \lesssim 3\mu$m. Good model fitting requires rather large inner holes, up about $5 - 20$ R_*, and mass accretion rates through the hole in excess of $\sim 10^{-6}$ $M_\odot\mathrm{yr}^{-1}$. These rather specialized models have been criticized by Hartmann, Kenyon & Calvet (1993), on grounds that the requirement of high accretion rates is inconsistent with an optically thin disk. In addition, a severe energy problem exists. Given the estimated accretion rates,

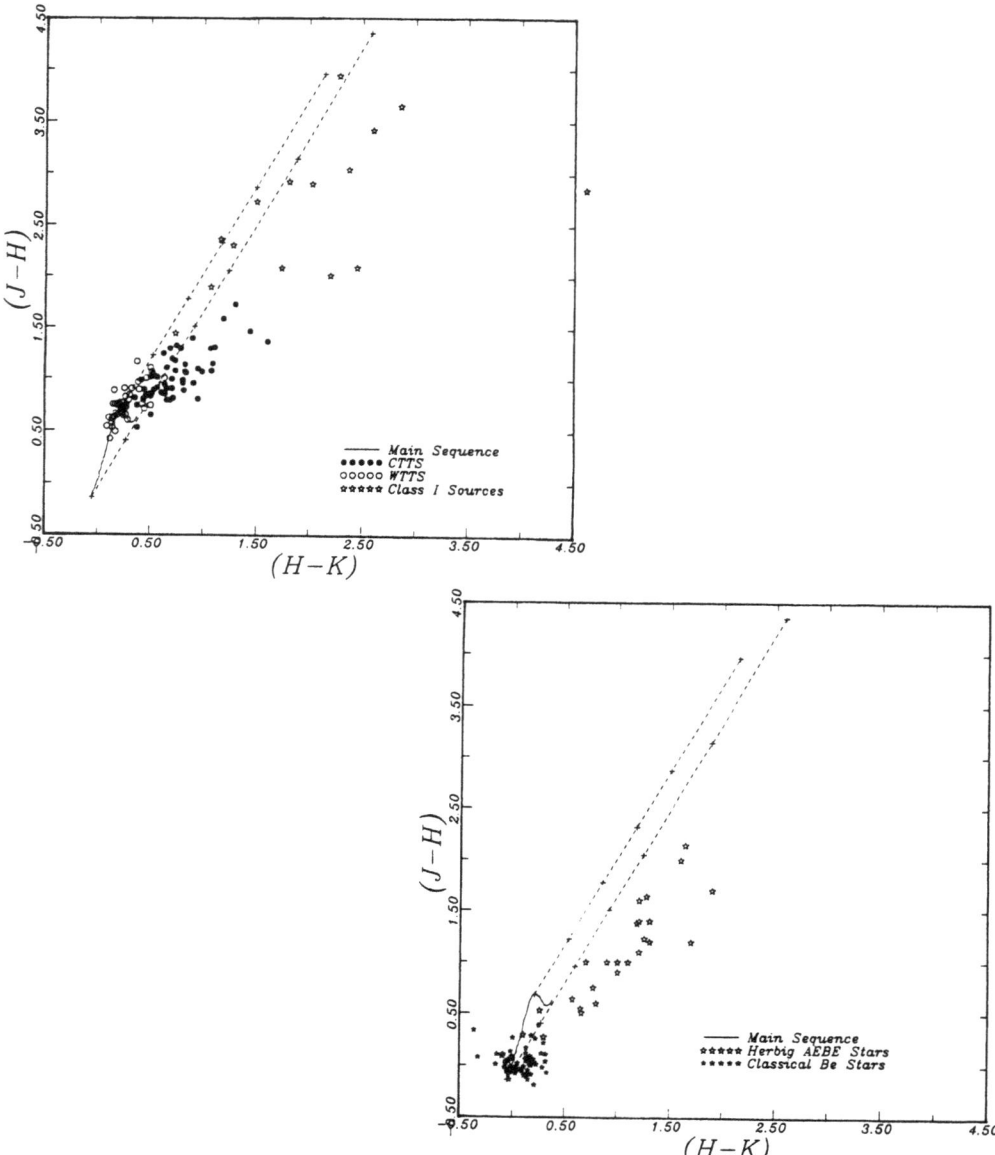

Figure 16: *Distribution of YSOs in the JHK color-color diagram. Low-mass YSOs in Taurus-Auriga are shown in the left panel, while the location of Herbig Ae/Be and classical Be stars is given in the right panel. The two dashed lines form the reddening band for normal stellar photospheres. Sources which fall outside this band have large intrinsic IR excess (from Lada & Adams 1992).*

the total luminosity released by a steady accretion shock, $GM_*\dot{M}/R_i$, where R_i is the inner disk radius, would exceed that of the central star itself. This radiation should appear as optically thin line or continuum radiation, either in the optical or in the UV. Thus far, there is no evidence of excess emission of that magnitude at those wavelengths. As an alternative, Hartmann *et al.* (1993) suggest that the excess comes from dusty envelopes, instead of circumstellar disks. As for the origin of the emission at ~ 3 μm, either transiently heated small grains at large distances from the star, or the presence of an embedded companion are considered as likely possibilities. However, the energy problem still remains an unsolved puzzle.

The presence of extended envelopes around Herbig Ae/Be stars has been directly probed by high spatial resolution observations in the far-infrared and at submillimeter wavelengths (Natta *et al.* 1992, 1993; Weintraub *et al.* 1992). These observations have shown that the size of the emitting region is far too large, typically between 5×10^3 and 8×10^4 AU, to be produced by an optically thick disk. By combining the prediction of radiation transfer models to the observations (the far-infrared scans and the SED from visual to the millemeter), it is found that the flux at wavelengths $\lambda \gtrsim 50\mu$m is dominated by the emission from the envelope. These structures are in all cases optically thin in the mid- and far-infrared ($\tau_{100} \lesssim 0.01$), but thick in the visual, where the extinction (absorption and scattering) is between 5 and 10 magnitudes. However, the effective extinction is smaller, suggesting that light is leaking or scattering out in anisotropic geometries. There is strong indication that inside the envelope a large fraction of the stellar radiation is already degraded to longer wavelengths, as it would be the case if a disk was present. However, the excess near- and mid-infrared luminosity, if generated internally would imply implausibly high accretion rates, $\dot{M}_{acc} \sim 10^{-3}$ $M_\odot$ yr^{-1}, that can hardly be accomodated by present disk accretion theories.

4 COLLECTIVE STAR FORMATION

The last Section of this chapter deals with the collective properties of star formation, as revealed by the study of young embedded star clusters. Star clusters are the smallest systems in which the processes that generate a distribution of stellar masses, the Initial Mass Function (IMF), can be studied observationally. An enormous boost in this field has been provided by the availability of large format near-infrared array detectors of high sensitivity that allow to completely sample the young stellar population over an entire star forming cloud. An account is given in this Section of the status of this rapidly developing field. Particular emphasis will be given to the interpretation of observed luminosity functions. Specific examples will be discussed, together with the description of theoretical models of the temporal evolution of a star forming cluster.

4.1 Embedded clusters and the Luminosity Function

The study of young clusters still deeply embedded in molecular clouds is particularly attractive, since it offers the unique opportunity to probe *in situ* the origin and development of the luminosity function of newly born stars, and to test the predictions of theoretical models, mainly developed for the formation of individual objects, on a statistically significant population. The determination of reliable luminosity functions (LFs) is the first step in the more ambitious process towards the knowledge of the IMF. A major goal of the studies of star formation is to understand the spectrum of masses with which stars are formed. The IMF plays a fundamental role in determining the observed properties of stellar systems and their evolution with time; it also provides the vital link between the local processes of star formation and the global properties of populations in galaxies.

The two basic functions to be determined are the Initial Luminosity Function (ILF), which gives the relative rates of appearance of main-sequence stars of specified luminosity, and the IMF, that measures the relative rates at different stellar masses. The traditional approach to determining the IMF has been to take the present-day LF in the solar neighbourhood, and infer the initial main-sequence mass function taking into account stellar lifetimes, galactic scale heights, and an assumed time dependence of the star formation rate in the Galaxy (Salpeter 1955; Scalo 1986). Although this derivation is subject to many uncertainties, two basic features of the IMF seem well established (e.g. Larson 1992). The first one, is that the typical stellar mass is of order 1 $M_\odot$; the second, that the IMF for relatively massive stars can be approximated by a simple power-law

$$\xi\,(logm_*) \;\propto\; m_*^{-\alpha} \;, \tag{16}$$

where $\xi(logm_*)$ is the IMF, and is related to the ILF, $\Phi(M_V)$, via

$$\Phi(M_V)dM_V \;=\; \xi(logm_*)dlog(m_*) \;. \tag{17}$$

In this equation, the ILF is given as a function of the absolute magnitude, M_V. In order to transform the observed M_V to m_*, one must make use of a mass-luminosity relation. For main-sequence stars, this relation is rather well established empirically, and has a power-law form $L_* \propto log\,M_V \propto m_*^\beta$ ($\beta = 3.75$). The spectral index α in eq. (16) was derived originally for field stars by Salpeter (1955), who found $\alpha = 1.35$. More recent estimates, based on field stars and open clusters, find a steeper value, $\alpha = 1.7 \pm 0.5$ (Scalo 1986). Within the uncertainties, the two estimates are still compatible. Since the index is less than unity, the power-law implies that more stellar mass is contained in low-mass stars than in high-mass stars. The power-law breaks down at low masses, and the IMF flattens or even turns over, at a mass $\sim 0.6\,M_\odot$. The reality of other features in the IMF below $0.3\,M_\odot$ remains uncertain; however, the suggested rise for very low masses, below about $0.1\,M_\odot$, attributed to the presence of brown dwarfs, seems at present highly unlikely (see chapter by Gilmore). At any rate, the important point is that most of the stars formed in our galaxy are low-mass stars with mass $\lesssim 2\,M_\odot$. Of course, it is of basic interest to know whether these properties of the locally derived IMF for a inhomogeneous sample are universal, both in space and in time, as it is often assumed.

In principle, the study of the stellar population in embedded clusters offers an alternative approach to the derivation of the IMF. Observationally, due to the extreme obscuration of circumstellar dust, it has been very difficult to find and survey systematically the youngest cluster in molecular clouds. Partially embedded clusters have been known for many years, and examples include the IR core in ρ Oph, NGC 2264, the Trapezium cluster, M17 (cf. Lada C. 1991), but their full content could not be sampled, especially at the low mass end because of the poor sensitivity and spatial resolution. This limitation has been overcome with the development of infrared arrays of high sensitivity, spatial resolution, and wide field coverage, and many embedded clusters of young stars have been discovered associated with galactic HII regions and OB stars.

The luminosity function is one of the more readily parameters that can be determined with an IR array camera. For embedded clusters, the observed LF is a direct measure of the ILF of the stellar population. However, the transformation of the ILF to an IMF is not at all straightforward. The main problem is that many cluster members are PMS objects, whose luminosity varies as a result of gravitational contraction toward the main-sequence. In addition, some sources are likely to be accreting protostars, whose mass is varying on time scales shorter than the cluster age, and there is no guarantee that the IMF correctly describes their distribution. The transfomation is hampered by the intrinsic time-dependent nature of the ILF. It is, then, inappropriate to convert LFs into stellar masses using time-independent main-sequence mass-luminosity relations and bolometric corrections. Despite these intrinsic difficulties, numerous studies aimed at deriving *monochromatic* and *bolometric* LFs have been already published (see reviews by Lada C. 1991 and Zinnecker *et al.* 1993). Of the two observed LFs, it is only the bolometric LF that really matters for the determination of the IMF. However, only in the case of nearby dark cloud complexes this task can be carried out observationally; in more distant clouds the poor linear spatial resolution of existing instruments at the longest wavelengths, and the higher degree of confusion from background stars, preclude the effort. The two situations present different problems, so they will be discussed separately.

4.2   NIR observations of embedded clusters

Typically, embedded clusters have small diameters, $\simeq 1$ pc, and host a population of several hundred young stars, whith ages estimated between $1 - 5 \times 10^6$ yr. The cluster members form a relatively homogeneous population that can provide information on the relative lifetimes of the embedded and PMS evolutionary phase, the star formation rate and efficiency (SFE), in addition to the LF. A cluster is defined as a group of physically related stars, whose stellar mass density exceeds the critical value for stability against tidal destruction by passing interstellar clouds, namely $1 M_\odot$ pc^{-3}. Whether embedded clusters will appear as bound or unbound objects after emergence from the parent molecular cloud depends on the magnitude of the internal motions of stars. While the criterion on the stellar density is not too difficult to estimate observationally, the dynamical state of an embedded cluster remains a mistery. The only indirect measure relies on the estimate of the star formation efficiency. Examples of known embedded clusters, in order of increasing distance from the Sun, include Ori-L1630 (d=480 pc, Lada E. *et al.* 1991), Ori-L1641 (d=480 pc, Strom *et al.* 1989a), the Trapezium cluster (d=480 pc, McCaughrean *et al.* 1993), S 106 (d=600 pc, Rayner 1993), NGC 2264 (d=800 pc, Lada C., Young & Greene 1993), LkHα 101 (d=800 pc; Barsony, Schombert & Kis-Halas 1991), the Vela GMC (d=900 pc, Liseau *et al.* 1993), NGC 6334 (d=1.7 kpc, Straw & Hyland 1989), and M17 (d=2.2 kpc, Lada C. *et al.* 1991). Observations and data analysis follow a standard procedure that is useful to briefly summarize.

Images of a star forming complex are taken at the standard infrared colors, J (1.25 μm), H (1.65 μm), and K (2.2 μm). A control field is also imaged in order to estimate and remove any stellar background contamination of the main survey region. Mosaic images in each of the IR colors are then constructed to determine the size and the extent of the observed cluster. Source extraction and photometry is performed by using standard image reduction processing routines. The distribution of the extracted sources provides information on the stellar density in the region, and the location of the embedded cluster. Stellar densities can be moderately high, 500-900 stars pc^{-3}, high, 4000 stars pc^{-3} for the densest cluster in L1630 (NGC 2024), or exceptionally high, exceeding 10^4 stars pc^{-3} in the central 1 arcminute core of the Trapezium Cluster. For comparison, the central core of the Rho Ophiuchi dark cloud, one of the most active star forming nearby dark clouds, has a number density of only 200 pc^{-3}, or twice as small as that of NGC 2024 for the same size scale. Some evidence exists for radial mass segregation, with brighter sources being more centrally located than the fainter ones. In the case of the Trapezium cluster, the segregation in luminosity is also a mass segregation, with the OB stars sitting in the center.

Having defined the stellar population, it is then possible to construct the cluster and control field luminosity functions. Usually, this is done in the K-band, where the effects of extinction are minimal. Of course, the two LFs should be very different, since background objects are overwhelmingly main-sequence stars with little emission at K band. The subtraction of the stellar background contamination leaves the LF of the embedded sources, that is more conveniently plotted as a cumulative K magnitude distribution. This histogram gives the log of the cumulative number of sources brighter than a given K-magnitude versus that magnitude (normalized to the total number of stars sources with magnitude less than the limiting magnitude). An example is shown in Figure 17 for the four clusters in L1630. In most cases studied so far, the cumulative K-LF is described by a power-law which monotonically decreases with luminosity to the completeness limit of the survey

$$\frac{dN}{d(logL_K)} \propto L_K^{-\gamma} \, , \qquad (18)$$

where L_K is the K band luminosity derived from the observed fluxes. Estimates of the index γ vary between 0.26, for very distant clusters, and 0.38, for the clusters in Orion. What is the significance of this determination, and how it can be used in relation to the index α of the IMF? The answer depends on how deep the surveys are made.

If the observations are sensitive only to early type stars, as for distant clusters, then it can be shown that the slope of the K-LF can be transformed directly into an exponent of the MF, using the fact that the wavelength K lies in the Rayleigh-Jeans domain (see Lada C. 1991). For example, in the case of the cluster in M 17, the observed $\gamma = 0.26$ implies a value of $\alpha = 1.35$, in very good

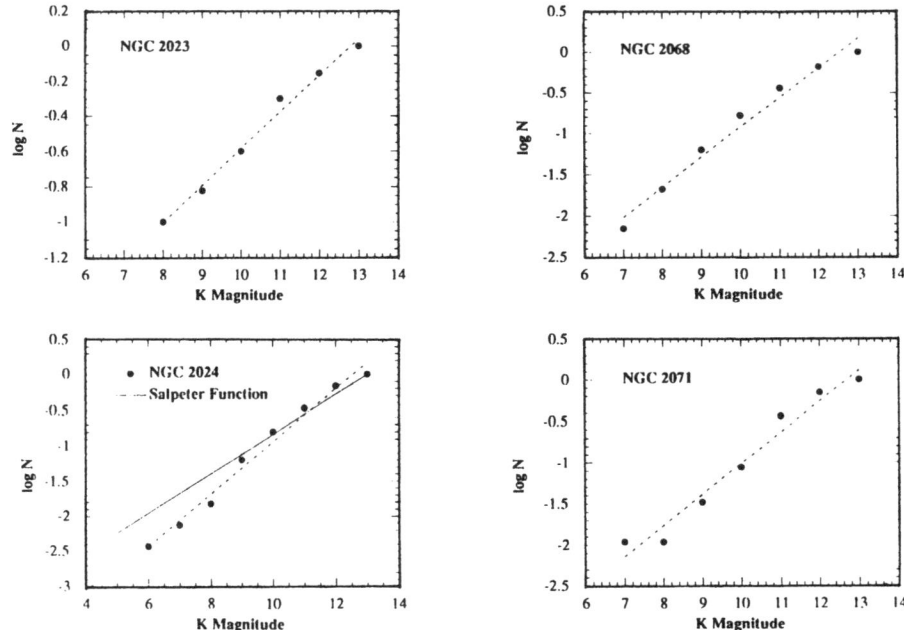

Figure 17: *Cumulative K magnitude distributions for the four embedded clusters in L1630. The dotted line gives the power-law index γ (from Lada E. et al. 1991).*

agreement with Salpeter's value. The stellar masses sampled in this cluster go from 8 to 30 $M_\odot$. As we have shown in the previous Section, these stars do *not* have a PMS phase, but are born directly on the ZAMS. Thus, the Rayleigh-Jeans approximation is appropriate. It is noteworthy that most cluster members show an unexpected excess emission at K, indicative of large amounts of circumstellar matter, and this might complicate the straightforward interpretation. In general, however, the high mass tail of the IMF in these clusters can be properly tested by the K-LF. Alternatively, if the survey is also sensitive to the lower luminosity stars, as for nearby clouds, the direct transformation cannot be applied, since most of the NIR emission comes from circumstellar matter and not from a photosphere. Evolutionary corrections must be made for an appropriate interpretation of the LF. The fact that the observed values of $\gamma \sim 0.4$ are larger than in the other case, can be attributed to the PMS nature of the embedded population (cf. Lada E. *et al.* 1991). Indeed, low-mass PMS stars have higher luminosity than main-sequence stars of the same mass, while higher mass stars have the same luminosity in both phases, so that the low-mass PMS population increases the proportion of high- to low-luminosity stars, making the LF steeper. Finally, the true PMS nature of the cluster members can also be ascertained by constructing NIR color-color diagrams.

Another important, and controversial, property of the K-LF is the presence

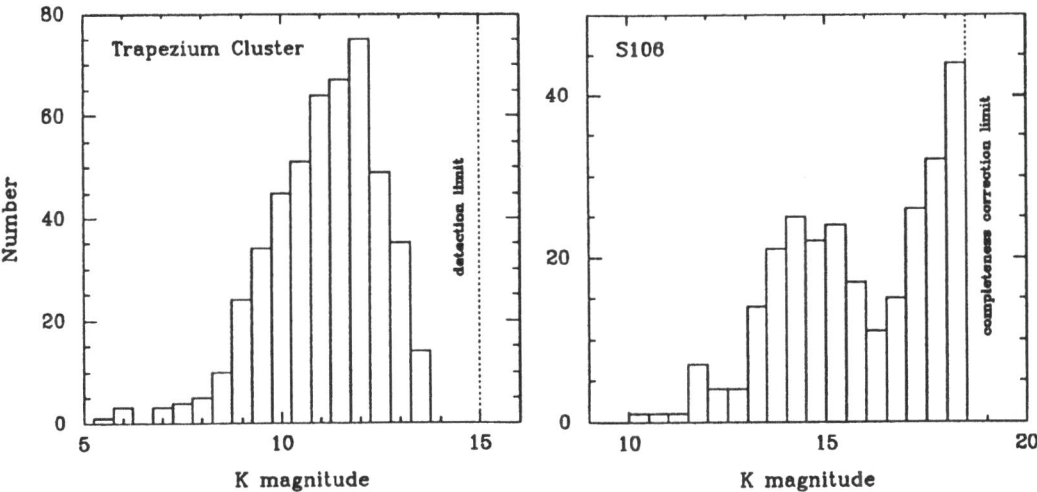

Figure 18: *Measured K luminosity functions for the Trapezium Cluster and the S 106 cluster. Both KLFs show a peak at $M_K \sim 14^m$. The KLFs have not been corrected for extinction (from McCaughrean & Zinnecker 1993).*

of peaks and turnovers, and their implication for the IMF. Figure 18 shows two examples of the K distribution function for the Trapezium Cluster and the S106 cluster, in which these features are particularly distinct. Note that often the observed LF flattens, or even sharply drops, near the completeness limit of the survey, reflecting an instrumental artifact and not an intrinsic property of the cluster population. However, in the case shown in the figure, the peaks appear well below the limiting magnitude, and correspond to $m_K = 12$ for the Trapezium and $m_K = 14 - 15$ for S106 (the rise at higher magnitudes is spurious and due to background contamination).

McCaughrean & Zinnecker (1993) have strongly cautioned against the over-simplistic interpretation of these features as evidence of actual features in the IMF. This procedure derives by the inappropriate application of power-law main-sequence mass-luminosity relations to a population of young stars. By using a simplified model of the evolution of a coeval cluster of PMS stars, they have shown that peaks and turnovers may artificially appear as a result of inflexions in the mass-K luminosity relation. McCaughrean & Zinnecker attribute this behavior to the slowing down effect of deuterium burning on the contraction of PMS stars, and the consequent departure from a pure power-law mass-K luminosity relation. This very aspect is reminescent of the same confusion that plagues the determination of the very low-mass tail of the field IMF, where spurious features in the mass-luminosity relation (due to opacity effects in that case) simulate a turn over in the IMF (Kroupa, Tout & Gilmore 1990). Despite the plausibility of the argument, we have seen before that deuterium burning initiates already during the accretion phase, and that its effect in the course of PMS evolution is greatly reduced. Better models that take into account the accretion phase, as well the PMS phase, are neccesary to clarify the importance of this phenomenon. The whole issue of the reality of the features in the K-LFs remains still open, but it is clear that single color LFs cannot constrain the properties of the stellar population of embedded clusters. It is necessary to derive bolometric LFs.

4.3   Bolometric Luminosity Functions

The necessary steps to obtain a reliable ILF, and possibly the IMF, are : *(i)* the construction of the spectral energy distribution of each cluster member in order to estimate the bolometric luminosity; *(ii)* the empirical classification of the SEDs according to their shapes; *(iii)* the transformation of the empirical scheme into an evolutionary sequence to determine the nature of each source; and *(iv)* the comparison of the observed LFs with time-dependent, theoretical LFs to estimate the age and mass. By analogy with the derivation for field stars, the IMF can then be computed via

$$\frac{dN}{dm_*} = \frac{dN}{dm_\lambda} \cdot \frac{dm_\lambda}{dM_{\rm bol}} \cdot \frac{dM_{\rm bol}}{dm_*} , \tag{19}$$

where the first term on the rhs represents the distribution function at a given wavelength; the second term is the bolometric correction; and the last one involves the time-dependent mass-luminosity relation. The determination of the individual SED yields the first two terms of eq. (19), and hence the bolometric luminosity. The estimate is not so simple, however, since the available spectral interval is usually limited at the shortest wavelengths ($\lambda \lesssim 1.25\mu$m) by extinction problems, and at the longest ($\lambda \gtrsim 60\mu$m) by source confusion. Extrapolations must then be made, that usually result in bolometric luminosities underestimated by, at most, a factor of two. Other important effects, such as cloud extinction, intracluster and foreground reddening, intrinsic variability of the sources, are often ignored.

To date, bolometric LFs have been presented for three regions only, the ρ Ophiuchi cluster (Wilking *et al.* 1989), Taururs-Auriga (Kenyon *et al.* 1990), and Chamaeleon I (Prusti, Whittet & Wesselius 1992). The resulting LFs are shown in Figure 19. Significant differences among the LFs exist, that probably reflect variations in the conditions for star formation in the individual clouds.

52

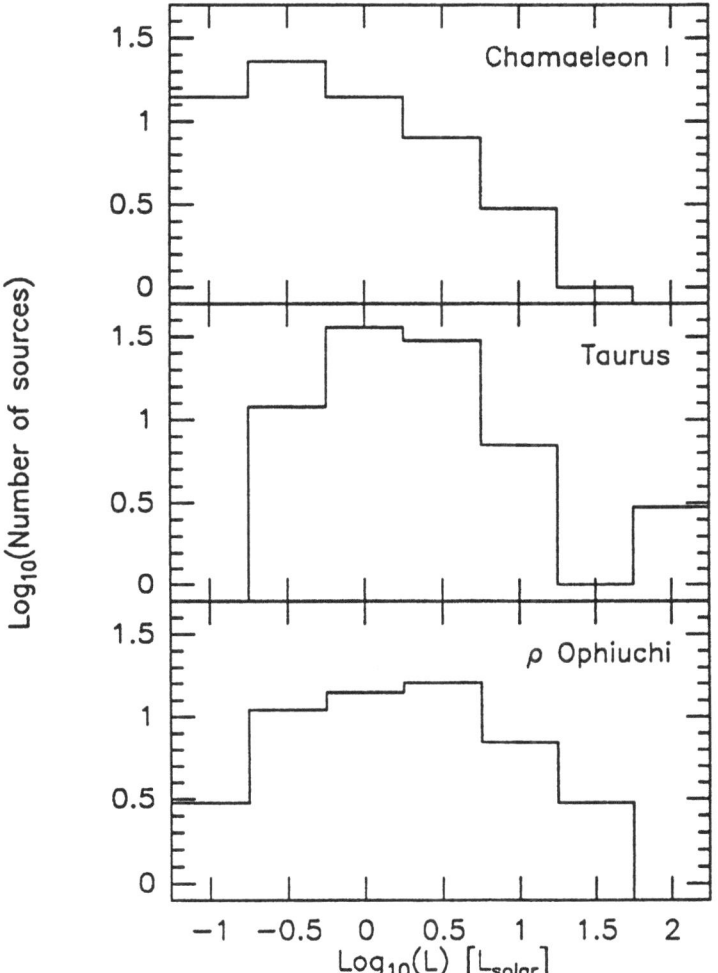

Figure 19: *Bolometric luminosity functions for three nearby star forming complexes (from Prusti et al. 1992).*

For example, there are about ten different star forming centers in Taurus-Auriga, scattered throughout the filamentary complex. In Chamaeleon I, star formation in concentrated in only three regions, while in Ophiucus most of the activity is limited to the core. Thus, the resulting LF is an average over inhomogeneous volumes. In addition, the slope of the LF, in the region where is most reliable, differs in the three cases. For all cases, however, the LFs are significantly steeper than for main-sequence stars. Similarly, the peaks of the LFs show a systematic shift towards higher luminosities, moving from Chamaeleon I to ρ Ophiuchi. This trend may reflect a difference in age. In Chamaeleon I, it is estimated that star formation started about 10^7 yr ago (Prusti *et al.* 1992); in Taurus, the rate of star formation has been approximately constant during the last several million years; while the ρ Oph cluster is thought to be the youngest one, less than one million years old.

A turnover in the low luminosity tail is also visible in all LFs. There has been a strong debate on the reality of this feature in the case of the central core of ρ Ophiuchi, because the first deep images at K band ($m_K^{lim} = 14.5$) by Rieke, Ashok & Boyle (1989) did not find other sources in addition to those seen at a lower sensitivity by Wilking *et al.* (1989), on which the LF shown in Fig. 19 is based. This finding prompted Rieke *et al.* to conclude that the turnover is indeed real, and argued that it would correspond to a dearth of stars with bolometric luminosity between $10^{-2} - 10^{-3}$ $L_\odot$. However, in a subsequent survey of comparable depth on a larger area by Barsony *et al.* (1989) more sources were discovered, in increasing number up to the detection limit. The issue has been definitively settled down very recently by the multi-color survey of Greene & Young (1992) on an area large enough to cover both previous surveys. The not too surprising conclusion is that the K-LF is significantly different in different regions of the survey area. Thus, previous studies had simply sampled two independent population of stars. Limited to the central core region, the new K-LF can be fitted with a power-law of index 0.43, not dissimilar from that found in more distant clouds.

A more significant difference between the LFs of these regions is the relative distribution of Class I and Class II sources. In ρ Ophiuchi, Class I objects dominate the intermediate luminosity range ($L_* \gtrsim 6$ $L_\odot$), while Class II sources are more numerous at low luminosities. The opposite is true both in Chamaeleon I and in Taurus-Auriga. The segregation of intermediate luminosity sources and the paucity of Class II sources cannot be accounted for by selection or confusion effects, but appear to be real. It is worth pointing out that such a deficit in the population of intermediate luminosity stars could result from the atypical behavior of PMS stars with mass between 2 and 4 $M_\odot$. As we have discussed in the previous Section, these stars begin the contraction phase as low-luminosity objetcs, and relax to the high-luminosity state that competes to their mass. The process is slow enough that is likely that, over a large population of young stars, a healthy fraction could be found in the low state.

This point underscores the importance of reliable theoretical LFs for the comparison with observed ones. Such models begin to being available. Particularly interesting is the approach followed by Fletcher & Stahler (1993), who have constructed time-dependent LFs incorporating both the protostellar and PMS phase. An important result of these simulations is the prediction of a very small population of protostars at any given time, even though the star formation process proceeds at a constant rate. This might explain the rather puzzling, and disappointing, lack of detection of embedded Class I sources, the putative protostars, in young clusters with luminosities corresponding to the expected accretion luminosity (e.g. Kenyon *et al.* 1990). These theoretical models are still preliminary and improvements are required before detailed comparison can be made. At the same time, an increase in the number of reliable bolometric LFs of other nearby star forming regions is obviulsy needed.

4.4   Conclusions

The study of embedded star clusters has become an integral part of the studies of star formation. Despite most of the fundamental questions about the origin of stellar masses and the IMF remain still largely unanswered, as it is clear from the partial conclusions given above, it should not be underestimated that before the advent of NIR arrays it was impossible even to approach these problems observationally. The first studies have shown unambiguously that in giant molecular clouds the most massive stars do not appear alone, but are surrounded by a cluster of low-mass stars. The idea of a bimodal distribution of stars has lost ground in face of this evidence. Whether star formation occurs preferentially in clusters or distributed throughout the cloud is another aspect that has been possible to tackle only with the use of infrared arrays. Unbiased surveys of molecular clouds have shown a healthy tendency for clustering (L1630 in Orion), but evidence of the opposite is also strong (L1641, also in Orion). Most of the surveys conducted so far have been done using arrays with 256×256 pixels. By the time this contribution is in the press, it is likely that the first arrays with 1024×1024 pixels will be already in operation. It is then expected that our knowledge of the embedded population will be expanded to include all galactic giant molecular clouds. A great future for IR astronomy is ahead of us.

5 REFERENCES

Adams, F.C., Lada, C.J., Shu, F.H.: 1987, *Astrophys. J.* **213**, 788.

Adams, F.C., Lada, C.J., Shu, F.H.: 1988, *Astrophys. J.* **326**, 865.

Adams, F.C., Ruden, S.P., Shu, F.H.: 1989, *Astrophys. J.* **347**, 959.

Adams, F.C., Lin, D.: 1993, in *Protostars and Planets III*, Eds. Levy, E.H., Lunine, J., and Matthews, M.S. University of Arizona Press. Tucson.

Alexander, D.R., Augason, G.C., Johnson, H.R.: 1989, *Astrophys. J.* **345**, 1014.

André, Ph., Ward-Thompson, D., Barsony, M.: 1993, *Astrophys. J.*, in press.

Appenzeller, I., Jankovics, I., Öestreicher, R.: 1984, *Astron. Astrophys.* **141**, 108.

Balachandran, S.: 1992, in IAU Coll. 137 *Inside the Stars*, Eds. Baglin, A., Weiss, W.W. PASP Conference Series, in press.

Barsony, M., Burton M.G., Russell, A.P.G., Carlstrom, J.E., Garden R.: 1989, *Astrophys. J.* **346**, L93.

Barsony, M., Schombert, J.M., Kis-Halas, K.: 1991, *Astrophys. J.* **379**, 221.

Basri, G., Bertout, C.: 1989, *Astrophys. J.* **341**, 340.

Bastien, P., Ménard, F.: 1990, *Astrophys. J.* **364**, 262.

Beckwith, S.V.W., Sargent, A.I., Chini, R.S., Güsten, R.: 1990, *Astron. J.* **99**, 924.

Beckwith, S.V.W., Sargent, A.I.: 1993, *Astrophys. J.* **402**, 280.

Beichman, C.A. et al.: 1984, *Astrophys. J.* **278**, L45.

Beichman, C.A., Myers, P.C., Emerson, J.P., Harris, S., Mathieu, R., Benson, P.J., Jennings, R.E.: 1986, *Astrophys. J.* **307**, 337.

Benson, P.J., Myers, P.C.: 1989, *Astrophys. J. Suppl. Ser.* **79**, 89.

Berrilli, F., Ceccarelli, C., Lorenzetti, C., Nisini, B., Saraceno, P., Strafella, F.: 1990, *Il Nuovo Cimento* **13**, 293.

Bertoldi, F., McKee, C.F.: 1990, *Astrophys. J.* **354**, 529.

Bertout, C., Regev, O.: 1992, *Astrophys. J.* **399**, L163.

Blaauw, A.W.: 1964, *Ann. Rev. Astron. Astrophys.* **2**, 213.

Blaauw, A.W.: 1991, in *The Physics of Star Formation and Early Stellar Evolution*, p. 125. Eds. Lada, C.J. and Kylafis, N.D. Kluwer Academic Publ. Dordrecht.

Bodenheimer, P.: 1966, *Astrophys. J.* **144**, 709.

Bodenheimer, P.: 1992, in *Star Formation in Stellar Systems*, p.3 Eds. Tenorio-Tagle, G., Prieto, M., and Sanchez, F. Cambridge University Press. Cambridge (UK).

Bok, B.J., Reilly, E.F.: 1947, *Astrophys. J.* **105**, 255.

Boss, A.P.: 1993, *Astrophys. J.*, in press.

Butner, H.M., Evans, N.J.,II, Lester, D.F., Levreault, R.M., Strom, S.E.: 1991, *Astrophys. J.* **376**, 636.

Butner, H.M., Natta, A., Evans, N.J.,II: 1993, *Astrophys. J.*, in press.

Cabrit, S., Edwards, S., Strom, S.E., Strom, K.M.: 1990, *Astrophys. J.* **354**, 687.

Calvet, N., Hartmann, L.W., Kenyon, S.J.: 1993, *Astrophys. J.* **402**, 623.

Cameron, A.G.W.: 1962, *Icarus 1*, 13.

Cesaroni, R., Walmsley, C.M., Churchwell, E.M.: 1992, *Astron. Astrophys.* **256**, 618.

Churchwell, E.M., Cesaroni, R., Walmsley, C.M.: 1990, *Astron. Astrophys. Suppl.* **83**, 119.

D'Antona, F.: 1991, Ed. of *The Problem of Lithium*, Mem. Soc. Astron. It., Vol. 62.

D'Antona, F.: 1992, in IAU Coll. 137 *Inside the Stars*, Eds. Baglin, A., Weiss, W.W. PASP Conference Series, in press.

D'Antona, F., Matteucci, F.: 1991, *Astron. Astrophys.* **248**, 62.

D'Antona, F., Mazzitelli, I.: 1993, *Astrophys. J.*, in press.

Draine, B.T., Lee, H.M.: 1984, *Astrophys. J.* **285**, 89.

Edwards, S., Cabrit, S., Strom, S.E., Heyer, I., Strom, K.M., Anderson, E.: 1987, *Astrophys. J.* **321**, 473.

Elmegreen, B.G.: 1991, in *The Physics of Star Formation and Early Stellar Evolution*, p. 35. Eds. Lada, C.J. and Kylafis, N.D. Kluwer Academic Publ. Dordrecht.

Elmegreen, B.G.: 1992, in *Star Formation in Stellar Systems*, p.383. Eds. Tenorio-Tagle, G., Prieto, M., and Sanchez, F. Cambridge University Press. Cambridge (UK).

Emerson, J.P.: 1987, in *Star Forming Regions*, p. 19. Eds. Peimbert, M. and Jugaku, J. Reidel Publishing Co. Dordrecht.

Evans, N.J.,II: 1991, in *Frontiers of Stellar Evolution*, p.45. Ed. Lambert D.L. (San Francisco: ASP).

Ezer, D., Cameron, A.G.W.: 1965, *Canadian J. Phys.* **14**, 1497.

Fletcher, A.B., Stahler, S.W.: 1993, *Astrophys. J.*, in press.

Franco, J.: 1992, in *Star Formation in Stellar Systems*, p. 515. Eds. Tenorio-Tagle, G., Prieto, M., and Sanchez, F. Cambridge University Press. Cambridge (UK).

Galli, D., Shu, F.H.: 1993a,b *Astrophys. J.*, in press.

Gatley, I., Merrill, K.M., Fowler, A.M., Tamura, M.: 1991, in *Astrophysics with Infrared Arrays*, p. 230. Ed. Elston, R. Astron. Soc. of the Pacif. Conf. Series N. 14.

Geiss, J., Reeves, H.: 1981, *Astron. Astrophys.* **93**, 189.

Greene, T.P., Young, E.T.: 1992, *Astrophys. J.* **395**, 516.

Grossman, A.N., Graboske, H.C.: 1971, *Astrophys. J.* **164**, 475.

Harris, S., Clegg, P., Hughes, J.: 1988, *Mon. Not. Roy. Astron. Soc.* **235**, 441.

Hartmann, L.W., Stauffer, J.R.: 1989, *Astron. J.* **97**, 873.

Hartmann, L.W., Kenyon, S.J.: 1990, *Astrophys. J.* **349**, 190.

Hartmann, L.W., Kenyon, S.J., Calvet, N.: 1993, *Astrophys. J.*, in press.

Hayashi, C.: 1961, *Pub. Astron. Soc. Jap.* **13**, 450.

Heiles, C, Goodman, A., McKee, C.F., Zweibel, E.G.: 1993, in *Protostars and Planets III*, Eds. Levy, E.H., Lunine, J., and Matthews, M.S. University of Arizona Press. Tucson.

Henyey, L.G., LeLevier, R., Levee, R.D.: 1955, *Pub. Astron. Soc. Pac.* **67**, 154.

Hillenbrand, L.A., Strom, S.E., Vrba, F.J., Keene, J.: 1992, *Astrophys. J.* **397**, 613.

Iben, I.: 1965, *Astrophys. J.* **141**, 993.

Keene, J., Masson, C.: 1990, *Astrophys. J.* **355**, 635.

Kenyon, S.J., Hartmann, L.W.: 1987, *Astrophys. J.* **323**, 714.

Kenyon, S.J., Hartmann, L.W., Strom, K.M., Strom, S.E.: 1990, *Astron. J.* **99**, 869.

Klein, R.I., Whitaker, R.W., Sandford, II,M.T.: 1985, in *Protostars and Planets II*, p.340. Eds. Black, D.C. and Matthews, M.S. University of Arizona Press. Tucson.

Kroupa, P., Tout, C.A., Gilmore, G.: 1990, *Mon. Not. Roy. Astron. Soc.* **244**, 76.

Kurucz, R.L.: 1991, in *Stellar Atmospheres*, Eds. Crivellari, L., Dubeny, I., Hummer, D.G. Kluwer Acad. Publ. (Dordrecht).

Lada, C.J.: 1985, *Ann. Rev. Astr. Ap.* **23**, 267.

Lada, C.J.: 1987, in *Star Forming Regions*, p. 1. Eds. Peimbert, M. and Jugaku, J. Reidel Publishing Co. Dordrecht.

Lada, C.J.: 1991, in *The Physics of Star Formation and Early Stellar Evolution*, p. 329. Eds. Lada, C.J. and Kylafis, N.D. Kluwer Academic Publ. Dordrecht.

Lada, C.J., DePoy, D.L., Merrill, K.M., Gatley, I.: 1991, *Astrophys. J.* **374**, 533.

Lada, C.J., Adams, F.C.: 1992, *Astrophys. J.* **393**, 278.

Lada, C.J., Young, E.T., Greene, T.P.: 1993, *Astrophys. J.*, in press.

Lada, E.A.: 1992, *Astrophys. J.* **393**, L25.

Lada, E.A., DePoy, D.L., Evans, N.J.,II, Gatley, I.: 1991, *Astrophys. J.* **371**, 171.

Lada, E.A., Strom, K.M., Myers, P.C.: 1993, in *Protostars and Planets III*, Eds. Levy, E.H., Lunine, J., and Matthews, M.S. University of Arizona Press. Tucson.

Ladd, E.F., Adams, F.C., Casey, S., Davidson, J.A., Fuller, G.A., Harper, D.A., Myers, P.C., Padman, R.: 1991, *Astrophys. J.* **366**, 203.

Larson, R.B.: 1969, *Mon. Not. Roy. Astron. Soc.* **145**, 405.

Larson, R.B.: 1972, *Mon. Not. Roy. Astron. Soc.* **157**, 121.

Larson, R.B.: 1982, *Mon. Not. Roy. Astron. Soc.* **200**, 159.

Larson, R.B.: 1992, *Mon. Not. Roy. Astron. Soc.* **256**, 641.

Leinert, Ch., Haas, M.: 1989, *Astrophys. J.* **342**, L39.

Levreault, R.M.: 1988, *Astrophys. J.* **330**, 910.

Liseau, R., Lorenzetti, D., Nisini, B., Spinoglio, L., Moneti, A.: 1993, *Astron. Astrophys.*, in press.

Lizano, S., Shu, F.H.: 1989, *Astrophys. J.* **342**, 834.

Lynden-Bell, D., Pringle, J.E.: 1974, *Mon. Not. Roy. Soc. Astron.* **168**, 603.

Magazzú, A., Rebolo, R., Pavlenko, Ya.V.: 1992, *Astrophys. J.* **392**, 159.

Mathis, J.S., Mezger, P.G., Panagia, N.: 1983, *Astron. Astrophys.* **128**, 212.

Mazzitelli, I.: 1989, in *Low Mass Star Formation and Pre–Main-Sequence Objects*, p.433. Ed. P. Reipurth ESO Conf. Proc. N. 33.

Mazzitelli, I., Moretti, M.: 1980, *Astrophys. J.* **235**, 955.

McCaughrean, M.J., Zinnecker, H.: 1993, *Astrophys. J.*, in press.

McCaughrean, M.J., Zinnecker, H., Aspin, C., McLean, I.S.: 1993, in preparation.

McCullough, P.R.: 1992, *Astrophys. J.* **390**, 213.

Menten, K.M., Serabyn, E., Güsten, R., Wilson, T.L.: 1987, *Astron. Astrophys.* **177**, L57.

Mercer-Smith, J.A., Cameron, A.G.W., Epstein, R.I.: 1984, *Astrophys. J.* **279**, 363.

Mestel, L., Spitzer, L.: 1956, *Mon. Not. Roy. Astron. Soc.* **116**, 505.

Mezger, P.G., Sievers, A.W., Zylka, R.: 1991, in *Fragmentation of Molecular Clouds and Star Formation*, p.245. Eds. Falgarone, E. et al. Reidel Publishing Co. Dordrecht.

Mitskevich, A.S.: 1993, *Astrophys. J.*, in press.

Mouschovias, T.Ch.: 1976, *Astrophys. J.* **206**, 753.

Myers, P.C., Benson, P.J.: 1983, *Astrophys. J.* **266**, 309.

Myers, P.C., Fuller, G.A.: 1992, *Astrophys. J.* **396**, 631.

Myers, P.C., Ladd, E.F., Fuller, G.A.: 1991, *Astrophys. J.* **372**, L95.

Myers, P.C., Fuller, G.A., Mathieu, R.D., Beichman, C.A., Benson, P.J., Schild, R.E., Emerson, J.P.: 1987, *Astrophys. J.* **319**, 340.

Natta, A.: 1993, *Astrophys. J.*, in press.

Natta, A., Palla, F., Butner, H.M., Evans, N.J.,II.: 1992, *Astrophys. J.* **391**, 805.

Natta, A., Palla, F., Butner, H.M., Evans, N.J.,II.: 1993, *Astrophys. J.*, in press.

Ohashi, N., Kawabe, R., Hayashi, M., Ishiguro, M.: 1991, *Astron. J.* **102**, 2054.

Palla, F., Stahler, S.W.: 1990, *Astrophys. J.* **360**, L47.

Palla, F., Stahler, S.W.: 1991, *Astrophys. J.* **375**, 288.

Palla, F., Stahler, S.W.: 1992, *Astrophys. J.* **392**, 667.

Palla, F., Stahler, S.W.: 1993, *Astrophys. J.*, in press.

Palla, F., Brand, J., Cesaroni, R., Comoretto, G., Felli, M.: 1991, *Astron.*

Astrophys. **246**, 249.

Parigi, G.: 1992, Thesis, Univ. of Florence.

Parker, N.D., Padman, R., Scott, P.F.: 1991, *Mon. Not. Roy. Astron. Soc.* **252**, 442.

Pinsonneault, M.H., Deliyannis, C.P., Demarque, P.: 1992, *Astrophys. J. Suppl. Ser.* **78**, 179.

Prusti, T., Adorf, H.-M., Meurs, E.A.: 1991, *Astron. Astrophys.* **254**, 421.

Prusti, T., Whittet, D.C.B., Wesselius, P.R.: 1992, *Mon. Not. Roy. Soc. Astron.* **254**, 361.

Rayner, J.T.: 1993, in preparation.

Rieke, G.H., Ashok, N.M., Boyle, R.P.: 1989, *Astrophys. J.* **339**, L71.

Rogers, F.J., Iglesias, C.A.: 1992, *Astrophys. J. Suppl. Ser.* **79**, 507.

Rucinski, S.M.: 1985, *Astron. J.* **90**, 2321.

Rudolph, A., Welch, W.J., Palmer, P., Dubrulle, H.: 1990, *Astrophys. J.* **363**, 528.

Rydgren, A.E., Zak, D.S.: 1987, *Pub. Astron. Soc. Pacific* **99**, 141.

Salpeter, E.E.: 1954, *Mem. Soc. Roy. Liège* **14**, 116.

Salpeter, E.E.: 1955, *Astrophys. J.* **121**, 161.

Scalo, J.M.: 1986, *Fundam. Cosmic Phys.* **11**, 1.

Shakura, N.J., Sunyaev, R.A.: 1973, *Astron. Astrophys.* **24**, 337.

Shu, F.H.: 1977, *Astrophys. J.* **214**, 488.

Shu, F.H., Adams, F.C., Lizano, S.: 1987, *Ann. Rev. Astr. Ap.* **25**, 23.

Shu, F.H., Najita, J., Galli, D., Ostriker, E., Lizano, S.: 1993, in *Protostars and Planets III*, Eds. Levy, E.H., Lunine, J., and Matthews, M.S. University of Arizona Press. Tucson.

Spitzer, L.: 1948, *Harvard Obs. Monogr.* **7**, 87.

Stahler, S.W.: 1983, *Astrophys. J.* **274**, 822.

Stahler, S.W.: 1988a, *Astrophys. J.* **332**, 804.

Stahler, S.W.: 1988b, *Pub. Astron. Soc. Pac.* **100**, 1474.

Stahler, S.W., Shu, F.H., Taam, R.E.: 1980, *Astrophys. J.* **302**, 590.

Straw, S.M., Hyland, A.R.: 1989, *Astrophys. J.* **340**, 318.

Strom, K.M., Margulis, M., Strom, S.E.: 1989a, *Astrophys. J.* **345**, L79.

Strom, K.M., Strom, S.E., Edwards, S., Cabrit, S., Skrutskie, M.F.: 1989b, *Astron. J.* **97**, 1451.

Sugitani, K., Fukui, Y., Ogura, K.: 1991, *Astrophys. J. Suppl. Ser.* **77**, 59.

Tatematsu, K., Umemoto, T., Kameya, O. et al.: 1993, *Astrophys. J.*, in press.

Tauber, J.A., Lis, D.C., Goldsmith, P.F.: 1993, *Astrophys. J.*, in press.

Tenorio-Tagle, G., Bodenheimer, P.: 1988, *Ann. Rev. Astron. Astrophys.* **26**, 145.

Terebey, S., Shu, F.H., Cassen, P.: 1984, *Astrophys. J.* **286**, 529.

Tscharnuter, W.M.: 1991, in *The Physics of Star Formation and Early Stellar Evolution*, p. 411. Eds. Lada, C.J. and Kylafis, N.D. Kluwer Academic

Publ. Dordrecht.

Umemoto, T., Tatematsu, K., Kameya, D. et al.: 1993, *Astrophys. J.*, in press.

Walker, C.K., Lada, C.J., Young, E.T., Maloney, P.R., Wilking, B.A.: 1986, *Astrophys. J.* **309**, L47.

Walker, C.K., Carlstrom, J.E., Bieging, J.H.: 1993, *Astrophys. J.* **402**, 655.

Weintraub, D.: 1990, *Astrophys. J. Suppl. Ser.* **74**, 575.

Weintraub, D.A., Kastner, J.H., Zuckerman, B., Gatley, I.: 1992, *Astrophys. J.* **391**, 784.

Wilking, B.A.: 1989, *Pub. Astron. Soc. Pacif.* **101**, 229.

Wilking, B.A., Lada, C.J., Young, E.T.: 1989, *Astrophys. J.* **340**, 823.

Wilner, D.J., Lada, C.J.: 1991, *Astron. J.* **102**, 1050.

Wood, D.O.S., Churchwell, E.M.: 1989, *Astrophys. J.* **340**, 265.

Wouterloot, J.G.A., Walmsley, C.M.: 1986, *Astron. Astrophys.* **168**, 237.

Wynn-Williams, C.G.: 1982, *Ann. Rev. Astr. Astrophys.* **20**, 587.

Yorke, H.W., Shustov, B.M.: 1981, *Astron. Astrophys.* **98**, 125.

Yun, J.L., Clemens, D.: 1990, *Astrophys. J.* **365**, L73.

Yun, J.L., Clemens, D.: 1992, *Astrophys. J.* **385**, L21.

Zhou, S.: 1992, *Astrophys. J.* **394**, 204.

Zhou, S., Evans, N.J.,II, Kömpe, C., Walmsley, C.M.: 1993, *Astrophys. J.*, in press.

Zinnecker, H., Bastien, P., Arcoragi, J.P., Yorke, H.W.: 1992, *Astron. Astrophys.* **265**, 726.

Zinnecker, H., Wilking, B.A.: 1992, in *Binaries as Tracers of Stellar Formation*, p.135 Eds. Duquennoy, A. and Mayor, M. Cambridge University Press. Cambridge (UK).

Zinnecker, H., McCaughrean, M.J., Wilking, B.A.: 1993, in *Protostars and Planets III*, Eds. Levy, E.H., Lunine, J., and Matthews, M.S. University of Arizona Press. Tucson.

Zuckerman, B., Evans, N.J.,II: 1974, *Astrophys. J.* **192**, L149.

Last Stages of Stellar Evolution

Prof. Dr. S.R. Pottasch

Kapteyn Laboratory
P.O. Box 800
9700 AV Groningen, The Netherlands

1   THEORETICAL EVOLUTION

In order to discuss the observations in a meaningful way, it is useful to first discuss the theory, because only then does one have a meaningful context in which to place the observations. One must note that this does not imply that the theory is well understood, because this is not necessarily the case. But this approach gives some insight as to what the key observations are, and by comparing the observations with the theory one can find the weakness in the theory which can be more thoroughly studied.

1.1   Review of the early evolution

The main sequence star transforms hydrogen into helium in the central regions of the star as its source of energy. This stage is usually referred to as hydrogen burning. As the hydrogen burns, the core gradually contracts and heats up, which produces an increase in the rate at which the hydrogen is burned. The increased burning rate closely offsets the diminished fuel supply, and the luminosity of the star does not change substantially. For the first 90% of its lifetime the star remains close to the main sequence, and may double its luminosity in this time. This is shown schematically in the H-R diagram, Fig. 1, as the motion from points A to B. The initial position on the main sequence is determined only by the mass of the star, as long as it consists mainly of hydrogen. Evolutionary tracks are shown for stars of 1.1 $M_\odot$ and 5 $M_\odot$. The former represent the 'low' mass stars and the latter represent 'intermediate' mass stars which will be presently more completely defined.

The most important stage in the evolution of a star away from the main sequence occurs after the central hydrogen is completely burned. The central regions then stop releasing nuclear energy and begin to contract, releasing gravitational energy slowly at first. As this happens, the region in which hydrogen is being burned gradually moves outward and a so-called hydrogen burning shell is produced. Since the only energy release in the central regions of the star now is gravitational potential energy, the luminosity in this region is very low. Therefore only a very small temperature gradient is needed to carry the energy outward and the star has an almost isothermal helium core (plus a small amount of heavy elements). The star now approaches point C in the H-R diagram. At point C the temperature gradient steepen in the stars so that over a large region sufficient energy can only be transported outward by mass motion or convection. The convection now extends from the outer envelope of the star to the regions where products of nuclear burning exist. This 'first dredge-up' will be discussed

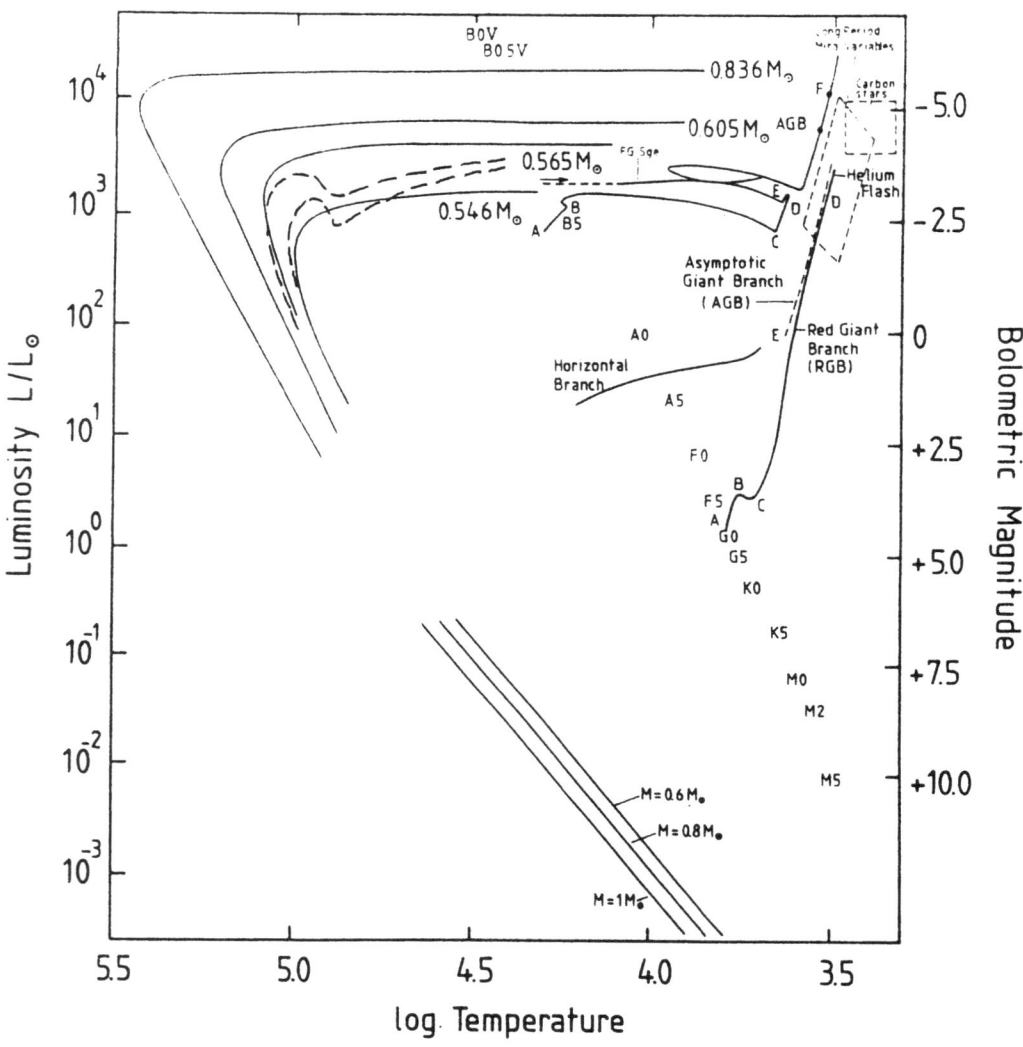

Figure 1: *The theoretical evolution of stars of 1.1 $M_\odot$ and 5 $M_\odot$ in the H-R diagram. The letters refer to stages described in the text. The spectral types on the zero age main sequence are shown. The results of Schönberner and co-workers for the evolutionary tracks of hydrogen and helium shell burning stars of 4 different core masses are shown as solid lines. The dashed lines are the evolutionary tracks of Vassiliades and Wood for stars of core mass 0.5478 $M_\odot$ and 0.554 $M_\odot$. The final theoretical tracks are of a completely degenerate star of 0.6, 0.8 and 1.0 $M_\odot$ as it cools. The observed evolution of FG Sge is also indicated.*

in more detail presently.

As the hydrogen burning shell moves outward, a dramatic change occurs when the isothermal core contains 10% to 13% of the mass of the star. The pressure gradient in the slowly contracting isothermal core is unable to support the outer regions of the star and the central regions now collapse rapidly, with a time scale which depends on the stellar mass.

At this point, it is useful to make the distinction between 'low' and 'intermediate' mass stars. For 'low' mass stars, i.e. stars less massive than $M \sim 2.25\ M_\odot$ the density in the core becomes so high that the electron gas no longer obeys the perfect gas law. The Pauli exclusion principle, which states that no more than two electrons with oppositely directed spin can be in a single energy state, now determines the properties of the electron gas which is said to be degenerate. In particular, the momentum of a particle is essentially determined by the Pauli exclusion principle rather than by the temperature of the gas. This means that the pressure and internal energy of the gas become essentially independent of the temperature. This temperature independence of the pressure has the following consequence. The contracting core increases in temperature due to the release of gravitational energy. It finally becomes so hot and dense that the helium in the core is ignited (helium is converted into carbon). This burning produces energy which further heats the core causing an increased energy production and a still higher temperature. But the matter does not expand because the pressure does not increase with increasing temperature in a degenerate gas. Since the matter does not expand (and cool), there is no reduction in the rate of energy release. The burning therefore continues in a runaway release of nuclear energy.

The process continues until the temperature becomes so high that the material becomes non-degenerate. At this stage the material will expand, but, because of the runaway release of energy, the expansion is likely to be explosive. This thermal runaway is called the 'helium core flash', which occurs only in 'low' mass stars, and occurs at point D in the figure.

The increase in luminosity between point C and point D reflects the increase of the hydrogen shell burning strength, necessary to balance the increasing gravity due to the increasing compactness of the degenerate core, which reaches central densities up to 10^6 gm cm^{-3}. The contraction of the core is accompanied by a dramatic atmospheric expansion which continues until the 'helium flash' begins. The track of 'low' mass stars from point C to D is known as the first giant branch (or red giant branch, RGB).

After the ignition of the helium core in 'low' mass stars, a rapid internal readjustment occurs, including contraction of the envelope. A temporary stability is achieved. The star is now on the 'horizontal branch', near point E on the diagram. The position of the star on the horizontal branch depends on its mass and chemical composition; the smaller the proportion of heavy elements (mostly carbon and oxygen), and the lower its mass, the farther it is to the left. In addition to the helium burning core, the star has a hydrogen burning shell. 'Intermediate' mass stars are then those stars more massive than 2.25 $M_{\odot}$, in which the electrons in the core do not become degenerate before helium is ignited. Such stars begin to burn helium shortly after reaching the red giant branch. At this stage the luminosity is a few times larger than the luminosity of the hydrogen burning phase on the main sequence. During most of the subsequent helium burning phase the star remains at roughly the same luminosity.

After a certain period, helium is exhausted in a sizeable fraction of the stellar core (E on the diagram). For 'low' mass the second ascent of the red giant branch then begins. The predominantly carbon-oxygen core, containing highly degenerate electrons, contracts and heats rapidly because no energy source is present. To keep pace with the contraction, the envelope expands to giant size as described above when hydrogen was first exhausted in the core. The sequence of giant stars formed in the first ascent is usually called the 'first red giant branch (RGB) stars', while the sequence now being formed with an inert carbon-oxygen core is referred to as the 'asymptotic giant branch (AGB'. The name applied to the "AGB stars" comes from the fact that the red giant branch in globular clusters (in which the most massive main sequence stars are less massive than the sun) is actually composed of two bands of stars that approach each other on the H-R diagram at high luminosity. The redder of the two bands contains the RGB stars with an electron-degenerate helium core, and the bluer consists of AGB stars. For 'intermediate' mass stars the name AGB has no morphological significance and is simply used to designate stars with a degenerate carbon-oxygen core.

Initially after the C-O core contraction and its rapid heating, the helium-burning shell narrows in mass and increases in luminosity. At the same time hydrogen burning in effect ceases temporarily. It is at this stage that for stars more massive than about 4 $M_{\odot}$ (point F on the diagram), a convection zone is formed which extends from the outer envelope almost to the helium burning shell. Matter that has undergone hydrogen burning can be brought to the surface by this convection. This important point will be discussed in the following sections.

1.2   Thermally pulsating AGB stars

The physical situation is now called the double-shell burning phase and is characterized by certain properties which are common to 'low' and 'intermediate' mass stars. The degenerate carbon-oxygen core is surrounded by a helium-burning shell which in turn is surrounded by a hydrogen-burning shell. The (non-degenerate) helium-burning shell is thermally unstable: a small increase in temperature results in an excess of energy release which cannot immediately be removed. This leads again to a thermonuclear runaway, reminiscent of the helium core flash, and is called the helium shell flash and is the beginning of a 'thermal pulse'. During a flash the helium luminosity increases by several orders of magnitude and the hydrogen burning is practically extinguished. Ultimately expansion and cooling turn off this thermonuclear runaway, but not before an important number of nuclear reactions take place. Not all of them concern us at present, but some consequences will be discussed in the next two sections.

After the thermonuclear runaway ends, the helium burning shell continues burning in a quiescent way for about 10 to 15% of the time between pulses. Expansion and cooling then extinguish the helium burning shell and hydrogen burning is renewed. It is the source of almost all the energy that escapes from the surface for the remainder of the pulse. When the mass of the helium-rich region left behind the advancing hydrogen burning shell reaches a critical value, the helium burning shell is again ignited and the whole process repeats itself and constitutes a second pulse. At the same time a convective region is established which extends from the helium-rich region all the way to the hydrogen-helium discontinuity. The interval between pulses depends on the mass of the C-O core and is about 6000 years for a core mass (M_c) of 0.88 $M_\odot$, 40,000 years for M_c = 0.7 $M_\odot$ and 100,000 years for M_c = 0.55 $M_\odot$. The time interval during which the surface luminosity exceeds the quiet pre-flash luminosity is of the order of 500 years and thus so small compared to the cycle time that the luminosity peak is generally unobservable .

Theoretically the AGB evolution could continue until the degenerate core reaches central densities sufficiently high for carbon ignition, which would lead to a supernova event. Empirically it is seen that this happens only for initial masses above 8 to 10 $M_\odot$. Apparently some other process, presumably loss of the outer envelope, terminates AGB evolution below these masses.

1.3   Uncertainty in early evolution

The rather complicated stellar evolution calculations require that simplified assumptions are made for physical processes which are only partially understood. It is sometimes necessary to ignore effects which are not of proven importance. The resulting model is then compared with observations of a star or sequence of stars and the model calculations are refined until there is agreement between calculation and observation. The good agreement between the stellar evolution models of stars with hydrogen burning cores with the observed main sequence stars indicates that the physics used in the calculations in this stage is adequate and no important effects have been ignored.

Globular cluster observations of the red giant branch (RGB) with a main sequence turnoff of the order of 1 $M_{\odot}$, compare well with stellar evolution calculations near the helium flash in a qualitative way. Quantitatively, however, there are discrepancies. For example, the predicted slope of the red giant branch does not quite agree with the observations in several clusters, and its position on the H-R diagram appears to be shifted from prediction on the basis of the observed atmospheric abundances.

At later stages of stellar evolution, through the helium core burning and beyond, it becomes more difficult to compare observed stars with the theoretical models. For example, Mira variables and carbon stars are low mass stars somewhere near the top of the AGB, but theoretical attempts to predict their absolute luminosity have not been successful. Probably the evolutionary calculations do not include all relevant physical processes. The uncertainties connected with convection and mass loss are two of the most important problems.

(a) Convection theory. A really satisfactory treatment of convective energy transport in stars is not yet available. Current astrophysical models make several assumptions which simplify the physics. First, the convective layer is taken to be thin compared to the scale height. Second, viscous dissipation is considered to be negligible. Third, the energy transport can be modelled by blobs of fluid travelling a mixing length, conserving entropy, and remaining in pressure equilibrium with the surrounding medium. All of these assumptions are very uncertain. The 'thin layer' hypothesis is almost always incorrect. Convection plays an important role in the later stages of evolution, but the present treatment is likely to lead to significant errors in estimating the extent of convective overshooting at the boundary of convection zones, and in the structure of low temperature convective envelopes.

(b) Mass loss. It is becoming increasingly clear that mass loss is the single most important phenomenon influencing the late stages of stellar evolution. No complete theory of mass loss from AGB stars exists at present. In general terms mass loss is thought to be a dual process involving levitation of matter above the photosphere by large amplitude radial pulsations followed by the formation of grains on which radiation pressure acts to drive the circumstellar material away from the star. Only semi-quantitative descriptions are available for use in evolution calculations and different calculations make use of different approximations. The quantitative understanding is actually so uncertain that a comparison with observations is used to judge how good the mass loss assumptions are.

Before discussing the observations two other topics related to the theory will be discussed: (1) Enrichment of certain elements, and (2) the core mass and its relation to the luminosity.

1.4   Enrichment of helium, nitrogen and carbon

Hydrogen burning produces helium in the core of the star. The nitrogen and carbon abundances are also increased, if the hydrogen burning occurs via the CNO cycle. In this process carbon is used as a catalyst, successively capturing protons as follows:

$$
\begin{array}{ll}
{}^{12}C + {}^{1}H \rightarrow {}^{13}N + \gamma & 1.3 \times 10^{7} \text{ yrs} \\
{}^{13}N \rightarrow {}^{13}C + e^{+} + \nu & 7 \text{ minutes} \\
{}^{13}C + {}^{1}H \rightarrow {}^{14}N + \gamma & 2.7 \times 10^{6} \text{ yrs} \\
{}^{14}N + {}^{1}H \rightarrow {}^{15}O + \gamma & 3.2 \times 10^{8} \text{ yrs} \\
{}^{15}O \rightarrow {}^{15}N + e^{+} + \nu & 1.3 \text{ minutes} \\
{}^{15}N + {}^{1}H \rightarrow {}^{12}C + {}^{4}He & 1.1 \times 10^{5} \text{ yrs}
\end{array}
$$

where the time is given that required for the reaction to take place in conditions representative of the center of a star. The relative amounts of carbon, nitrogen and oxygen depend on the rates of the individual reactions, which are such that the abundance of nitrogen increases substantially. During the main sequence stage a star converts most of its initial ${}^{12}C$ into ${}^{14}N$ within the inner half of its mass.

As the star approaches the red giant branch for its first ascent (point C in Fig. 1), envelope convection extends inward to the boundary of the region of nearly total conversion of C to N. As it extends beyond the boundary, ${}^{12}C$ is convected inward and ${}^{14}N$ is convected outward. By the time convection has extended inward almost as far as the hydrogen burning shell, the surface abundance of ^{12}C has dropped by about 30%. At the same time the ^{14}N surface abundance has roughly doubled, assuming that the initial abundance is solar. This important change in surface composition is sometimes referred to as the 'first dredge-up', and it occurs in both low and intermediate mass stars.

In stars more massive than 4 $M_\odot$, a second mixing or dredge-up occurs on the AGB. As discussed in the previous section, a convective zone is formed when the helium burning shell increases in strength and the hydrogen burning temporarily stops (point F in Fig. 1). The convection extends from the atmosphere almost to the location of the outward advancing helium burning shell. Matter that has previously been completely processed by hydrogen burning via the full CNO cycle is mixed with relatively unprocessed envelope material. The result is an increase in surface ^{4}He and ^{14}N, and a reduction in surface ^{12}C and ^{16}O. The surface ^{14}N ultimately exceeds the ^{12}C abundance and the surface ^{16}O is reduced only slightly. The He may be increased by about 40%.

A final change in surface composition occurs in intermediate mass stars during the thermal pulses (helium shell flash). Not only ^{4}He and ^{14}N are affected in this 'third dredge-up'. Now ^{12}C, which is created in helium burning, is brought to the surface, as well as 's-process elements'. These are elements which have been formed by successive capture of neutrons, beginning with ^{56}Fe. The reactions producing the s-process elements have a much larger cross section than fusion reactions between charged particles, since there is no Coulomb barrier between the incoming neutron and the charged nucleus. However the rate of red-giant evolution is relatively slow, with an average interval between neutron captures for a given nucleus of many years, so that enough time is available for beta-decay processes to occur between neutron captures. The beta-decay creates a stable nucleus when an unstable nucleus was originally formed by neutron capture. Hence the name 's-process', implying slow. All heavier elements can be built by this process, up to lead. The elements particularly recognizable in red-giants are Ba, Zr, Y, La and Tc. The latter element, technetium, has no stable isotope, and yet it is often observed in red giants. Since the observed isotope has a half life of about 10^5 years, it must have been formed and brought to the surface within this time interval. The existence of Tc and abnormal abundances of other 's-processes' elements in the stellar atmosphere is direct evidence that the products of nuclear burning in the interior have been carried to the surface of the star.

1.5 Core mass-luminosity relation

The early evolutionary models of Paczynski already showed that when a carbon-oxygen core exists, and it is surrounded by an inner helium burning shell and an outer hydrogen burning shell, there is a relation between the total luminosity and the mass of the core. This was later substantiated by Wood and Zarro, and by Boothroyd and Sackmann. The theoretical basis for this relation is rather vague, although it has been given by Kippenhahn when only a hydrogen burning shell is present. The relation given by Boothroyd and Sackmann is given in the following table:

M_{core}	Luminosity
0.51 $M_\odot$	2000 $L_\odot$
0.535 $M_\odot$	4000 $L_\odot$
0.57 $M_\odot$	6000 $L_\odot$
0.615 $M_\odot$	8000 $L_\odot$
0.65 $M_\odot$	10000 $L_\odot$
0.68 $M_\odot$	12000 $L_\odot$

where the luminosity range covers most of the observed post AGB range. The relationship is quite uncertain at low luminosities (and core masses).

1.6 Post AGB evolution

It is now generally accepted (Schönberner; Wood and Vassiliadis) that strong mass loss is occurring in AGB stars. At the same time thermal pulses periodically take place (intervals are typically 5×10^4 to 10^5 years for low mass stars) as the helium shell ignites violently and burns up the helium produced by the hydrogen shell since the last phase of helium shell activity. At the same time there is a long term increase in the core mass (due to the burning) and consequently the luminosity. There is also a decrease in the envelope mass. In the last one or two pulses the mass loss increases, the envelope is to a large extent ejected and the core with a very thin envelope begins to become visible. The envelope ejected in this way is presumably a planetary nebula and the stellar core is its central star.

Because the mass loss processes are so poorly understood, the quantitative descriptions of this evolution are uncertain. In Fig. 1 tracks are shown as calculated by Schönberner and coworkers for stars of various core masses which leave the AGB burning hydrogen and continue burning hydrogen until it is exhausted

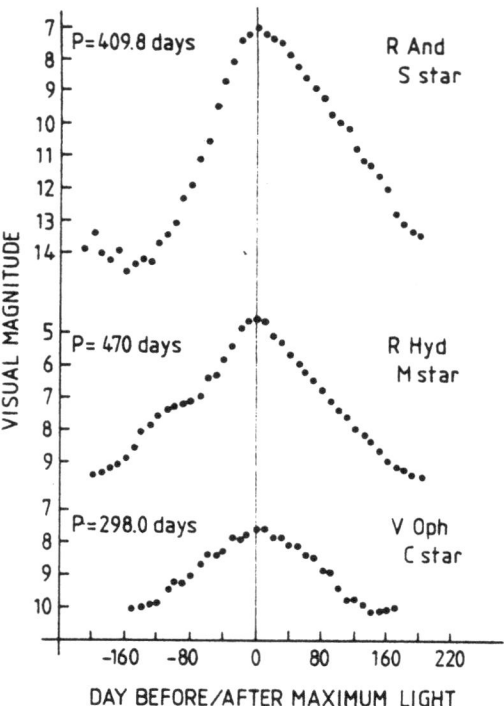

Figure 2: *The mean light curves of three typical Mira variables are shown. Some light curves are symmetric, while others have a steeper rise than decline. There are also those which have a shoulder in the curve, either on the rising or declining part. While the illustrated stars are representatives of the M, S and C spectra, the curves are not characteristic of the different spectra. For example, the curve shown for the S star R And is exactly the same shape as is seen in many M stars (e.g. o Ceti). The curves shown are mean curves: the different cycles do not repeat each other exactly. In fact sometimes large departures from the mean are observed.*

and cooling tracks begin. In the same figure tracks are shown as calculated by Vassiliadis and Wood for stars which leave the AGB burning helium (as a result of a helium shell flash) and which re-ignite the hydrogen burning shell at high luminosity and high stellar temperature. Note that the different authors use different mass loss rates so that a detailed comparison is difficult. Nevertheless, the global properties are probably correct. Vassiliadis and Wood expect that 'at least 40% (of the observed PN) are helium burners'.

2 OBSERVATIONS

The difficulty in summarizing the observations is that it is sometimes uncertain as to whether a given object, or class of objects, belongs to the late stages of stellar evolution. Comparison with expectation observations are an important way of learning. We begin with the best studied objects.

2.1 The cool giants

This term is used to describe the stars whose effective temperature is less than 4000 K, and often less than 3000 K, and whose luminosity is approximately 10^3 to $10^4 L_\odot$. These giants are roughly divided into three spectral types, each type covering the same range of temperature, but having different spectra. They are called M, S and C type. The M stars are characterized by strong absorption bands of TiO and VO in the red part of the spectrum. In the S stars the TiO (and VO) are replaced by absorption bands of ZrO, and to a lesser extent LaO and YO. It is striking that Zr, Y and La are s-process elements. In the C (for carbon) stars none of the metal oxide molecules are important, and the strong absorption in these stars is caused primarily by C_2, CN and CH molecules. Since the same temperatures (and pressures) occur in each of the three spectral sequences, the difference is almost certainly due to changes in abundance. In particular, the C star atmospheres must have an excess of carbon with respect to oxygen. On the other hand, the M star atmosphere must have more oxygen than carbon. The spectra of the S stars indicate that the heavy metal abundances are enhanced, especially the s-process elements; at the same time there must be a reduction in the amount of oxygen, so that the ratio O/C is only slightly higher than unity. There are a small number of stars with spectra intermediate between two of the groups.

The relative number of stars of each type is summarized in Table I. Close to the sun 96% of the red giants are M type, 3% are C type and a very small number are S type. In the galactic bulge, the region closer to the galactic center, the situation is more extreme: more than 99% are M type and only about 0.2% are C type. Yet in the Magellanic Clouds the proportions are entirely different: there are more C stars than M stars.

In the Large Magellanic Cloud the difference is not so pronounced, but in the Small Magellanic Cloud there are 20 times as many C stars as M stars. Since the Clouds and the galactic bulge do not have very large abundance differences, if follows that the ratio of C to M stars is quite sensitive to small differences in

Table I: *Relative numbers of M, S and C cool giant stars*

Type	Local	Galactic Center	Large Magellanic Cloud	Small Magellanic Cloud
M	96%	99%	40%	5%
S	1.3%	0.1%		
C	3%	0.2%	60%	95%

abundance.

All cool giants vary their light output with time. The lower temperature stars are more likely to show such instabilities. If variability is defined as showing luminosity changes greater than 0.4 magnitudes in the visual, then for the spectral classes M2 to M5, 9% of the stars are variable, for M6 20% and for M7 30% are variable. These objects are divided into 3 classes: the irregular, the semi-regular, and the Mira variables. The stars in the latter class, named after the prototype Mira whose variations have been studied for over 300 years, have a definite period of usually between 100 and 1000 days (sometimes higher) and a light amplitude range of at least 2.5 magnitudes between minimum and maximum in the visual. Typical light curves for three types of Mira variables are shown in Fig. 2. For all the light curves emission lines are clearly seen at some phase. The Miras will be discussed in more detail in the next sections. The semi-regular variables have a smaller amplitude light variation (less than 2.5 magnitudes) and do not usually show emission lines. This group of stars may be considered an extension of the Mira sequence on the short period side. The irregular variables do not have a fixed period of variation, nor do they have emission lines. Their light variation is the smallest of the three groups.

In the discussion at the beginning of this section of the relative numbers of M, S and C stars, no mention was made of variability. In fact the surveys on which Table I are based generally do not distinguish whether a star is variable or not. The different kinds of variables (or non-variables) are not distinguished by particular spectral types. For example, some Mira variables are M stars, others are C stars and S stars. In the solar neighbourhood, about 50% of the total number of M and S variables are Mira stars, while about 30% of the C variables are Miras.

2.2   The Mira Variables

For the cool giants it is sometimes difficult to establish which are on the red giant branch (RGB) and which are on the AGB (see Fig. 1). It is now generally accepted that the Mira variables are on the AGB. The reasons for this are: (1) their luminosity extends to higher values than the other giant stars; these high values are only predicted for AGB stars, (2) their atmospheres are unstable as can be seen from the strong light variations and the very high rates of mass loss observed.

2.2.1   Light curves, periods, absolute magnitudes and mass loss

The Miras exhibit a variety of light curves. Three examples are given in Fig. 2. Sometimes the light curves are symmetric, sometimes the rise time is substantially shorter than the decay time. Generally, it is not possible to distinguish M. S and C stars on the basis of the light curves. The light curves are periodic, but one cycle is not an exact replica of the preceeding one. In fact, rather large differences are not predictable. The light curves shown in the figure are average values over a large number of cycles. The periods were found to range from about 100 days to 500 days, based only on photographic and visual observations. Infrared observations which permit measurements of stars normally hidden in their own dust, have shown that longer periods exist, certainly up to 1600 days and perhaps longer. In the visible, the difference in the light emitted at maximum and minimum can be very large: 8 magnitudes (a factor of more than 1000) is not uncommon. In the infrared this difference is much smaller. Since most of the energy is emitted in the infrared, the total bolometric magnitude variation is also considerably smaller, usually not more than 1 mag (a factor of 2.5) for the longer periods and as low as 0.2 mag for the shortest periods.

The distance to galactic Mira stars is not well known in most cases. A trigonometric parallax is available for Mira itself establishing the absolute magnitude as $M_{BOL} = -3.8$. Further, when a Mira is in a globular cluster, its distance can be determined. Mira's in the galactic bulge and in the LMC have also been extensively observed recently. Using these distances, absolute brightness can be found. These are shown in Table II and are taken from a recent review by Whitelock.

There are certainly large individual variations and these values should only be considered as indicative. Approximate values of the effective temperature near maximum light are also given in the table. The temperature is somewhat

Table II: *Absolute bolometric magnitude and effective temperature for Mira variables*

Period (days)	125	200	300	400	500
M_{BOL}	−3.4	−4.0	−4.5	−4.9	−5.2
T_{eff}	3000	2800	2650	2500	2350

lower at minimum light. The effective temperatures show a pronounced decrease towards larger periods; this decrease certainly continues towards periods longer than shown in the table.

There is a reason that the longer period Miras are more easily observed in the infrared. While it is true that the longer period stars have low temperatures and are therefore less bright in the visual, this is not the most important effect. A clue to the problem is found in the fact that most Miras show excess radiation in the 2 μm to 20 μm wavelength range, compared to that normally expected from stars of this temperature. This excess is interpreted as radiation from a dust shell surrounding the Mira. The dust shell absorbs radiation from the star and re-radiates in the infrared. Direct evidence for this shell is provided by interferometric measurements of Mira itself. At 11.1 μm the emitting surface has a diameter of 0″.7, or at a distance of 77 pc, a radius of about 6000 $R_\odot$ and a temperature of about 500 K. This dust shell is probably formed by a constant outflow of matter from the star. The material either contains dust, or more likely, the dust condenses in the outflowing matter when the cooling has reduced the temperature below a critical temperature (probably 700 K to 1000 K). This dust shell may be a quasi-steady phenomenon whose radius is determined by the rate of mass flow and the properties of the dust particles.

In this case the flow of matter is such that a substantial amount of radiation is emitted in the infrared, but the shell remains optically thin in the visible wavelengths so that the star is easily observed. Apparently for the longer period Miras the rate of mass flow is so great that the dust component becomes thick enough to hide the star from view, at least at visible wavelengths. The stars can then only be observed in the infrared. The infrared radiation will be discussed in detail presently.

2.2.2 Spatial distribution, kinematics and local space density of Miras

Mira variables become much more numerous towards the galactic center. Because of the extinction it is difficult to quantify this statement, which is based on various observations. The strongest evidence for the concentration is the measurement of the surface density of Miras in three 'windows' near the galactic center. These 'windows' are regions of much lower extinction which makes it possible to see quite far through the galactic plane. The windows are located at $\ell = 0°9$, $b = -3°9$; $\ell = 1°4$, $b = -2°6$; and $\ell = 4°2$, $b = -5°1$. There is a rapid decline in the sky surface density of Miras with increasing angular distance from the galactic center in these windows. Extinction cannot play a role because the furthest window is the one with the lowest extinction. This rapid decline is similar to the decline in surface density of planetary nebulae form the galactic center. The total number of Miras in the bulge (within 2 kpc of the center) has been recently estimated by Whitelock as at least 2×10^4. Only about 100 of these have luminosities greater than 8000 $L_\odot$ and periods greater than 500 days. The distribution of Mira periods is approximately the same in these three windows as for the nearby Miras.

Radial velocities have also been measured for a selected sample of Miras (with periods ranging from 147 to 345 days) in these windows near the galactic center. When the velocities are compared to those of other stars near the center, the average velocity of the Miras with respect to the local standard of rest can be obtained. This velocity is not significantly different from zero. But the velocity dispersion is very high: 112 ± 17 km s^{-1}.

The kinematics of nearby Miras have also been studied as a function of period. The velocity dispersion of the stars can be compared with known groups of stars. The velocity dispersion of the short period Miras differs from that of the longer period stars. The short periods are associated with an intermediate population II component (not as extreme as globular clusters), while the bulk of the Miras with periods between 250 and 450 days are associated with younger population II or disk population. A general conclusion from these kinematic studies, combined with the occurrence of some short period Miras in globular clusters, is the strong suggestion that the masses of the Mira progenitors vary from about 1 $M_\odot$ for those Miras with periods near 200 days to about 2 $M_\odot$ for periods of 450 days. Since these arguments are statistical other values of mass for individual objects are certainly possible.

No precise masses are known for individual Mira stars. Both o Ceti and X Oph are visual binary stars with rather long orbital periods. Because only a part of the orbit has been measured, the mass determination is not very accurate. The resultant masses for the Mira variables are 1 $M_\odot$ in both cases, with an uncertainty of about 50%.

The space density can be determined by counting the observed Mira variables once the absolute magnitude and individual extinction is known. There are about 250 Miras per kpc^{-3} in the solar neighbourhood, most of which have periods between 200 and 600 days. The scale height perpendicular to the galactic plane is about 300 pc. The formation rate depends on the lifetime as a Mira. There is no observational method to determine the lifetime so that the theoretical determination must be relied upon. Whitelock has recently estimated the lifetime as at least 10^5 years. This leads to a birthrate of less than 2.5×10^{-12} pc^{-3} yr^{-1}.

2.2.3 Mass loss

The presence of a dust shell and its formation as a consequence of cooling and dust condensation has been discussed above. It is not certain what causes the initial motion of the gas and dust. It is possible that the large radiation field from the star which impinges on the dust and is partially absorbed by it causes the dust to be accelerated after its formation. The dust is collisionally coupled to the gas, so that the gas would be set in motion as well. To describe this quantitatively, the acceleration of the gas of density n (cm^{-3}) at a point r (cm) distant from the star of luminosity L (erg s^{-1}), can be set equal to the absorption by grains of density n_d (cm^{-3}), cross section σ_d (cm^2) and absorption efficiency as a function of wavelength, $Q(\lambda)$:

$$nm\frac{dv}{dt} = Q(\lambda)n_d\sigma_d\frac{L}{4\pi r^2 c}, \qquad \text{gm cm}^{-2}\text{ s}^{-2} \qquad (1)$$

where m is the mass of an atom (gm), c is the velocity of light (cm s^{-1}) and v is the velocity of the shell (cm s^{-1}). The mass loss rate $\dot{M}$ can be written

$$\dot{M} = 4\pi r^2 vnm, \qquad \text{gm s}^{-1} \qquad (2)$$

where the optical depth τ_d of the dust is defined as

$$d\tau_d = Q(\lambda)n_d\sigma_d dr. \qquad (3)$$

Equation (2) can then be written

$$\dot{M}\frac{dv}{d\tau_d} = \frac{L}{c}. \qquad \text{gm cm s}^{-2} \qquad (4)$$

If the initial velocity was small at the place of dust formation, and if the final (terminal) velocity is denoted as v_T, integration of eq. (3) gives

$$\dot{M} = \frac{L\tau_d}{v_T c} \text{ gm s}^{-1} = 2 \times 10^{-8} \frac{\tau_d L}{v_T} \qquad M_\odot \text{ yr}^{-1} \qquad (5)$$

On the right hand side of the equation L is given in units of $L_\odot$ and v_T in km s^{-1}. L can be determined (Table II) and v_T can also be measured as will be described in this section and the next. The optical depth τ_d is an average value for different wavelengths weighted according to the distribution of the radiation with frequency and the response of the dust to this radiation. As an approximation, the value of the peak flux from 2 μm to 5 μm can be used. But it is difficult to determine τ_d from eq. (3), not only because the dust properties are not well understood, but also because the size of the region is uncertain. It is nonetheless clear that in the longer period Miras the dust shell causes considerable absorption, so that a value of about unity or somewhat higher for the optical depth τ_d is appropriate. If it were much higher no radiation would be visible.

The values of mass loss derived in this way are given for six Miras in the 4th column of Table III. They are very approximate values, and are based on the assumption that radiation pressure on dust drives the mass loss.

There are other indications of mass loss. One of them is the measurement of the emission lines of CO at millimeter wavelengths from regions surrounding the Mira star. Often ^{12}CO and ^{13}CO can be observed; this is important because ^{12}CO is often so abundant that the observed line is saturated (optically deep). The measurements are very similar to those of CO surrounding the planetary nebula NGC 7027. Just as for that case, the mass loss can be estimated by using a simplified model. First the CO mass is determined, usually from the ^{13}CO line. Then a ratio of CO to H is estimated. This step is very uncertain both because the element abundances are not well known, and because the fraction of carbon and oxygen which are in the form of CO must be estimated. Knowing the mass, the mass loss is found by dividing by the time since the assumed constant mass loss began. The time is found by dividing the radius (found from the observed angular radius and the distance) by the expansion velocity. This last quantity is obtained from the observed line profile. It is clear that mass loss rates determined in this way are quite uncertain. Some values of mass loss and terminal velocity derived in this way are shown in columns 7 and 6 of Table III for eleven Mira variables. The expansion velocities range from 4 to 28 km s^{-1} and the mass loss rates from 2×10^{-7} to 2×10^{-5} $M_\odot$ yr^{-1}. There is a rough agreement between these values and those obtained from eq. 5. There seems to be a correlation between the period of the Mira and both the expansion velocity and the mass loss in the sense that the longer the period, the higher the other

81

Table III: *Mass loss rates for Mira variables determined form CO emission lines and from dust shell*

star	period (days)	spectral type	mass loss eq. (5) $M_\odot$ yr^{-1}	dist. pc	CO exp. vel. km s^{-1}	CO mass loss $M_\odot$ yr^{-1}	OH exp. vel. km s^{-1}
R Leo	313	M6–9	2×10^{-6}	300	7	8×10^{-7}	5
o Ceti	332	M6–9		77	4	2×10^{-7}	
S CMi	335	M6–8		580	18	6×10^{-6}	
R LMi	372	M7–8	3×10^{-6}	400	6	1×10^{-6}	5
X Cyg	407	S7–10	3×10^{-6}	100	8	2×10^{-7}	
R And	410	S6		300	15	2×10^{-6}	
V Cyg	421	C7		600	14	2×10^{-5}	
R Cas	431	M6–8	1×10^{-6}	200	11	6×10^{-7}	9.5
IK Tau	460	M6–10	5×10^{-6}	250	28	4×10^{-6}	19
W Aql	489	S		470	20	2×10^{-5}	
IRC+10011	655	M 8	1.5×10^{-5}	500	24	2×10^{-5}	19

quantities. Higher values of mass loss are typically what are expected from a 'pre-planetary' progenitor.

2.2.4 Mass of Mira variables and pulsation theory

From the fact that the Mira variations are periodic, and from the form of the light curve, it is generally agreed that the atmosphere of the Mira is pulsating. No completely adequate theory for the pulsation in Miras is available at present. General consideration from pulsation theory show that there is a pulsation 'constant' Q for a given type of star (e.g. red giant). The value of Q is defined as

$$Q = P(M/M_\odot)^{1/2}(R/R_\odot)^{-3/2} \qquad \text{days} \qquad (6)$$

where P is the period in days, M the mass of the star in units of solar mass and R its radius expressed in solar radii. Q is not absolutely constant, but is a slowly varying function; there is a distinct value of Q for each mode of pulsation. It is the object of the pulsation theory to predict the value of Q.

If the value of Q is known from theory and if the period is observed and the radius is known from photometry and the distance of the star, the mass of the Mira can be determined. But in practice this is difficult because the radius is difficult to determine. Furthermore, the value of Q is uncertain, principally because it is not completely certain whether the Mira is pulsating in its fundamental mode or its first overtone mode. The more accepted conclusion that Miras are pulsating in their fundamental mode leads to estimates of their mass between 1.5 and 2 $M_\odot$.

2.3 The OH/IR stars

The first Mira was discovered around 1600 and several thousand have been discovered since. But a small, highly significant subgroup escaped detection until some twenty years ago. These are the reddest examples. In the first infrared survey, the IRC, several new, very red variables were found, that later turned out to be Miras of long periods (e.g. IK Tau and IRC+10011). Such objects are very red and cool. This does not reflect the photospheric temperature but the presence of cool circumstellar material – more specifically the presence of dust grains with temperatures of a few hundred kelvin. More extreme cases have since been found in large numbers by IRAS, that searched deeper and at longer wavelengths. The most extreme cases are the "OH/IR stars": objects with a mass loss so rapid that the thick shell makes the star invisible. The circumstellar material transforms the stellar light into infrared radiation: the object is a pure infrared source with a peak in the spectrum around 10 to 30 microns. The photosphere of the OH/IR stars have never been seen, but there are three good reasons to consider OH/IR stars as a subclass of the Mira variables: (1) Almost all are long period, large amplitude variables: periods range from 500 to 2000 days and bolometric amplitudes of 2 magnitudes are no exception. The much larger periods can be explained if one assumes that the mass of the star is now much smaller than when it was on the main sequence; most of the hydrogen envelope has already been ejected. (2) For a small group of OH/IR stars with the largest pulsation amplitudes it has been possible to measure distances directly (by a method which will be described below). In combination with (measured) total fluxes one can calculate stellar luminosities, which are indeed in the range expected for AGB stars (1000 to 60,000 $L_\odot$). (3) The kinematical properties of OH/IR stars and their galactic distribution are the same as those of the planetary nebulae- and of the Mira stars.

Because the OH emission is observed at radio frequencies it is not absorbed by dust. Thus these objects can be seen even if they are surrounded by dust shells or are occurring in regions of the galaxy where a great deal of extinction takes place, e.g. near the galactic center. Furthermore, the OH emission from

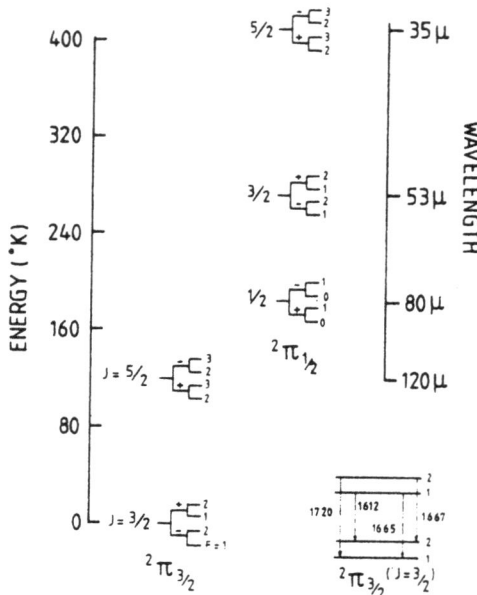

Figure 3: *The rotational levels of OH which can decay radiatively directly to the ground state ($^2\pi_{3/2}$). F is the total angular momentum, including nuclear spin. The strong maser transitions are detected in the ground state, which is plotted in enlarged form on the right side of the diagram. The number near each transition denotes its frequency in MHz.*

the stars has a unique signature and so is easily recognizable. The galactic distribution of these objects can therefore be more easily studied.

2.3.1 OH maser emission

Before discussing the objects themselves, the OH emission will be described. A diagram of some of the important rotational energy levels of this radical is shown in Fig. 3. The observed transitions are all formed in the ground state, which is split into four sublevels. In normal conditions the radiative rates between the sublevels are such that the intensities expected from the 1665 and 1667 MHz lines (main lines) are considerably greater than in the 1612 and 1720 MHz lines. But in the OH/IR stars, the 1612 MHz transition is usually much stronger than the main lines. This is one of the indications that a maser effect is present. The other is that the strength of the line is much greater than expected on the basis of collisional or radiative excitations.

A maser is the result of stimulated (or induced) emission, i.e. an incident photon interacting with the molecule causes the emission of a second photon.

For stimulated emission to occur, the ratio of the population of the upper level of the transition to that of the lower level must be greater than that expected from the Boltzmann relation (this is referred to as a population inversion). There must therefore be some way of pumping the maser, i.e. exciting the upper level without causing an increase in the population of the lower level. The mechanism probably responsible for the pumping in the OH/IR stars, is absorption of 35 μm and/or 53 μm photons by the ground levels of OH, populating the $^2\pi_{1/2}$ J = 5/2 and 3/2 levels (see Fig. 3). This is because the star at the center and its associated dust emit very strongly in the far infrared. From these levels cascade occurs via the $^2\pi_{1/2}$ J = 1/2 level to (preferentially) the upper level of the 1612 MHz transition. Different conditions will cause different overpopulation; in fact when the optical depth in the dust is low, masering is observed to occur in the main lines.

Thus the strong radiation in the maser line is due to the predominance of stimulated emission relative to spontaneous emission. When the stimulated emission becomes so strong that the induced de-excitation depletes the upper level of the maser line as fast as the other 'ordinary' loss rates (e.g. collisions), the population of the masering level becomes self-controlled by the radiation it produces and the maser is 'saturated'. It has then achieved its maximum efficiency and every pumping event leads to roughly one emitted photon. Most of the 1612 MHz masers to be discussed are 'saturated'.

2.3.2   Observed OH emission

The OH maser lines from late type stars have a characteristic profile consisting of two intense peaks separated in velocity by about 20 to 50 km s^{-1}. An example is shown in Fig. 4. Note that the extreme velocities of the profile of each peak show a steeper fall-off than the inside. When the radial velocity of the star itself can be measured, either using the optical spectra or in an SiO maser (which is formed in the photosphere of the star), it is found to lie precisely at the central velocity between the two spikes. That the radial velocity of the star lies midway between the OH peaks can be explained in terms of a shell expanding from the star. In a radially expanding shell, molecules at different directions of the radius have velocities which are 'pointing' in different directions and therefore cannot reabsorb radiation to cause stimulated emission. Strong amplification is therefore only possible along radial directions, and each segment of the shell can emit strong maser radiation only along a given radius, towards and away from the center. An observer at an arbitrary location will therefore detect radiation from only two regions which correspond to the intersection of the shell with the line of sight. The 'blue' and 'red' shifted components, corresponding to

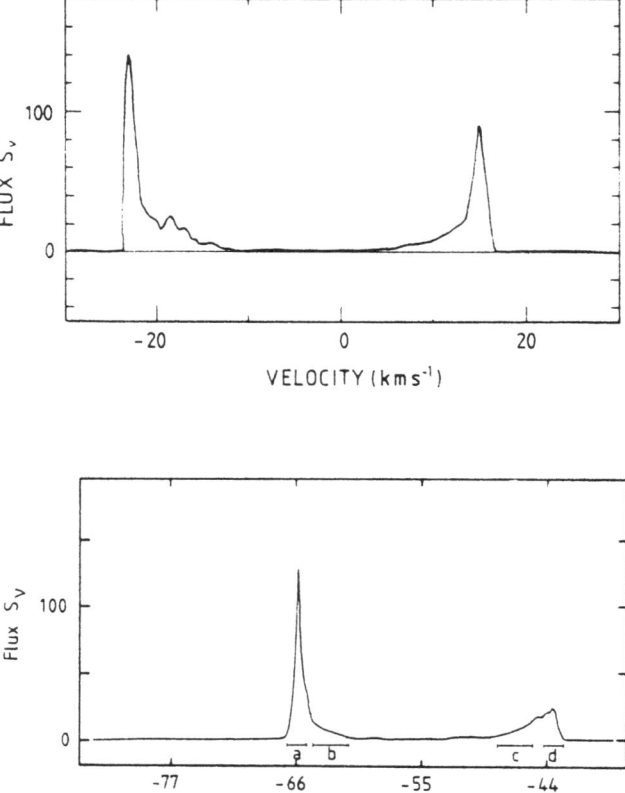

Figure 4: *Two examples of an observed 1612 MHz OH maser line. The line is almost always double and each profile is steeper on the outside than on the inside. This is interpreted in terms of an expanding shell.*
(a) The object OH 344.9+0.0;
(b) The source OH 127.8−0.0 for which long baseline interferometric maps have been made.

the front and back of the shell respectively if the shell is expanding, should therefore be separated by twice the velocity of this expansion. The slower fall-off of emission on the inside of the profile is caused by an apparent (geometrical) velocity gradient in the shell and the effect of thermal motions which cause radiation from parts of the shell farther from the point of intersection of the shell with the line of sight to be seen.

An interesting direct confirmation of this picture can be made. Because the extreme velocities (the peaks) come from parts of the shell close to the point of intersection of the shell with a line of sight through the center, and less extreme velocities (the slower fall-off) come from parts of the shell farther away from this point, a map of an OH/IR source made at the velocities of the peaks should be

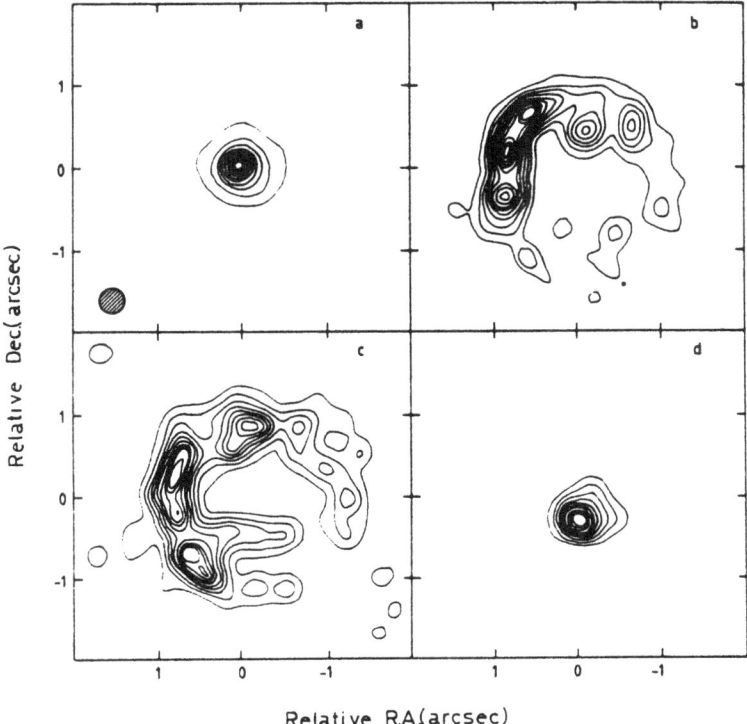

Figure 5: *Maps of the spatial distribution of the OH emission from OH 127.8–0.0. In each map the integrated emission in a small velocity interval a to d respectively is given. The velocity intervals are indicated in Fig. 4(b). In each map the contour interval is 10% of the peak integrated emission in that velocity interval, and the lowest contour is about 10% (from Booth, R.S., Kus, A.J., Norris, R.P. and Porter, N.D. 1981, Nature* **290**, *382).*

considerably smaller than a map made at less extreme velocities. An example, produced by long baseline interferometry techniques is shown in Fig. 5. Part (a) and (d) of this figure correspond to the peak velocities of the 'fall-off'. The differences in size are quite dramatic, as expected from the above picture.

The intensity of the OH emission usually varies with time. If the central star is an observable Mira, the OH variation follows the variation of infrared light from the Mira with almost no phase delay. This confirms that the infrared radiation is the pumping mechanism for the population inversion. A plot of the intensity of the OH emission as a function of time when no optical counterpart is known permits the determination of a period. In this way periods have been found for many 'stars' which are not observable in visible light. These periods are mostly longer than from the observable Miras, extending to almost 2000 days. This is a factor of three higher than the longest optically observable Mira period. Infrared observations have been made of selected OH masers which

Table IV: *OH/IR sources with measured shell sizes (Distances are from van Langevelde et al., Astron. Astrophys.* **239**, *193, 1990)*

Star	Period	V_{exp}	Radius	Ang. Diam.	Dist.	Mass loss	Lumi-nosity
	(days)	km s^{-1}	10^{16} cm	sec	kpc	$M_\odot$ yr^{-1}	$L_\odot$
R Aql	280	6.5		3.1	0.45	8×10^{-7}	
U Ori	372	4.		≤1	0.3	3×10^{-7}	
RR Aql	394	6.5		1.0	0.5	5×10^{-7}	
WX Ser	425	7.5	4.4	0.8	1.5	1×10^{-6}	
OH 44.8−2.3	530	15.9	2.2		1.1	5×10^{-6}	4300
OH 20.7+0.1	1670	18.2	6.3	1.7	5.0	7×10^{-6}	5900
OH 104.9+2.4	1465	14.9	4.2	2.9	2.3	5×10^{-6}	4300
OH 127.9-0.0	1650	11.0	5.4	3.1	2.9	1×10^{-4}	57000
OH 39.7+1.5	1420	16.5	3.0	4.0	1.0	1×10^{-5}	8400
OH 32.0−0.5	1440	20.6	9.5	1.1	11.3	7×10^{-5}	67000
OH 32.8-0.3	1550	16.4	10.3	2.7	5.0	3×10^{-5}	23000
OH 26.5+0.6	1600	14.1	4.9	4.5	1.3	2×10^{-5}	18000
OH 21.5+0.5	1800	18.8	12.0	1.9	7.3	7×10^{-5}	70000

have no visible counterpart. They are often detected with the same period as found in the OH. They also appear to be an extension of the Mira sequence in temperature, since the infrared colors indicate still lower temperatures with an increase in period.

There must be a small phase delay between the stellar light variation and each of the OH peaks. The phase delay will be greatest between the two OH peaks since they are formed in small regions at opposite sides of the expanding shell. The delay is simply the extra time that it takes for the light emitted from the farthest side of the shell to reach the observer, compared to the light emitted by the nearest side. This measurement can be immediately interpreted as the diameter of the shell. This diameter (radius) is given in the 4th column of Table IV, and depends only on the measured phase delay; it does not require a knowledge of the distance. In fact, the distance can be determined from the radius if the angular radius is known. Measured OH angular diameters shown in the 5th column of the table are probably smaller than the actual value because only the front and back of the shell are seen in the masered OH emission. The distance given in column 6 of the table is based on the assumption that the angular size can be measured. Since this is difficult, the distance is uncertain, but it agrees reasonably well for those OH sources which are visible Miras, and whose distances are determined from the empirical period-luminosity relation calibrated for Miras of known distance (e.g. known cluster membership). The phase delay measurement is probably the only reliable way of estimating the distance of the OH sources.

In Fig. 6 the expansion velocity of the shell, as determined from OH measurements, is plotted against the period. A relation is clearly seen for periods below 700 days. For larger periods the velocity seems to remain constant. The mass loss rate is shown as a function of the period in Fig. 7 using the values listed in Tables III and IV. For the shorter periods, less than 400 days, the mass loss is low. It increases with increasing period to values of 10^{-5} $M_\odot$ yr^{-1} at periods of 500 days or more.

In summary, the following result is suggested by the variations in radius, expansion velocity and mass loss for the different objects in Tables III and IV. It seems that Mira variables and OH/IR stars form an evolutionary sequence in which the Miras are the earlier stage. Mass loss is still low and the expelled gas is ejected at a very low velocity. With time, the period increases due to an increase in the stellar radius, the stellar temperature decreases and the mass loss rate increases. Extinction then becomes more pronounced, either because more dust is being ejected from the star or because conditions are favorable for condensation in the shell at this time. Optical observations become difficult at this stage. The velocity of ejection and mass loss rate both increase initially and then level off at values $v_{exp} \simeq 20$ km s^{-1} and $\dot{M} \simeq 2$ to 3×10^{-5} $M_\odot$ yr^{-1}. These values of mass loss are sufficient to produce a planetary nebula in a time of 5×10^3 years.

The scatter seen in the relations between radius, expansion velocity and mass loss listed in Tables III and IV is probably real, indicating that the situation is more complex than described above. The mass of the star involved is expected to play a role, in the sense that both the expansion velocity and the mass loss rate are higher for higher mass stars. Because individual masses cannot be determined, this result is only statistically significant.

2.3.3   Distribution and kinematics of OH/IR maser sources

After the initial discovery of this type of OH source, the association with Mira variables and infrared stars was established. Initial follow-up work concentrated on Mira variables. Gradually more emphasis was placed on those sources without optical identification, especially after it was realized that these objects are often

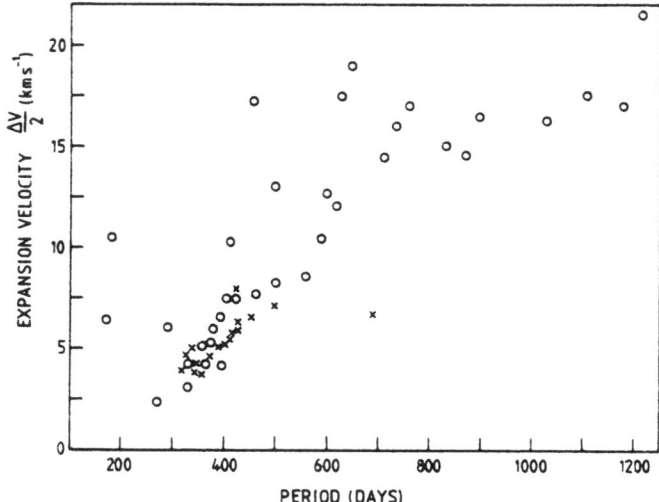

Figure 6: *The expansion velocity of the OH shell is plotted against the period of light variation. Most of the stars with periods less than 570 days are optically observed Mira variables while only one of the others is a classically known Mira. The circles are type II OH masers (seen predominantly at 1612 MHz) while the crosses are type I or main line masers.*

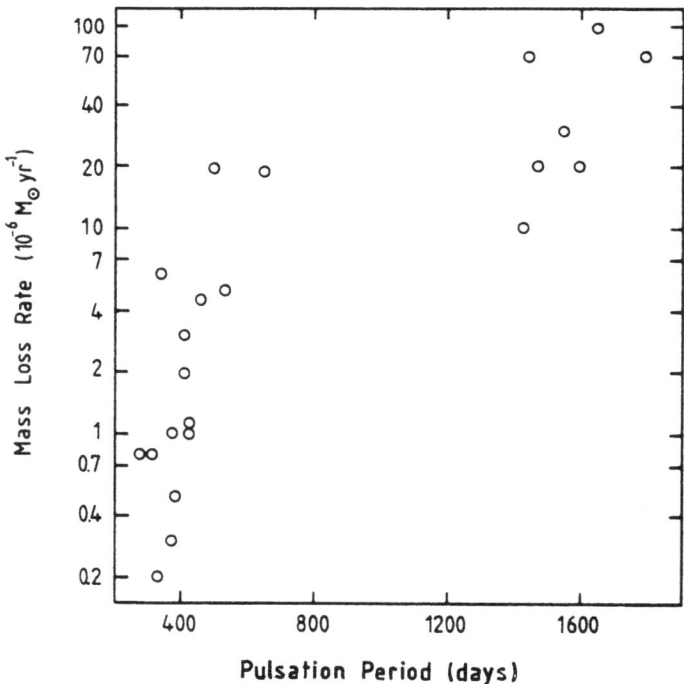

Figure 7: *The mass loss rates are shown as a function of the period of light variation (either optical, infrared or OH). Mass loss rates are derived from CO measurements and from the assumption of radiative acceleration of dust (eq. 5).*

intrinsically more luminous OH sources. Since these radio emitters were thought to be concentrated towards the galactic plane, surveys were generally limited to the region near the plane. This was thus a self-fulfilling prophecy. Only much later, when Te Lintel Hekkert made a survey of all sources found in the IRAS catalogue with far infrared colors similar to OH/IR stars, was it realized that the concentration to the galactic plane exists but is not really so pronounced as for population I stars. It is actually rather similar to that of planetary nebulae. What is also similar to planetary nebulae is the strong concentration of these objects in the direction of the galactic bulge. Fig. 8 shows a velocity-galactic longitude plot for OH/IR stars where the concentration toward the galactic bulge is evident, as well as the large range in radial velocity for those OH/IR stars associated with the galactic bulge. As with planetary nebulae, this can be interpreted, at least in part, as a spherical population with a high velocity dispersion: in this case about 125 ± 15 km s^{-1} in the line of sight. The Miras close to the galactic centre have a similarly high velocity dispersion: 112 ± 17 km s^{-1}. Thus like planetary nebulae, this group has population II (globular cluster) characteristics. Also like planetary nebulae, there is superimposed on this group a younger population, as testified by their concentration to the galactic plane and their adherence to general galactic rotation.

If this younger population of OH/IR sources is divided according to expansion velocities into two groups, those with expansion velocities equal or greater than 15 km s^{-1} and those with expansion less than 15 km s^{-1}, a small difference in the kinematic properties of the two groups becomes noticeable. A plot similar to Fig. 8 for each of the groups shows that the group with the smaller expansion velocity has statistically the largest dispersion in its radial velocity. It is therefore the older of the two groups, originating from progenitor stars of somewhat lower mass.

2.3.4   The birthrate of OH/IR stars

The age of OH/IR stars, i.e. the length of time spent in the OH/IR stage, can be computed in the same manner as is done for planetary nebulae: the size of the largest objects is divided by their expansion velocity. This leads to an age of 4×10^3 years to 10^4 years, which is somewhat shorter than the age in the planetary nebula stage. The total number of OH/IR stars in the galaxy is rather more difficult to determine. About 2000 have already been detected, and an assessment of the selection effects involved leads to the conclusion that the total number must be about 20,000, quite similar to the number of planetary nebulae. Thus 2 to 4 OH/IR stars are formed per year in the galaxy, compared to 1 to 2 planetary nebulae per year. But the planetary nebulae live somewhat

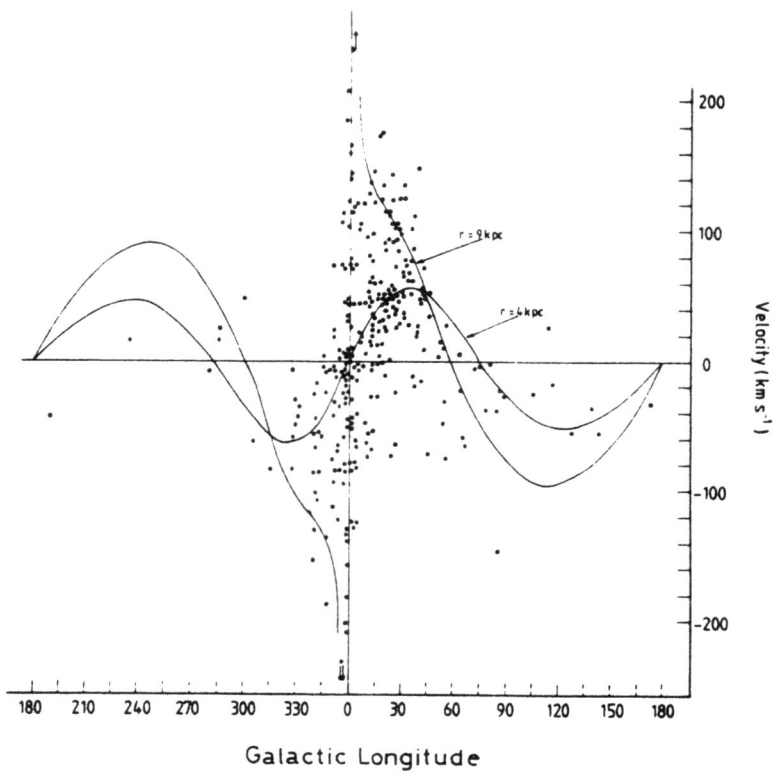

Figure 8: *The distribution of known OH/IR maser sources with galactic longitude plotted against the observed radial velocity of the source after correcting for the solar motion (LSR). The latitude is limited to* $|b| \leq 10°$. *This distribution should be compared with that of the planetary nebulae given in Fig. 11. The solid curves are the predicted velocities due to circular orbits in the galactic plane at two distances from the sun.*

longer, thus their total number is about the same.

3 FROM AGB TO PLANETARY NEBULAE

As has already been discussed, the stars on the AGB have a high density carbon-oxygen core surrounded by helium and by hydrogen burning shells and a very extensive low density atmosphere. These atmospheres are unstable or becoming unstable. The rate of mass loss from the atmosphere now becomes very high for reasons not completely understood in detail, but connected to the unstable atmosphere. When the material emitted from the star is some distance from the star (10 to 15 stellar radii is suggested in the literature) it becomes cool enough so that some of the heavy elements, possibly in the form of molecules (silicates, silicon carbide, iron oxide, graphite, polyaromatic hydrocarbons, etc.) 'condense out' to become 'dust' which strongly absorbs the radiation of the central star, especially in the optical and ultraviolet part of the spectrum. As a consequence the star becomes 'invisible' in these wavelengths and (to a lesser extent) in the near infrared as well. It is only visible in the far infrared and the radio (mm and cm) wavelength range. Since most of the energy is in the far infrared it is clear that a concentrated observational study of this stage of evolution only began after the IRAS observations were available. Here the use of the IRAS data will be described.

3.1 The IRAS color-color diagram

This has proved to be a useful tool to study the properties of various astronomical objects. For the present purpose a diagram of the IRAS broad band 12 μm/25 μm vs the 60 μm/25 μm flux density is most useful. Different ways of making this plot are found in the literature. An example of such a plot is given in Fig. 9a. The symbol [x] is defined as -2.5 $(\log[F(x)])$. This plot was made by selecting optically bright known sources of various types and plotting their IRAS flux ratios. The types of sources shown are: late type stars, especially Mira variables (M), OH/IR stars (S), planetary nebulae (O), HII regions (H) and galaxies (X). There is seen to be very little overlap between these various kinds of sources, in spite of the fact that each of the groups shows a considerable spread. This suggests that it may be useful to isolate certain regions of this diagram as 'belonging' to a special type of object. It may also prove possible to trace evolution in this diagram. For example, it is clear that the OH/IR stars form in a rough way a band on the diagram extending from the lower left (close to the position of the late type stars) to the region where the planetary nebulae begin to be found.

This same kind of diagram has been plotted schematically by van der Veen

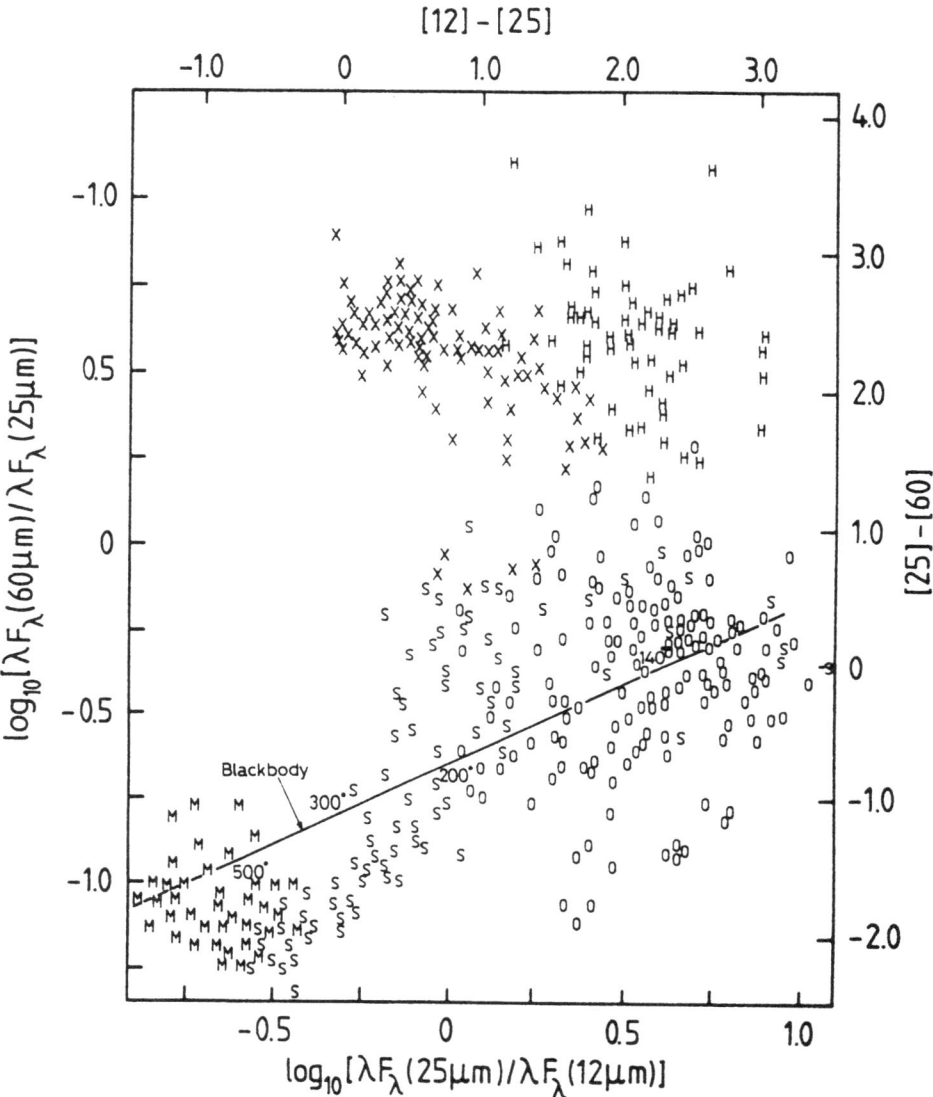

Figure 9: *(a) The IRAS color-color diagram. The symbols are as follows: (M) M stars and Mira variables; (s) OH/IR stars; (O) Planetary nebulae; (H) HII regions; (x) Galaxies. The position of a blackbody in this diagram is shown by the solid line on which various temperatures have been listed.*

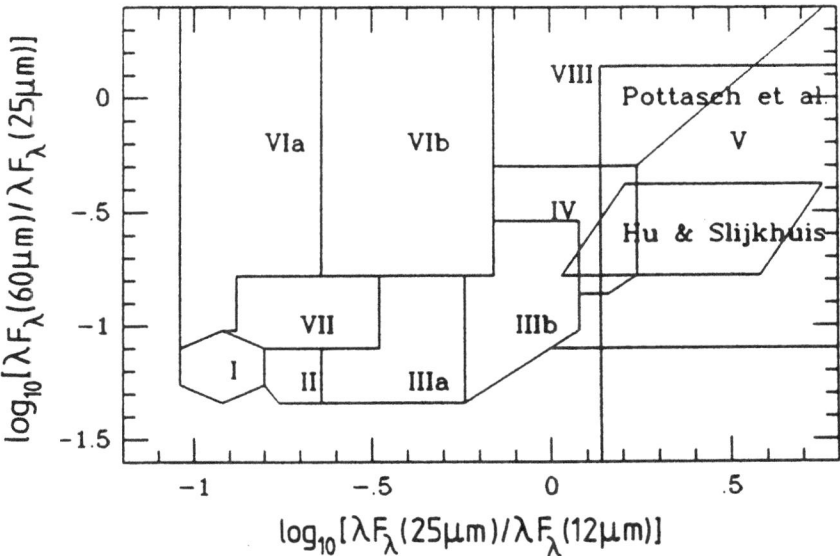

Figure 9: *(b) The IRAS color-color diagram with the different regions corresponding to various classes of objects (van der Veen and Habing, 1989). The color criteria used by Hu and Slijkhuis is shown by a trapezium and the rather relaxed criteria used by Pottasch et al. is shown by a bigger rectangle.*

and Habing and is shown in Fig. 9b. It covers only a part of the total region shown in Fig. 9a, since these authors were especially interested in AGB evolution. They suggest that regions I, II, III and IV form an evolutionary sequence of oxygen rich stars with increasingly thicker and more opaque circumstellar shells. The last three of these four regions contain a substantial number of variable stars. These regions contain the OH/IR stars discussed above. Van der Veen and Habing state that region V is where planetary nebulae occur, but reference to Fig. 9a shows that many occur in regions IV and VIII as well.

In general, the color-color diagram must contain clues as to the evolution of stars with circumstellar envelopes. The radiation of carbon rich dust is expected to differ from that of oxygen dust and the contribution of the central star will vary according to the thickness of the shell. But care is required in interpreting the position of otherwise unknown objects. For example, as can be seen in Fig. 9a, region VIII contains not only OH/IR stars and planetary nebulae, but the (completely unrelated) Seyfert galaxies as well.

3.2 The LRS spectra

The IRAS satellite also made low resolution spectra of each source on the sky in the spectral region from 8 to 22 μm. The resolution is about $\lambda/\Delta\lambda \approx 20\text{--}30$. The brightest sources (about 5400) are published by the IRAS Science Team. Many other spectra are available if one goes to the original data base.

Planetary nebulae are usually easily recognizable from the spectra because of the ionic emission lines which are strong in spite of the low resolution. The other features in the spectra which are important include the silicate features at 10 and 18 μm. These are almost certainly formed in a circumstellar envelope; the features are in emission if the envelope is not very thick. When the envelope becomes very thick these features appear in absorption. Thus models of the envelope can be made which reproduce the observed spectra quite well. These models confirm the evolution discussed in the previous section.

The silicate features are confined to oxygen rich envelopes. Sometimes a feature at 11.3 μm is seen, which is identified as due to silicate carbide, indicating a carbon rich envelope. Sometimes the silicate feature is seen in the LRS spectrum of a carbon star, presumably indicating that carbon has only been brought to the surface of the star after the ejection of the envelope. These spectra provide clues to a rather complex evolution which is not yet completely understood.

3.3 Classes of objects suggested to be post AGB stars

1. Non-variable OH/IR stars

The OH/IR stars previously discussed are thought to be on the AGB. They were found to pulsate like Mira stars and are expected to be a somewhat more evolved stage, with an increased mass loss rate and usually a longer period. Their far infrared colors are also similar but appearing to evolve toward lower temperature dust shells. But, as can be seen from Fig. 9a, some OH/IR stars have far infrared colors which overlap with those of planetary nebulae. In fact, 12 objects are now known which are both OH/IR stars and planetary nebulae at the same time, i.e. they show OH emission at 1612 MHz and show either emission lines characteristic of low excitation PN or free-free continuum emission from an ionized medium (Zijlstra et al., 1989). These objects clearly have already evolved off the AGB. It is expected that other OH/IR stars are also in the post AGB phase, especially those which no longer show Mira-like variations but have a continuous energy output.

2. High latitude 'supergiants'

There is a surprising coincidence of IRAS sources which fall in the region of the planetary nebulae on the IRAS color-color diagram, with bright supergiant stars of spectral type between K and B and especially spectral type F. Many of these stars had long been suspected of being peculiar because of their galactic distribution. They are often found at (very) high latitudes which is strange because supergiants are expected to be massive stars which evolve very quickly. They should therefore be very young objects which are confined to regions close to the galactic plane. Because the precise positional coincidence, the suggestion was made (Parthasarathy and Pottasch, 1986) that these objects are post-AGB stars with low gravity atmospheres which mimic supergiant spectra. Subsequent detailed analysis of some of these objects has strengthened this interpretation. Especially interesting is the lack of metals, especially iron, in their atmospheres. These elements are thought to have 'condensed out' to form the dust which is responsible for the far infrared radiation. At present, these stars are generally accepted to be post-AGB objects. It is still puzzling why object which have an extensive dust shell, and emit more than half of their energy in the far infrared, are easily visible in the optical region. In fact they appear to have little extinction in this spectral region.

3. Young planetary nebulae

Various authors have selected objects found by the IRAS in color-color diagram where planetary nebulae are found for further examination from the ground (the limits of the region examined are sometimes slightly different). A large survey has been made in the radio continuum (6 cm and 3 cm) by Pottasch and co-workers. If the objects are found to emit at these wavelengths it is almost certainly due to free-free emission from ionized gas. The object is thus a (young) planetary nebula. Of the more than 1000 objects observed in this way, about 20 to 25% were found to be detectable and thus PN. Subsequent optical spectra of some of these objects have confirmed this classification, but many are optically invisible.

Using the same IRAS selection criteria, optical spectra can be directly obtained for those objects which can be found on optical survey plates (SERC or Palomar plates). Various groups are doing this in the hope of discovering interesting objects (Manchado and García-Lario; Kwok and co-workers). Hu and Slijkhuis have attempted to investigate a complete sample of these objects, but to keep the sample size to reasonable proportions they had to use very stringent criteria, both in the position on the color-color diagram as well as only investigating the brightest objects. Of the 42 sources they investigated, 23 have no optical counterpart, while 17 are identified as bright stars with spectral types F, G and M (usually classified as supergiants).

Sometimes extremely interesting objects are found. Both García-Lario et al. (1993) and Parthasarathy et al. (1993) have found objects which are now planetary nebulae but have only become so in recent years. One is the bright star SAO 244567 which is now an O8 star surrounded by a small (1 arc sec diameter) planetary nebula recognizable both by the forbidden line radiation and the radio continuum. But in 1970 it was classified as a B3 star and the only line emission seen was Hα emission. The star itself is rapidly becoming fainter.

4. Objects with deep 9.7μm absorption

Especially Kwok and co-workers have searched for post-AGB objects among IRAS point sources with deep, flat bottomed 9.7μm silicate absorption in the LRS spectra. Such objects have thick circumstellar shells and are thus good candidates for the post-AGB. This has proved to be a fruitful selection criterion.

5. R Cor Bor and RV Tau stars

R Cor Bor stars are supergiants characterized by periodic dimming and extremely hydrogen deficient, carbon rich helium envelopes. The dimming is thought to be the result of obscuration by dust removed from the star by mass loss. The peculiar composition is attributed either to essentially complete ejection of the hydrogen-rich envelope or full mixing and burning of the envelope hydrogen.

In the IRAS data Gillett et al. (1986) have found a large extended shell around R Cor Bor itself -very suggestive of "a fossil hydrogen envelope", and thus of a previous existence as a RGB or AGB star.

RV Tau stars are luminous pulsating stars (typical periods 75 days) of spectral type F, G and K, estimated to have a luminosity of about 10^3 L$_\odot$. They have OH emitting shells indicating low outflow velocities ($\sim$10 km s^{-1}). This is consistent with the temperatures and luminosities expected from the less massive post-AGB stars. Jura (1986) analyzing the IRAS data suggests that they have significantly decreased in mass loss rate over the past 500 years, which is consistent with their moving form the AGB to the post AGB stage. Their birthrate, however, is low; at the most only 10% of the planetary nebulae could be formed from these stars.

6. Miscellaneous objects

The OH source OH 231.8+4.2 ('Calabash nebula') is a bipolar nebula with two lobes expanding at high velocity, about 100 km s^{-1}. It resembles very closely to a planetary nebula except that the central star is of M type. Four other OH maser sources are known which resemble the calabash nebula. It is thought that these objects are just leaving the AGB. IRAS 09371+1212 ('Frosty Leo') was discovered because of its extremely strong 60 μm IRAS flux. It has been detected in the CO line with low velocity, proving that it is a galactic source in spite of its high galactic latitude. The strong 45 μm emission band due to H_2O ice suggests that it is a post-AGB star.

Many other individual objects have been suggested to be post-AGB stars. Apparently, a variety of scenarios are possible in this as yet poorly understood stage.

4 PLANETARY NEBULAE

The theoretical expectation for the appearance of central star of rapidly increasing temperature at roughly constant luminosity appears to be realized in planetary nebulae. Some of the observed characteristics of PN are given below.

4.1 Spatial distribution and kinematics

It has long been known that planetary nebulae are distributed along the plane of the galaxy with a peak in the direction of the galactic center (Minkowski, 1965; Pottasch, 1984). This distribution can be reconsidered in the light of recent 'cleaning' of the list of PN, especially by Acker et al. (1987, 1990). These authors took low resolution spectra of almost all suggested PN and were able to remove several hundred objects which were mainly M stars and symbiotic stars, although galaxies, emission line stars, HII regions and even plate flaws have also been a source of confusion.

The distribution of the 'clean' list of PN is shown in Fig. 10. This histogram is divided into three parts. The upper diagram shows the distribution in galactic longitude of all 391 nebulae which are within 3° of the plane; the lower diagram gives the distribution of the 216 nebulae with are farther than 10° from the galactic plane. Several conclusions can be drawn from these figures. First, there is a clear excess of nebulae within 10° and 15° of the direction to the galactic center. This excess is only marginally seen at latitudes greater than 10° from the plane. We may conclude that 80 to 90% of the PN seen within 10° latitude of the galactic center are actually within a distance of 1.3 kpc from the center.

Secondly, while 1004 PN are within 10° of the galactic plane, their average distance from the plane is about 5° to 6°. This is considerably higher than HII

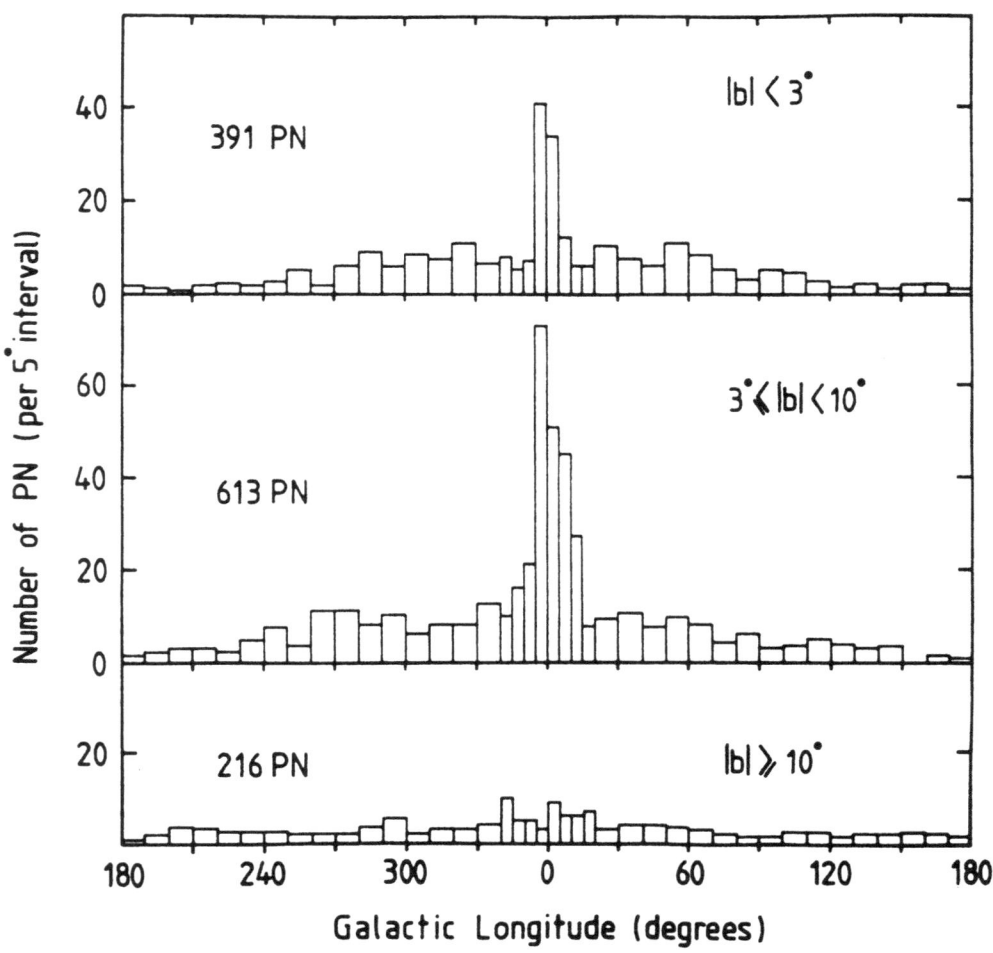

Figure 10: *(a) Distribution in galactic longitude of all PN within 3° of the galactic plane.*
(b) Distribution of all PN between 3° and 10° of the galactic plane.
(c) Distribution of all PN farther than 10° of the galactic plane.

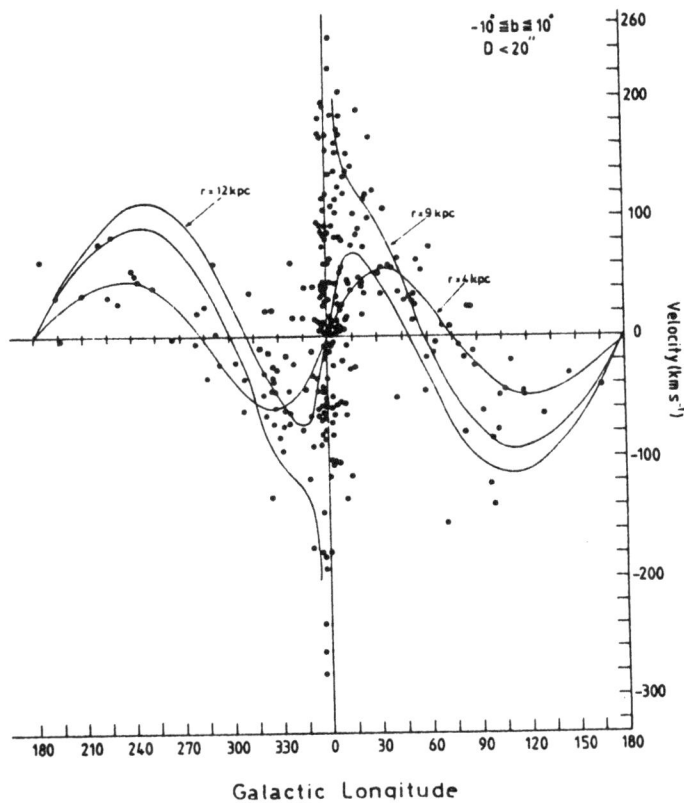

Figure 11: *The radial velocities (LSR) of planetary nebulae whose diameters are less than 20″ and which lie within 10° of the galactic plane, are plotted against the galactic longitude at which the nebula is found. The solid curves are the predictions of expected radial velocities assuming circular orbits at various distances from the sun.*

regions, which, except for the closest, are almost all within 0°.5 of the plane.

Radial velocities are known for over 400 nebulae, about 40% of the total number. They are shown in Fig. 11 where the radial velocities are corrected for the solar motion with respect to the average velocity of the stars in the local neighbourhood, i.e. reduced to the local standard of rest. They are plotted as a function of galactic longitude as was done for the OH/IR stars in Fig. 8. The resemblance between the two figures is striking indicating the close connection between the two classes of objects. Again, in the galactic bulge the retrograde orbits are seen with the very high velocity dispersion of 140 km s^{-1}. Not only does this resemble the OH/IR stars and Mira variables but the globular clusters and RR Lyrae stars as well.

An explanation is now in order. In the above it has been shown that many nebulae near the galactic center closely resemble old, extreme Population II objects in their spatial and kinematic properties. Earlier it had been shown that many nebulae are confined to the galactic plane and are therefore associated with younger Population I objects. Paradoxically both are true. A group of stars or other objects can be assigned to a population class (I, II or intermediate) only if they are all the same age. Planetary nebulae progenitors can have greatly different ages. A star of 1 $M_\odot$ will take 10^{10} years to reach the planetary nebula stage, whereas a star of 5 $M_\odot$ will become a nebula at a much younger age. Thus there are old (Population II) nebulae and much younger (intermediate Population I) nebulae.

4.2 Distances

Distances to PN are fundamental to any good physical discussion of the properties of both the nebula and the central star. Unfortunately it has remained difficult to determine distances to individual PN because most are too far away for presently known parallax measurements to be reliable. Work is in progress both from the ground (Pier et al., 1993) and from space (the Hipparcos satellite). Recent work by Pier et al. (1993) give trigonometric parallaxes for seven PN. Among them are NGC 6720, 6853, PW 1 and S 216 for which distances are obtained of 500, 360, 320 and 120 pc respectively. There are now several methods available for determining distances to a limited sample of nebulae. These are:

4.2.1 Binary stars and clusters

There are five binary PN nuclei with distance determinations found from the spectral type of the secondary star: NGC 246, 1514, 2346, 3132 and A35. The distance to K648 is determined with the assumption that it belongs to the globular cluster M15; that to NGC 2818 with the assumption that it belongs to the open cluster of the same name. There is also a bizarre PN which in all likelihood belongs to the globular cluster M22 (Cohen and Gillett, 1989).

4.2.2 21 cm absorption line

Gathier et al. (1986a) have derived distances for 12 PN using 21 cm line measurements. In this method the absorption lines are measured against the background of the PN continuum emission, and of a background radio-bright source, usually a galaxy, which is situated very close to the PN in the plane of the sky. Since each of the spiral arms in our galaxy will produce a 21 cm absorption line at a known velocity, it can immediately be seen which of the spiral arms are between us and the PN and which are beyond the PN. Since the distance to a spiral arm in a particular direction is known, the distance to the PN can be approximately determined.

This method has the drawback that it can only be applied to PN close to the galactic plane. The method will work if a more distant absorption can be seen in the background source than in the PN. The average uncertainty of this determination is half the distance between spiral arms. The distances to spiral arms are usually determined from HII region contained within them, and are consequently uncertain.

4.2.3   Visual extinction

A plot can be made of interstellar extinction versus distance for stars which are in the 'same' direction in the sky as the PN. The PN can be placed on this diagram if its extinction can be determined. This is usually found from Balmer line ratios or radio/Hβ measurements. The PN distance can then be read from the diagram.

The limitations of this method are severe because of the patchiness of the extinction. The first study was made by Lutz (1973). The most extensive study was made by Gathier et al. (1986b) who restricted the fields around the PN to angular radii of less than 0.3 degree to limit the patchiness. Furthermore 5 color photometry was done to determine the normal colors and extinction of the field stars. In spite of these precautions, some of the fields showed a large amount of scatter in the extinction diagram. Distances to 12 PN were determined by these authors. More recent work has been done by Weinberger and collaborators (e.g. Weinberger and Ziener, 1988).

4.2.4   Stellar atmosphere analysis

Very high resolution spectra of central stars are obtained and compared to non-LTE model atmospheres. This comparison is made with the profiles of H lines, and where possible, HeI and HeII lines. The wings of the Balmer lines yield the gravity, the He^+/He^{++} ionization ratio determines the effective temperature and the absolute strength of the HeII lines give the helium abundance. For hotter objects where no HeI lines are observed, the temperature is obtained from the profiles of the H and HeII lines. From the temperature and gravity obtained in this way combined with the visual magnitude of the central star and an assumed core mass-luminosity relation, the core mass and the PN distance can be derived.

This type of analysis is very promising, but is at present difficult. It is carried out essentially by only one group (see Mendez et al., 1988, and Kudritzki and Mendez, 1989, for a description of the method and for detailed results). The difficulty is that the use of non-LTE models requires the precise knowledge (and inclusion) of all physical properties of the stellar atmosphere, including the radiation field at all frequencies at every point in the atmosphere. Since this is at present impossible an estimate of the uncertainties due to the various assumptions introduced must be made. According to Kudritzki and Mendez, the most important approximations at present are: a) no metal absorption was included in the models; b) wind blanketing has been neglected; c) atmospheric extension and deviations from hydrostatic equilibrium due to stellar winds have been neglected.

4.2.5 Expansion distances

By comparing PN expansion velocities, determined spectroscopically, with measured angular expansion distance estimates can be obtained (e.g. Liller et al., 1966). Optically this is done by comparing the positions of knots, filaments, etc. taken on photographs at different epochs. The first epoch photographs were usually taken at the end of last century and different photographic emulsions with different spectral response from the present epoch plates have been used. Thus different mixtures of emission lines are being compared. Since the angular expansion expected is in any case very small, it is now thought that the results may be unreliable and should not be used.

This is not true for angular expansion measured with the VLA in radio frequencies. These results are reasonably reliable, but until now only two nebula have been measured: NGC 7027, for which Terzian (1992) using an 8 year baseline finds a distance of 1100±300 pc, and BD +303639 for which Terzian (1992) finds a distance of 3300±1100 pc. Earlier Masson (1986, 1989a,b) found a value of 900 pc for NGC 7027 using a baseline of only two years. Terzian (1992) reports that NGC 6210, 6572, 2392 and 3242 are now being analysed and results should be available soon. The greatest uncertainty will occur in determining the correct expansion velocity to use both because of a possible non-spherically symmetric velocity field, and because the velocity of the ionization front differs from the material velocity.

4.3   Central star temperatures

Various methods are now available for determining the central star 'temperature'. This word is put in quotation marks because several of the methods to be discussed required that the frequency distribution of the energy be known. Often a blackbody distribution is assumed, but other assumptions can be made. The most important methods follow.

4.3.1   'Zanstra' temperatures

This rather ingenious method for determining the temperature of any hot star which is surrounded by a tenuous nebula was first discussed by Zanstra in 1931 and applied by him to determine the temperature of the central stars of planetary nebulae. Zanstra himself, in his later years, referred to it as a "cheap way to do space research", because it enables one to count the number of photons which can ionize hydrogen. These photons are shortward of λ 912 Å and do not penetrate the atmosphere, hence the reference to 'space research'.

The necessary condition for this method to work is that the optical depth of the hydrogen ionizing radiation, the Lyman continuum (L_c), be greater than unity, which means that all ionizing radiation is absorbed by the nebula. For each absorption an atom is ionized. After a certain time the electron will be captured by a proton. There now exist two possibilities: 1) the electron falls directly to the first (ground) level, or 2) the electron falls into one of the higher levels. In the first case an L_c -quantum is emitted and the process is repeated. In the second case the electron has to make a series of cascade transitions, thus emitting quanta in the subordinate series which escape from the nebula since the nebula is largely transparent at the frequencies of these series. In planetary nebulae the radiation field is so dilute, and the material density so low, that this chain of cascade transitions goes uninterrupted in the vast majority of cases. The last link in the cascade is a transition to the first level, accompanied by the emission of some quantum in the Lyman series. Now two things can happen.

Assume that the electron is in the second level, after a transition of the atom from some higher state (discrete or continuous) to the second and, hence, by the emission of one, and only one, quantum in the Balmer series or Balmer continuum. These Balmer quanta escape from the nebula. Afterwards a $2 \rightarrow 1$ transition takes place with emission of one Lα-quantum. However, it has been assumed that the optical depth of the nebula in the frequencies of the Lyman lines, including Lα, is very high. Therefore, this quantum, after travelling a

105

short distance inside the nebula, will be absorbed by some neutral hydrogen atom in the ground state. The atom is excited to the second state and again, due to the lack of collisions, it will make a spontaneous transition to the ground state after a very short time (of the order of 10^{-8} sec), emitting a Lα-quantum. Thus, under nebular conditions the Lα-quanta can only be modified after many scattering, e.g. by dust absorption or two quantum emission.

Assume that the electron has been captured in the third state. This capture is accompanied by the emission of a quantum beyond the Paschen limit, which escapes from the nebula. Now the electron has two possibilities: it can go either directly to the first level with emission of a Lβ-quantum, or first to the second level and then to the first with emission of two quanta, Hα and Lα. In the first case, the emitted Lβ-quantum will be absorbed again due to the large optical depth of the nebula in the frequencies of the Lyman lines, and later on, it will again excite an atom to the third state. This process will continue until the second possibility occurs. Then the Hα-quantum leaves the nebula and the Lα-quantum remains and is scattered many times. The same reasoning can be applied to the cases where, instead of a Lβ-quantum, one has Lγ-quanta, Lδ-quanta, and so forth.

Thus, one Lα-quantum and one Balmer quantum result from each L$_c$ quantum absorbed in the nebula. If the optical depth τ_c of the nebula is of the order of unity or higher, it will absorb all the Lc-quanta emitted by the star per unit time. By measuring the total number of Balmer quanta emitted by the nebula, the number of ionizing photons emitted in the central star may be determined. In practice, it is necessary only to measure one Balmer transition (Hβ is usually used) since the ratio of that transition to the total Balmer emission can be predicted. This ratio has a weak dependence on the nebular temperature T_e, which is always sufficiently well known.

Once the number of ionizing photons emitted by the central star has been determined, a temperature can be determined by comparing this number with the number of photons emitted in the visual (as found from the magnitude of the star, for example). This step requires knowing what the continuous spectrum of the star really is as a function of temperature. Either a model atmosphere or an assumption of blackbody radiation may be used. These two possibilities will be discussed in more detail presently.

4.3.2 'Stoy' temperatures

This method of determining the temperature is named after the South-African astronomer who first discussed it. It is more descriptively called the 'Energy Balance' method.

Consider the ratio of the sum of the energy in the 'forbidden' lines to the energy emitted in one of the hydrogen lines, for example, Hβ. The 'forbidden' lines (a term used to designate all the transitions, both forbidden and permitted, formed in the nebula by electron collisions) are usually optically thin so that they escape from the nebula, and are the most important source of energy loss. Energy input in the nebula takes place by ionization, primarily of hydrogen and helium. The ionizing photon gives up that part of its energy required for ionization; the energy it had in excess of this amount goes into the motion of the ejected electron. This electron then shares its energy with the other particles by elastic collisions. This is the mechanism which heats the nebula.

Thus the ratio of the 'forbidden lines' F(FL) to Hβ, F(Hβ) is essentially the average value of excess energy of an ionizing photon. The higher the temperature of the star, the higher is the value of this ratio. This is the essence of the 'Stoy' temperature calculation. The ratio F(FL) to F(Hβ) is a purely observational quantity, requiring knowledge of the entire spectrum. The [OII] lines in the visible are often the most important but the CIII] lines at λ 1909 Å and the CIV lines at λ 1550 Å are also important and require a knowledge of the ultraviolet spectrum. For low excitation nebulae the infrared line of [NeII] at 12.8 μm can be very important.

The theoretical value of F(FL)/F(Hβ) depends not only on the temperature but also on the optical depth in the nebula shortward of λ 912 Å. The optical depth is not of such critical importance as in the Zanstra method, where not 'registering' a photon affects only the inferred number of ionizing photons, while leaving the other part of the ratio, the number of visual photons, unchanged. In the 'Stoy' method, not 'registering' a photon affects both the 'forbidden' line flux and the Hβ flux, but in slightly different ways.

The results of detailed calculations are given in Fig. 12, where the predicted values of F(FL)/F(Hβ) are given as a function of the effective stellar temperature. The solid curves correspond to a stellar blackbody radiation field. Several cases are distinguished. In case II it is assumed that the nebula is optically thin to radiation between λ 912 Å and λ 228 Å, and thick to radiation shortward of λ 228 Å. Case III assumes that the nebula is thick to all radiation. The case in which the nebula is thin to all radiation shortward of λ 912 Å is not plotted, largely because this case is probably not applicable to any nebulae except the very largest. Each of these cases is subdivided according to the extent of the helium ionization in the nebula. Case (a) assumes $He^+/H = 0.05$, $He^\circ/H = 0.05$; case (b) assumes $He^+/H = 0.03$, $He^\circ/H = 0.07$. Lower values of helium ionization may also be distinguished but are not shown in the figure because they usually occur when the stellar temperature is less than 10^5 K; at such temperature the small amount of very high energy radiation minimizes the effect of the helium ionization. Below $T = 7 \times 10^4$ K all cases assuming blackbody radiation are almost indistinguishable from each other, so that only a single curve is shown.

4.3.3 Stellar atmosphere analysis

This is the only method which is completely based on observation of the spectrum of the central star. High resolution spectra of the star are taken and the line profiles are compared with the prediction of a grid of non-LTE model atmospheres. In principle this is an extremely interesting way of obtaining the temperature. Because the model atmosphere plays an essential role, the surface gravity can be obtained as well.

The calculation of a non-LTE model atmosphere is a difficult and time consuming problem which at present requires that simplifying assumptions be made in order to obtain a solution. First of all only the elements hydrogen and helium are explicitly taken into account in calculating the atmospheric absorption. Secondly, in Mendez et al. (1988), McCarthy et al. (1990) and earlier work by this group, a plane parallel atmosphere in hydrostatic equilibrium is assumed. Gabler et al. (1991) have suggested using spherically extended models including radiation driven winds. This has not yet been done in detail. Gabler et al. (1991) report a rough calculation which suggests that these effects are important (see also Mendez et al., 1992). These authors state that these additional effects will not change their derived T_{eff} because the absorption lines are formed deep in the atmosphere where winds are unimportant, and only features formed at much higher layers, such as the He^+ ionizing continuum, will be affected.

4.3.4 Temperatures from nebular models

Nebular models are commonly used to determine nebular abundances. They are in less common use to determine the stellar temperature. One of the reasons for this is that one does not know which stellar atmosphere model to assume. A

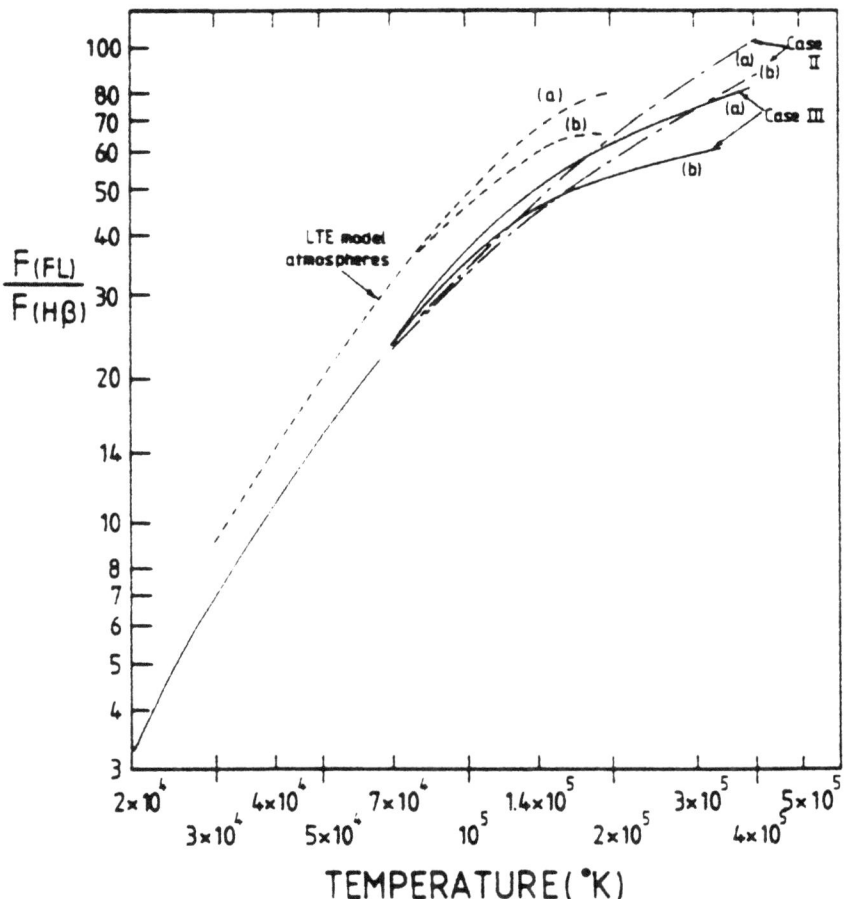

Figure 12: *Determination of the Stoy temperature. The ratio of the 'forbidden' line flux F(FL) to the Hβ flux F(Hβ) is plotted against the stellar temperature. The solid curve case (III) assumes blackbody stellar radiation, with the nebula optically deep shortward of λ 912 Å. The curve (case II) assumes blackbody stellar radiation, with the nebula optically thin between λ 912 Å and λ 228 Å and optically deep shortward of λ 228 Å. The dashed curve assumes model atmosphere (Hummer-Mihalas) radiation, with the nebula optically deep shortward of λ 912 Å. The curves (a) assume a nebula with $He^+/H = 0.05$ and $He^\circ/H = 0.05$, while curves (b) assume $He^+/H = 0.03$, $He^\circ/H = 0.07$. The differences between the different assumptions (and a blackbody radiation field) are small below $T = 2 \times 10^5$ K, and are negligible below $T = 7 \times 10^4$ K so that only a single curve is shown.*

second reason is that the resultant spectrum is more sensitive to the abundance used (the derivation of which is also one of the functions of the model) than to the precise value of the temperature derived. This is especially true for the higher temperatures. The insensitivity of the resultant spectrum to the temperature, while difficult for the temperature determination, is the reason that abundances can be derived with quite good accuracy.

This method is nevertheless in use, especially for the Magellanic Cloud PN. Here the star cannot be observed separately, so that Zanstra temperatures can only be determined when the star is bright enough so that its continuous emission can be easily separated from the nebular continuum. Stellar atmosphere analyses are of course impossible because the required high resolution stellar spectra cannot be obtained. Energy balance temperatures are also uncertain, because the faintness of the nebulae make ultraviolet and (far) infrared observations difficult or impossible.

4.3.5   Evaluation of the temperature determination

Sometimes the various methods give very consistent results. For example, all the methods agree to within 5% in assigning a temperature of 36,000 K to the central star of IC 418. But sometimes the various methods strongly disagree. The central star of NGC 2392 is perhaps the worst example. In the former case it appears that the star radiates approximately as a blackbody. In the latter case the stellar radiation appears to be far from a blackbody. Thus the present situation is not yet satisfactory. Optical depth effects play some role, but it is not yet clear how large their role is. Departure of the stellar radiation field from a blackbody also appears to play a role, but it may be a very important effect in some central stars and much less important in others. Present stellar atmosphere models are not yet able to make this distinction. It does not appear to be directly related to the stellar mass, since the central star of NGC 2392 (where the effect is large) has about the same mass as the central star of IC 418 (where the effect is small) (Mendez et al., 1988). Adam and Köppen (1985) have suggested that it is the stellar wind that enhances the radiation field in the far ultraviolet. The calculations of Gabler et al. (1991) point in the same direction. More detailed calculations including the structure and physical processes in the wind are necessary.

In the meantime it is wise not to be too dogmatic about the temperature. Energy balance temperatures are probably a reasonable approximation in many cases, but only when ultraviolet and infrared lines can be included.

110

4.4 Luminosity of the central stars

Once the stellar temperature has been determined, the luminosity may be discussed. Since the results will be presented mainly for the galactic bulge and Magellanic Cloud PN, the problems specific to these groups will be discussed.

It is desirable to eliminate the uncertainty concerning the optical depth of the nebula to hydrogen ionizing radiation as much as possible. To this end, the energy balance temperature should be used as much as possible, because its value is not strongly affected by whether the nebula is thick or thin. If, in addition, the visual magnitude can be observed, the stellar radius can be found, because the derivation of this quantity is not very dependent on reasonable departures from blackbody radiation.

This method depends on being able to measure the visual magnitude with reasonable accuracy. Since the lower temperature stars generally have the larger radii and thus brighter magnitude, the method is usually limited to stellar temperatures less than 100,000 K. Even for the Magellanic Clouds, Dopita and Meatheringham (1991a,b) claim to have measured the stellar flux by subtracting the nebular continuum from the total measured continuum for PN where the stellar continuum is a large fraction of the total continuum. This is very difficult because of the crowding in the Magellanic Clouds which causes other (field) stars to be present in the diaphragm. Because these authors used long-slit CCD spectra some of the contamination can be eliminated. Nevertheless, these authors show a reasonable correlation between their derived Zanstra temperature and that which they found from nebular modelling, which supports their claim.

If the stellar radius can be found from the visual magnitude, then the luminosity can be determined directly. If the visual magnitude is not measured then the luminosity can be found only for those nebulae which are known to be optically thick. In addition to the methods already described to determine whether or not a nebula is optically thick both Kaler and Jacoby (1990, 1991b), and Dopita and Meatheringham (1991a,b) use the strength of the low excitation lines of OII λ 3727 and NII λ 6584 as an important indicator. If these lines are relatively strong in PN whose central star temperatures are high, one concludes that very little of the ionizing radiation reaches the outer boundary of the nebula (a value of λ 3727/Hβ >1 to 1.5 is often used (e.g. Kaler and Jacoby, 1990) to indicate complete absorption of hydrogen ionizing photons).

Luminosities have been determined by Pottasch and Acker (1989) for about 30 PN, assumed to be optically thick, using Zanstra temperatures. A much larger sample has been studied by Ratag (1992), this time using energy balance temperatures. For the 30 PN in common, the agreement is quite good. Because Ratag et al. include the effect of far infrared radiation as measured by IRAS, their luminosities are slightly higher. A histogram of the distribution of luminosities of those PN whose temperatures are low is shown in the top diagram of Fig. 13. Low temperatures in this case refer to PN where the HeII line λ 4686 is absent or weak (λ 4686/Hβ < 0.08). The actual central star temperature is then usually less than 75000 K. Most of the PN fall in the range between 1000 L$_\odot$ and 6300 L$_\odot$ with 5 higher values ranging to 20000 L$_\odot$. The distance to the galactic bulge has been assumed to be 7.8 kpc in these determinations.

Luminosities have been determined in the Magellanic Clouds by Monk et al (1988), Kaler and Jacoby (1991a,b) and Dopita and Meatheringham (1991a,b). Their methods of determining temperature have already been discussed. The luminosities have been determined with the assumption that the nebulae are optically thick using methods described above. Dopita and Meatheringham occasionally find a PN to be optically thin but still feel able to assign a luminosity to these central stars. We have ignored these values in Fig. 13, where the central diagram gives the distribution for 34 PN in both clouds whose central stars have low temperatures. The distances used were 51 kpc to the LMC and 57.5 kpc to the SMC.

The distribution of the luminosities is very similar to the galactic bulge. The peak luminosity in both cases occurs between 4000 and 5000 L$_\odot$. If any significant difference between the two samples exists, it is that there are relatively fewer PN in the Magellanic Clouds with luminosities between 1000 and 2000 L$_\odot$. This could be a selection effect. A plot of the Hβ flux of the two samples of PN shows that there are relatively more objects further from the peak value in the galactic bulge sample.

The scale on the top of Fig. 13 gives the core mass, using the core mass-luminosity relation of (a) Schönberner (1981), (b) Boothroyd and Sackmann (1988), and (c) Wood and Zarro (1981). As can be seen the range of core mass is considerably narrower if the relation of (a) is used (0.54 M$_\odot$ to 0.57 M$_\odot$). It is at least 3 times wider if the relation (b) is used, and (c) gives an intermediate range. There is no strong evidence for central stars of luminosities greater than 12000 to 15000 L$_\odot$, corresponding to core masses of 0.7 M$_\odot$.

At the bottom of Fig. 13, the luminosities obtained from the central star parameters found from the stellar atmosphere analysis (Mendez et al., 1988; Gabler et al., 1991) for stars in the same temperature range are shown. The existence of only high luminosity central stars in this sample is certainly in part a selection effect and indicates that any attempts to obtain a distance scale for

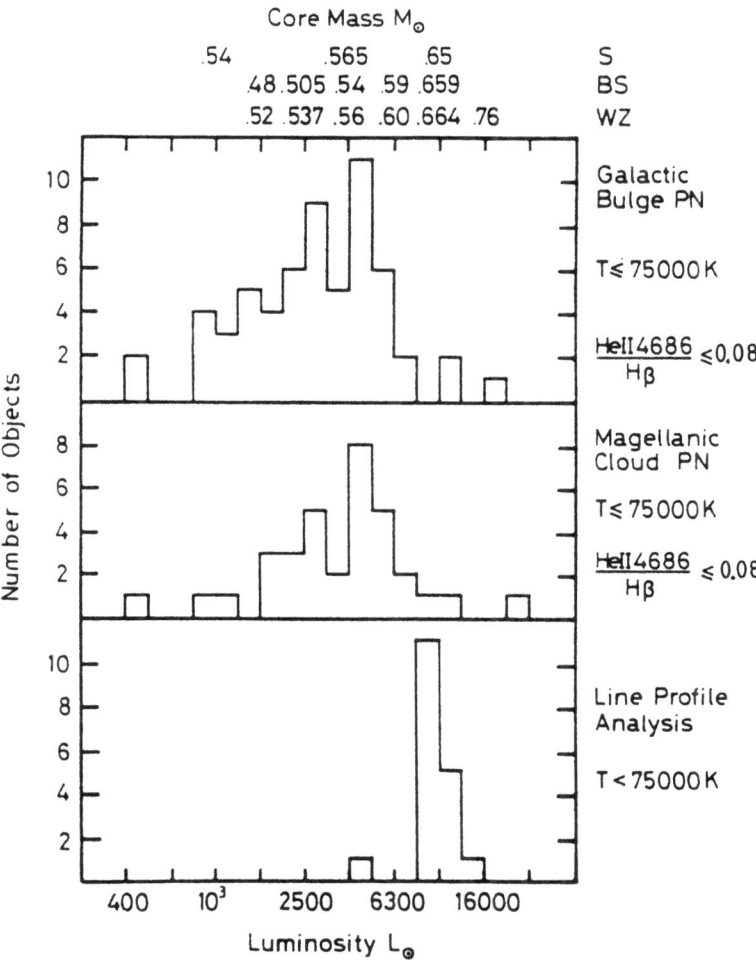

Figure 13: *Histogram of the central star luminosity when its temperature is less than 75000 K (and/or the nebular HeII line λ 4686 is weak) and the nebula is optically deep. The upper diagram refers to the galactic bulge (Ratag 1992) at a distance of 7.8 kpc. The middle diagram is for the Magellanic Clouds (Kaler and Jacoby, 1991; Dopita and Meatheringham, 1991a,b) assuming the distance of the LMC is 51 kpc and the SMC is 57.6 kpc. The lower diagram is from the work of Mendez et al. (1988) corrected somewhat for stellar winds (Gabler et al., 1991). The top scale gives the core mass from the luminosity according to (S) Schönberner (1981), (BS) Boothroyd and Sackmann (1988) and (WZ) Wood and Zarro (1981).*

PN from these results will lead to much too large distances.

The luminosities of central stars with temperatures above 75000 K have not been plotted in Fig. 13, because many of these stars are on the cooling tracks and are expected to have, on average, lower luminosity. This is actually found. The results of Dopita and Meatheringham (1991a,b) show luminosities distributed between 250 $L_\odot$ and 10^4 $L_\odot$ in the Magellanic Clouds. Kaler and Jacoby (1990) find luminosities between 50 and 6000 $L_\odot$. But the difference on the low side is a selection effect due to the inclusion of fainter PN in the latter sample. The luminosity range for galactic bulge PN is 180 to 8000 $L_\odot$ (Ratag 1992) which is consistent with the luminosity range in the Clouds.

4.5   Number and formation rate of planetary nebulae in the galaxy

A recent summary of the formation rate of PN, and their total number in the galaxy, has been given by Phillips (1989). He finds that there are about 30000 PN in the galaxy, and that they are being formed at the rate of 1.3 PN year^{-1}. As Phillips points out, these values are very uncertain because the distances are so uncertain. Indeed, various estimates of these quantities in the past decade are usually within a factor of 3 of these values and occasionally there are greater differences.

Let us try to estimate the numbers and formation rate in the galactic bulge, because, as has already been discussed, these nebulae are easy to identify and their distance is reasonably well known. There are approximately 283 optically observed PN within 10° of the galactic center which probably belong to the galactic bulge population. This is about 25% of all PN known. There are of course more PN which belong to the galactic bulge population but which have not yet been observed because of one of the following reasons.

1. They lie in regions of very high extinction and cannot therefore be seen

2. As nebulae expand, their surface brightness becomes lower and they are more and more difficult to discover

3. Intrinsically low luminosity nebulae are difficult to find at the distance of the galactic bulge

Taking these points into account there are 750 PN in the galactic bulge with diameters less than 6″, or radii less than 3.5×10^{17} cm if the bulge is at a distance of 7.8 kpc. Assume that the average expansion velocity is 15 km s^{-1}. This is less than the usual value of 20 to 30 km s^{-1}, but there are indications that the galactic bulge PN have lower velocities (Pottasch and Acker, 1989; Sahu et al., unpublished). The observable lifetime of a galactic bulge PN is then 8000 years. The formation rate of PN in the galactic bulge is thus about 10^{-1} PN/yr. This number may be increased (by about 30%) to take into account the new PN found from the IRAS colors described in the previous section. This will increase the formation rate to 1.3×10^{-1} PN/yr within 10° of the galactic center. From their distribution about 1/3 of these nebulae are within 3° (or 400 pc) of the center. Using the mass model of the inner galaxy given by Lindquist et al. (1992) there is a mass of about 1.5 to 3 $\times 10^9$ $M_\odot$ within 3° of the center. The formation rate per $M_\odot$ is then 1.5 to 3 $\times 10^{-11}$ PN $M_\odot^{-1}$ yr^{-1}. If the mass of the galaxy is 1.4×10^{11} $M_\odot$, and if the formation rate per $M_\odot$ and the observable lifetime are constant, which is uncertain, then there are 20000 to 40000 PN in the galaxy.

The scale height of PN in the neighborhood of the Sun has been examined by Zijlstra and Pottasch (1991). The method they use is different from many previous works (cited in Phillips, 1989) in that it makes no use of distance scales and little use of individual distances. It is based on observations of the number of PN at various latitudes, which appears to be a complete sample above latitudes of 30°–40°. In this way they find a scale height of 250±50 pc if an exponential disk is assumed. The reason given by these authors that earlier investigations gave too low values is that they were not corrected for Malmquist bias.

The discussion of Zijlstra and Pottasch (1991) leads to a total number of PN in the galaxy of 23,000, in reasonable agreement with both the number given above and the value given by Phillips cited at the beginning of the section. There is also reasonable agreement in the rate of formation in the Galaxy. The major part of the uncertainty can be traced to the different lifetimes used. Phillips uses a value of 2 $\times 10^4$ years, which may be a reasonable estimate for the more nearby PN but is too high for the galactic bulge objects. The lifetime is related to the expansion velocity which appears to be higher in the nearby PN (as already mentioned) but the selection effects related to determinations of average velocity have never been investigated. Thus the assumptions made in extrapolating the galactic bulge rates (and numbers) to total rates in the galaxy (constant formation rates and lifetimes) are probably not justified.

Since it is difficult to better analyze the nearby PN because of the uncertainty in their distance, one has turned to examining the statistics of PN in nearby galaxies. Because M31 has been well studied and is similar to our galaxy it is often used. Recently Peimbert (1990), making use of PN densities per unit luminosity in M31 found by Ciardullo et al. (1989) derived the total number of PN in our Galaxy to be 7200. There are two reasons why this number may be too low however. First, the PN densities are transformed into a birthrate using a mean PN lifetime of 25000 years, which, as we have seen above, probably only applies to a fraction of the PN. Secondly, the numbers given by Ciardullo et al include only PN which are within a factor 10 of the brightest PN in M31. Since PN can be another factor of 10 to 100 fainter and the extrapolation to fainter PN is very difficult to make, the total numbers could easily be underestimated. Thirdly they include only PN with reasonably strong [OIII] lines, because their measurements are made only in this line. Thus they are probably not counting PN with central star temperatures less than 35–40000 K. Furthermore Ciardullo et al. specifically say that their numbers are lower limits because no account was taken of dust extinction.

At present it appears difficult to give accurate values for numbers of PN: there are at least 10^3 PN in the galactic bulge and between 10^4 and 4×10^4 PN in the rest of the galaxy. About 1.3×10^{-1} PN are being formed per year in the galactic bulge and between 0.5 and 2 PN yr^{-1} are formed in the rest of the galaxy.

5   TRANSITION TO THE WHITE DWARF

It is now commonly accepted that white dwarfs are formed through post-AGB evolution with the ejection of a planetary nebula (e.g. Liebert, 1989). The evidence for this is both direct and statistical. The direct evidence is the fact that in old (large, low surface brightness) PN the central star often is a white dwarf or a star related to a white dwarf. In the hydrogen rich central stars both that of Abell 7 and NGC 7293 have white dwarf characteristics and values of log g $\simeq$ 7. One of the hottest known DA white dwarfs, PG 0950+139, is now known to be the central star of an extended planetary nebula of low surface brightness, which has a compact nebula at the center, indicating recent mass loss (Liebert et al., 1989). In the helium rich PN central star sequence, white dwarf candidates are also found. The high surface gravity of the central star of Jn1 indicates that it must be one, and the pulsating nucleus of K1–16 and possibly VV 47 show their relation to the PG 1159 stars which are hot stars (10^5 K) with surface gravities close to white dwarf values (log g $\simeq$ 7). A detailed discussion of these stars is given by Napiwotzski and Schönberner (1991).

The link between PN central stars and white dwarfs is clear and is expected. As the PN central star loses its hydrogen and/or helium rich envelope a star will remain which no longer can support nuclear processes and will slowly cool to a white dwarf. It is therefore interesting to study the statistics of these objects to see if the main path to white dwarf formation is through the PN stage or whether there may be other paths to white dwarf formation.

The comparison is difficult because white dwarfs, being faint, are only measured in a reasonably complete way to distances of the order of 100 pc, while it is doubtful if there is even one PN this close. The best determination of the white dwarf scale height is 275±50 pc (Boyle, 1989) which is in reasonable agreement with the value of 250 pc given for PN by Zijlstra and Pottasch (1991), but, as discussed previously, smaller values for PN also exist in the literature.

Fleming et al. (1986) using the Palomar Green survey find a local space density of DA white dwarfs of $0.49\pm0.05 \times 10^{-3}$ (pc)$^{-3}$ down in a magnitude $M_v = 12.75$. Liebert et al. (1988) give a value of 3×10^{-3} (pc)$^{-3}$ as the local space density of all observed degenerate dwarfs. Iben and Laughlin (1989) point out that there is a significant uncertainty to this number. In any case this number cannot be directly compared with PN densities and must first be transformed into œ[Bœ[Ba formation rate for white dwarfs. To do so the rate of cooling of the white dwarfs must be used. Iben and Laughlin then obtain a birthrate for white dwarfs of 1.25×10^{-12} pc^{-3} yr^{-1} in the local neighborhood. (A reevaluation by Weidemann (1991) gives a value in the range 1.4–2.3×10^{-12} pc^{-3} yr^{-1}, including binary stars.) If the local mass density is 0.1 to 0.2 $M_\odot$ pc^{-3} (Gilmore, 1989) then the local birthrate is between 6.3 and 12.5×10^{-12} $M_\odot^{-1}$ yr^{-1}. This can be extended to the entire galaxy: if the galaxy has a mass of 1.4×10^{11} $M_\odot$ and the birthrate of white dwarfs is everywhere the same then between 0.9 and 1.8 white dwarfs/yr are formed. This may be compared to our estimate at the end of the previous section that between 0.5 and 2 PN yr^{-1} are formed in the galaxy. This rather good agreement, if taken literally would mean that most white dwarf are preceded by a PN phase. At present the errors in both the formation rates are large enough that this result should only be taken as an indication, but it appears that most white dwarfs pass through the PN stage.

The mass distribution of white dwarfs is reasonably well known (Shipman, 1989; Weidemann and Koster, 1984; McMahan, 1989; Bergeron et al., 1992). It is sharply peaked with a mean mass between 0.56 and 0.58 $M_\odot$. The extreme values extend from less than 0.4 $M_\odot$ to about 1.2 $M_\odot$ (1.05 $M_\odot$ for Sirius B). The distribution is sharper on the low mass side. A controversy has developed as to whether the PN central stars can reproduce this distribution. Since the core masses are determined from the luminosities, the questions of distance and nebular optical depth play a role. Even more important is the exact form of the core mass-luminosity relation. The present situation is as follows. There is reasonable agreement that the central stars have core masses which show a peak at approximately 0.6 $M_\odot$ and have a core mass fall-off on the high mass side which seems to approximate that of the white dwarfs except that central stars above 0.7 to 0.8 $M_\odot$ are very hard to identify individually, perhaps because they evolve too quickly to be seen (but they may be observable and our determination of the luminosity could be at fault). The controversy is more concerned with the low core mass end. Using the core-mass-luminosity relation of Schönberner (1981) there are no central stars below 0.54 $M_\odot$. One must therefore conclude that the low mass white dwarfs (about 30% of the total) are formed without going through the PN state. On the other hand, if the relation of Jeffery (1988) is used, the low mass end of the central star distribution will reproduce the low mass white dwarf distribution reasonably well. This requires that evolution times for low mass stars must be faster than present calculations (e.g. Schönberner, 1981) allow, perhaps because of higher envelope mass loss rates. It is also possible that close binary evolution may play a role in producing both low mass central stars and low mass white dwarfs.

References

I. Theory
A general description is given by

Pottasch, S.R.: 1984, *Planetary Nebulae*, chap. IX and X, Kluwer Academic Publishers, Dordrecht (the Netherlands).

Much more detailed summaries are given by

Iben, I. Jr., Renzini, A.: 1983, *Ann. Rev. Astron. Astr.* **21**, 271.

Iben, I. Jr.: 1985, *Q. J. R. Astr. Soc.* **26**, 1.

Weidemann, V., Schönberner, D.: 1990, *From Miras to PN*, p. 3. Eds. Mennessier, M.O., Omont, A. Editions Frontières.

The core-mass vs. luminosity relation has been discussed by

Paczynski, B.: 1970, *Acta Astr.* **20**, 47.

Wood, P.R., Zarro, D.M.: 1981, *Astrophys. J.* **247**, 247.

Boothroyd, A.I., Sackmann, I.-J.: 1988, *Astrophys. J.* **328**, 641.

Evolution of hydrogen and helium shell burning cores is discussed by

Schönberner, D.: 1981, *Astron. Astrophys.* **103**, 119.

Blöcker, T., Schönberner, D.: 1991, *Astron. Astrophys.* **240**, L11; **244**, L43.

Vassiliades, E., Wood, P.R.: 1992, *Astrophys. J.* in press.

II. Observations of AGB stars

Many review papers can be found in the workshop proceedings *From Miras to PN*, eds. Mennessier, M.O., Omont, A. Editions Frontières, 1990. Other important reviews are:

Whitelock, P.A.: 1990, *ASP Conf. Ser.* **11**; 1992, *ASP Conf. Ser.* **30**.

Whitelock, P.A., Feast, M.W., Catchpole, R.M.: 1991, *Mon. Not. Roy. Astr. Soc.* **248**, 276.
Whitelock, P.A., Feast, M.W.: 1993, in *Planetary Nebuale*, IAU Symp. 155. Eds. Weinberger, R., Acker, A. Kluwer Academic Publ., Dordrecht.

A good discussion of the OH/IR stars is found in the thesis of te Lintel Hekkert, P. 1990 (University of Leiden) and the summary of Habing, H.J. in the workshop cited above.

III. From AGB to PN

The IRAS color-color diagram is discussed by

van der Veen, W., Habing, J.J.: 1988, *Astron. Astrophys.* **195**, 125.

Pottasch, S.R.: 1990, in *From Miras to PN*, p. 381. Eds. Mennessier, M.O., Omont, A. Editions Frontières.
Lewis, B.M.: 1989, *Astrophys. J.* **338**, 234.

The catalogue of IRAS low resolution spectra can be found in: *Astron. Astrophys. Suppl.* **65**, 607, 1986.

Other references:

García-Lario, P., Manchado, A., Sahu, K.C. and Pottasch, S.R.: 1993, *Astron. Astrophys.* **267**, L11.
Gillett, F.C., Backman, D.E., Beichman, C. and Neugebauer, G.: 1986, *Astrophys. J.* **310**, 842.
Hrovnak, B.J., Kwok, S., Volk, K.M.: 1989, *Astrophys. J.* **346**, 265.

Jura, M.: 1986, *Astrophys. J.* **303**, 327.

Kwok, S.: 1993, in *Planetary Nebuale*, IAU Symp. 155. Eds. Weinberger, R., Acker, A. Kluwer Academic Publ., Dordrecht.

Parthasarathy, M., Pottasch, S.R. : 1986, *Astron. Astrophys.* **154**, L16.

Parthasarathy, M., García-Lario, P., Pottasch, S.R., Manchado, A., Clavel, J., de Martino, D., Van de Steene and Sahu, K.C.: 1993, *Astron. Astrophys.* 267, L19.

Slijkhuis, S.: 1992, Thesis, University of Amsterdam.

Waters, L.B.F.M., Sahu, K.: 1993, in *Planetary Nebuale*, IAU Symp. 155. Eds. Weinberger, R., Acker, A. Kluwer Academic Publ., Dordrecht.

Zijlstra, A.A., te dintel Hekkert, P., Pottasch, S.R., Caswell, J.L., Ratag, M. and Mabing, J.: 1989, *Astron. Astrophys.* **217**, 157.

IV. Planetary nebulae

A general description can be found in the book of the same title by Pottasch, S.R.: 1984, referred to above.

Other references:

Acker, A., Chopinet, M., Pottasch, S.R. and Stenholm, B.: 1987, *Astron. Astrophys. Suppl.* **71**, 163.

Adam, J. and Köppen, J.: 1985, *Astron. Astrophys.* **142**, 461.

Ciardullo, R., Jacoby, G.H., Ford, H.C. and Neill, J.D.: 1989, *Astrophys. J.* **339**, 53.

Cohen, J.G., Gillett, F.C.: 1989, *Astrophys. J.* **346**, 803.

Dopita, M.A., Meatheringham, S.J.: 1991a, *Astrophys. J.* **367**, 115; 1991b *Astrophys. J.* **377**, 480.

Gabler, R., Kudritzki, R.P. and Méndez, R.H.: 1991, *Astron. Astrophys.* **245**, 587.

Gathier, R., Pottasch, S.R. and Goss, W.M.: 1986a, *Astron. Astrophys.* **157**, 191.

Gathier, R., Pottasch, S.R. and Pel, J.W.: 1986b, *Astron. Astrophys.* **157**, 171.

Kaler, J.B., Jacoby, G.H.: 1990, *Astrophys. J.* **362**, 491.

Kaler, J.B., Jacoby, G.H.: 1991a, *Astrophys. J.* **372**, 215; 1991b, *Astrophys. J.* **382**, 134.

Kudritzki, R.P., Mendez, R.H. 1989, in *Planetary Nebulae*, IAU Symp. 131, p. 273. Eds. Torres-Peimbert, S., Kluwer Academic Publ., Dordrecht.

Liller, M.H., Welther, B.L. and Liller W.: 1966, *Astrophys. J.* **144**, 280.

Lindquist, M., Habing, H.J. and Winnberg, A.: 1992, *Astron. Astrophys.* **259**, 118.

Lutz, J.H.: 1973, *Astrophys. J.* **181**, 135.

Masson, C.R.: 1986, *Astrophys. J.* **302**, L27.

Masson, C.R.: 1989a, *Astrophys. J.* **336**, 294; 1989b, *Astrophys. J.* **346**, 243.

McCarthy, J.K., Mould, J.R., Méndez, R.H., Kudritzki, R.P., Husfeld, D., Herrero, A. and Groth, H.G.: 1990, *Astrophys. J.* **351**, 230.

Méndez, R.H., Kudritzki, R.P., Herrero, A., Husfeld, D. and Groth H.G.: 1988, *Astron. Astrophys.* **190**, 113.

Méndez, R.H., Kudritzki, R.P. and Herrero, A.: 1992, *Astron. Astrophys.* **260**, 329.

Minkowski, R.: 1965, in *Galactic Structure*, p. 321. Eds. Blaauw, A., Schmidt, M.

Monk, D.J., Barlow, M.J. and Clegg, E.S.: 1988, *Mon. Not. Roy. Astr. Soc.* **234**, 583.

Peimbert, M.: 1990, *Rev. Mex. Astron. Astrof.* **20**, 119.

Phillips, J.P.: 1989, in *Planetary Nebulae*, IAU Symp. 131, p. 425. Ed. Torres-Peimbert, S., Kluwer Academic Publ., Dordrecht.

Pier et al: 1993, in *Planetary Nebulae*, IAU Symp. 155. Eds. Weinberger, R., Acker, A. Kluwer Academic Publ., Dordrecht.

Pottasch, S.R., Acker, A.: 1989, *Astron. Astrophys.* **221**, 123.

Ratag, M.A.: 1992, Thesis, University of Groningen.

Schönberner, D.: 1981, *Astron. Astrophys.* **103**, 119.

Terzian, Y.: 1993, in *Planetary Nebuale*, IAU Symp. 155. Eds. Weinberger, R., Acker, A. Kluwer Academic Publ., Dordrecht.

Weinberger, R. and Zeiner, R.: 1988, *Astron. Astrophys* **191**, 297.

Zijlstra, A.A., Pottasch, S.R.: 1991, *Astron. Astrophys.* **216**, 245.

V. Transition to the white dwarf.

Several reviews are available: In IAU Symp. 131 (1989) both J. Liebert and

121

H.L. Shipman review various aspects of white dwarfs and their evolution. This is brought up to date by Liebert in IAU Symp. 155 (1993). The conference proceedings 'White Dwarfs' (eds. Vauclair, G., Sion, E.), 1991, have various articles of which those by Weidemann and Napiwotzski and Schönberner are referred to in the text:

Other references:

Boyle, B.J.: 1989, *Mon. Not. Roy. Astr. Soc.* **240**, 533.

Bergeron, P., Saffer, R.A. and Liebert, J.: 1992, *Astrophys. J.* **394**, 228.

Fleming, T.A., Liebert, J. and Green R.F.: 1986, *Astrophys. J.* **308**, 176.

Gilmore, G.: 1989, Saas-Fee Course, University of Geneva.

Iben, I.Jr., Laughlin, G.: 1989, *Astrophys. J.* **341**, 312.

Jeffery, C.S.: 1988, *Mon. Not. Roy. Astr. Soc* **235**, 1287.

Liebert, J., Dahn, C.C. and Monet, D.G.: 1988, *Astrophys. J.* **332**, 891.

Lindquist, M., Habing, H.J. and Winnberg, A.: 1992, *Astron. Astrophys.* **259**, 118.
McMahan, R.K.: 1989, *Astrophys. J.* **336**, 409.

Weidemann, V., Koester, D.: 1984, *Astron. Astrophys.* **132**, 195.

THE MILKY WAY GALAXY AND THE GALACTIC CENTRE

Gerard Gilmore

Institute of Astronomy
Madingley Road
Cambridge CB3 0HA, UK

1 INTRODUCTION

It is widely though erroneously believed that one can see the Milky Way Galaxy. In fact, one's image of the Milky Way depends more on how one looks at it than on what is available to be seen. For reasons which are related to population biology more than to astrophysics, our eyes are optimised to detect the peak energy output from thermal sources with a surface temperature near 6000K. Thus, unless such an object is typical of the entire contents of the Galaxy, there is no reason why we should be able to see by eye a representative part of whatever may be out there. If we had X-ray or UV sensitive eyes we would 'see' only hotter objects, if infrared or microwave eyes only cooler objects.

No single section of the electro-magnetic spectrum provides the 'best' view of the Galaxy. Rather, all views are complementary. However, some views are certainly more representative than are others. The most fundamental must be a view of the entire contents of the Galaxy. Such a view would require access to a universal property of matter, which was independent of the state of that matter. This is provided by gravity, since all matter, by definition, has mass. Mass generates the gravitational potential, which in turns defines the size and the shape of the Galaxy. While the most reliable and comprehensive, such a view is also the hardest to derive. Nonetheless, we will repeatedly return to the gravitational picture of the Galaxy in these lectures.

Complementary and relatively readily available views of much of that part of the ordinary baryonic mass whose state has been identified can be provided by the sum of optical and near infrared studies. Much of the mass of the Milky Way is in stars, the more massive of which are visually luminous. Lower mass stars, those objects hiding behind interstellar extinction, and much of the inter-stellar medium (ISM), are most readily observed in the near infra-red.

The importance of infrared astronomy for study of obscured objects, a property which is of particular significance when studying the central regions of the Galaxy and regions of current or recent star formation, is illustrated in Table 1. This tabulates representative values for the wavelength dependence of extinction, illustrating the relative transparency of interstellar dust at wavelengths just a little longer than those to which our eyes are sensitive.

Infra-red astronomy provides the closest and greatest complementary match to optical astronomy, while at the same time extending the source temperature range available for study to those lower temperatures at which many known astrophysical sources are to be found, and at which a substantial part of the higher energy radiation is re-emitted.

TABLE 1

Optical/IR Interstellar Extinction

Photometric Passband	Wavelength μm	Extinction (magnitudes)
U	0.36	1.56
B	0.44	1.33
V	0.55	1.00
R	0.64	0.78
I	0.79	0.59
J	1.25	0.28
H	1.65	0.17
K	2.2	0.11
L	3.5	0.06
M	4.8	0.02
N	10	0.05

1.1   What is there to be seen?

Perhaps the most basic question one should ask about the Galaxy is *What is it made of?* The answer of course, is matter. Mostly. This leads to two supplementary questions, *Where is the matter?* and *In what form is the matter?*. Supplementary questions to which we shall return are *Do massive black holes matter?* and *Do brown dwarfs matter?*. The answers to those questions are *probably not* and *probably not* as we shall see. The availability of both answers is in large part thanks to infrared astronomy.

We now recall the basic evidence concerning the large scale mass distribution in the Galaxy. The most direct evidence comes from the Galactic rotation curve. HI 21cm rotation curves in external spirals show behaviour which is crudely characterised as a flat rotation curve. In fact very few rotation curves are flat, and systematic differences exists between the rotation curves of galaxies of different sizes, surface brightnesses and Hubble types (Casertano & van Albada 1990). Nonetheless, to an adequate approximation for present purposes, disk galaxy rotation curves may be characterized as 'flat', *ie* as having an approximately constant circular velocity with radius, to radii far outside the radius at which luminous matter is readily detectable. A combination of HI and direct stellar dynamical studies in the outer Milky Way (Fich & Tremaine 1991; Arnold & Gilmore 1992) suggests similarly that the rotation curve of the Milky Way remains approximately flat to at least 50kpc from the centre, and possibly to considerably greater distances.

The significance of flat rotation curves is illustrated by the simple force-balance relation involving the circular velocity V(R) at some radius R from the centre of the Galaxy and the enclosed mass interior to that radius, M(R):

$$V^2(R) = G\frac{M(R)}{R} \tag{1}$$

A flat rotation curve is the statement that

$$V(R) \approx const \tag{2}$$

so that

$$M(R) \propto R \tag{3}$$

That is, the integral mass of the galaxy increases linearly with increasing distance. Since the volume element in interval dR goes as R^3, this suggests that the mass density of total mass in a galaxy is distributed as $\rho_{mass}(R) \propto R^{-2}$.

Galaxy luminosity profiles behave quite differently, generally decreasing as an exponential with a scale length of a few kpc, or (equivalently) as a power law with index ~ 3, so that $\rho_{light}(R) \propto R^{-3}$. Thus the total luminosity of a galaxy is convergent, and is strongly centrally concentrated. The total mass is divergent when last seen, and is hardly, if at all, concentrated to the luminous central regions. As yet, no single photon has been detected which certainly has emanated from the mass which dominates the outer parts of galaxies. It is well named 'dark matter'. The same mass, as well as a variety of other things, is also often called 'missing mass'. This name is less appropriate, since the only property of the mass which is unambiguously determined is its presence. It is an understanding of the mass which is missing, not the mass itself.

The fundamental difference between the spatial distributions of mass and of light is the most important conclusion of studies of the mass distribution in galaxies. It is rarely appreciated. At its most simple, this distinction between mass and light proves that any attempted explanation of the nature of the dark matter which involves scaling identified mass must be wrong. Such attempted explanations are legion, with speculation that a substantial mass in very low mass 'stars', which are of too low mass to burn hydrogen, and so are very under luminous, may accompany the more readily visible stellar mass function. While this may be true – we will discuss later in these lectures the status of such investigations – any such mass by definition is distributed like the luminous

127

part of a galaxy. The dark mass is distributed differently. Thus, it cannot be explained by scaling the luminous mass in any way.

Identification of the luminous mass in the galaxy, and determination of the true number of very low mass objects associated with the stellar initial mass function remain subjects of considerable astrophysical importance. Complementary approaches to these studies involve detailed determinations of the mass distribution in the Galactic disk, analysis of the stellar mass function for objects near to the minimum mass for hydrogen burning, and direct IR surveys. The first two of these approaches will be covered in detail in these lectures, while the last is discussed elsewhere in this volume.

The best reason why one expects brown dwarfs to exist is really plausibility. Stars are known with masses covering the whole range from $\sim 100 \mathcal{M}_\odot$ to $\sim 0.1 \mathcal{M}_\odot$, while planets, comets, people and dirt extend the mass range of known objects over the range $\sim 0.001 \mathcal{M}_\odot$ down to near zero. The masses defining the lower end to the stellar mass regime and the upper end to the regime of known planets are determined by physics which has no known relationship to whatever is the physics determining stellar and planetary formation. Thus there is no *a priori* reason to expect that the small fractional mass range in which it is maximally difficult for us to detect single objects should happen to correspond to a mass range in which physics has problems making things. There is however no valid *a priori* reason to expect that physics is maximally efficient at making things in that region either. This assumption is adopted surprisingly often, in spite of its implausibility. Thus, one expects brown dwarfs and massive planets to exist, but *a priori* they should be neither rare nor copiously overabundant compared to the numbers of objects of slightly different masses.

1.2   Mass to Light Ratios

When discussing mass distributions one often makes use of mass to light ratios in solar units, $M/L = (M/\mathcal{M}_\odot)(\mathcal{L}_\odot/L)$. When using solar visual passband units some expectations based on the best calculations currently available can be noted. For the immediate Solar Neighbourhood direct studies of the stellar mass function, which will be detailed further below, show that the local mass density in stars more massive that $0.35 \mathcal{M}_\odot$ is $0.05 \mathcal{M}_\odot \mathrm{pc}^{-3}$. For stars and brown dwarfs less massive than this the corresponding value is $0.008 \mathcal{M}_\odot \mathrm{pc}^{-3}$. The luminosity density can be calculated from the known stars near the Sun, and is

$$j = 0.063(\pm 0.006)\mathcal{L}_\odot \mathrm{pc}^{-3}. \tag{4}$$

Thus, the local mass to light ratio in stars and substellar mass objects is $M/L_{\odot,v} \sim$ 1. When restricting the sample of stars considered to avoid the rather rare luminous stars which live only in the central Galactic Plane, so as to provide a value representative of the Galactic disk on larger scales, one derives $M/L_{\odot,v} \sim 3$. This and some other values, to set this value in context, are noted in Table 2.

TABLE 2

MASS to LIGHT Ratios, Solar Visual units

Cosmology	$\Omega = 1$	$M/L \sim 1500$
Standard Big Bang	$\Omega_B \lesssim 0.05$	$M/L \lesssim 75$
Milky Way	to 50kpc	$M/L \sim 10$
Milky Way	local disk	$M/L \sim 3$
Globular clusters	luminous part	$M/L \sim 2.5$

These values illustrate the complexity of definition required when discussing mass distributions in astrophysics. The disagreement by a factor of order 20 between the mass to light ratio required for simple solution of the cosmological horizon and flatness problems ($M/L \sim 1500$) and that derived from the Standard Big Bang model ($M/L \lesssim 75$) is the basic 'dark matter' problem. If both these numbers are correct, most mass in the Universe is non-baryonic, and of otherwise unknown properties. The second 'dark matter' problem is the disagreement by another factor of order twenty between the baryonic mass content of the Universe deduced from the Standard Big Bang model and the baryonic matter actually identified in galaxies. If this disagreement is real, then most of the ordinary baryonic matter in the universe is distributed in some state which efficiently defies detection. However, since this mass is baryonic, it can couple to photons, and in principle be detected. It can also cool, and may find its way into galaxies. It might even make up the dark halos of galaxies like the Milky Way. So long as it is indeed dark. Hence the interest in under-luminous matter, such as brown dwarfs, inside galaxies. There is always the possibility that brown dwarfs represent the toe of a crouching giant, in mass distribution terms, rather than the body of a midget.

What is loosely called 'dark matter' has that name just because it is not a prominent source of photons. Intrinsically low luminosity sources are easiest to detect when they are close, so our immediate neighbourhood is one of the places where one can most readily derive tight limits on the luminosity of any such matter. Additionally, stellar dynamics, and hence determination of both the total gravitational potential and the identified mass contributing to that potential, can be carried out with greatest precision with the high quality data available uniquely in the Solar neighbourhood. This is the explanation of the very considerable interest in the last few years in studying dark matter near the Sun.

Although any local dark mass need not have any relation to the dark matter of the Galactic halo, there is always the possibility that it might provide a valuable clue to the location and state of the unidentified baryons deduced from the Standard Big Bang Model. As it has turned out, there is no significant evidence for dark matter associated with the Galactic disk, so that the evidence provides a limit on, rather than a clue towards, the nature of the halo dark matter, and the location of the missing baryons. In spite of that, our part of the Milky Way remains one of the best places to look for brown dwarfs, while the infrared is the most appropriate wavelength region. Such studies, as we will see later, are of more general interest than just as a mass counter.

1.3 Galactic Structure: Why Bother?

There is more to the study of Galactic structure than its relevance for dark matter studies – it continues to have considerable intrinsic interest. The spatial distribution of stars in the Galaxy is related, through the gravitational potential Φ, to their kinematics. Specifically, the scale length of the spatial distribution is determined by the combination of the total energy of the stellar orbits and the properties, primarily the spatial gradient, of the gravitational potential. The shape of the spatial distribution allows one to determine the relative amounts of angular momentum (rotational), and pressure (stellar velocity anisotropy) balance to the potential gradients.

To be more explicit, the total orbital energy and angular momentum of the gas which will become a star depend on the maximum distance from the centre of the proto-Galaxy which it ever reached, the angular momentum of its orbit at that time, the depth of the potential well (generated by both dark and luminous mass) through which it fell, the fraction of the total orbital energy which was dissipated, the amount of viscous transport of angular momentum before star formation, and the subsequent dynamical evolution of the stellar orbit.

That is, the present kinematic properties of old stars in the solar neighbourhood are determined in part by the initial conditions in the proto-Galaxy, and in part by the physics of galaxy formation. Hence, determination of the large scale structural and kinematic properties of the Galaxy can help to determine both the detailed physics of galaxy formation and the distribution of gravitating mass, both identified and dark, in the Galaxy. As well as satisfying our inherent curiosity about our neighbourhood, such studies provide the most promising opportunity to understand the fundamental features of the formation, evolution, and structure of galaxies in general. The dominant processes are summarised in Table 3.

The important feature of the Table is not its completeness or its extent, but the extent to which all aspects of the present structure of the Galaxy are interrelated. One cannot reliably deduce the important properties of the Galaxy from a subset of the data treated in isolation. Rather, a wide range of astrophysical constraints need to be included at all times.

A specific example of the importance of ensuring that the widest possible astrophysical context is utilised can be seen in the history of the 'Kapteyn Universe', which is discussed in Chapter 2 of the book 'The Milky Way as a Galaxy' (Gilmore *et al.* 1990). The lessons of that major research project are still valid.

TABLE 3: PHYSICS OF THE MILKY WAY GALAXY

Type of Data		Model Function and Physics
Kinematics	$\Longleftrightarrow$	Dynamics
radial velocities		phase space distribution function
proper motions		spatial distributions
		$\Rightarrow$ gravitational potential
		dissipational history
Chemical Abundances	$\Longleftrightarrow$	Chemical Evolution
line strengths,		star formation history
photometry		ISM history
		$\Rightarrow$ stellar initial mass function
		gas flows, dissipation, SFR
Star Counts	$\Longleftrightarrow$	Galactic Structure
colour magnitude data		spatial distribution function
surface brightness		luminosity functions
		$\Rightarrow$ stellar IMF, binarism,
		dissipation, SFR, ...

2   GALACTIC STRUCTURE: THE BUILDING BLOCKS

In these lectures the major structural features of the Milky Way Galaxy – the thin disk, the thick disk, the bulge and the stellar halo – are described, with particular emphasis on the relationship between the central bulge and the other components of what we believe to be a typical large spiral galaxy.

Many of the important physical processes in Galactic formation and evolution noted in Table 3 above are relevant to each of the present components of the Galaxy, emphasising the inter-relationships between all the apparently diverse types of data. In fact, the distribution of stars in the Galaxy in coordinate space, in kinematic space, and in chemical abundance space, are intimately related. A meaningful understanding of any sub-part of stellar phase space presupposes consideration of all other parts.

Although we have just been discussing the Galaxy in terms of discrete building blocks. It is very important to remember that the relationships between these units are our major topic of research in Galactic structure. The discreteness or otherwise of the stellar populations in the Galaxy is our primary evidence for the past history of Galactic evolution. Thus it is important not to *assume* a *physical* discreteness behind a division which is made mostly for convenience and historical reasons. With that caveat however, it is very often convenient to discuss the Galaxy in terms of components. These are typically the central bulge, the stellar halo (often called also the stellar spheroid), the thick disk, the thin disk, the dark halo, and the interstellar medium (ISM).

The first requirement is to decide how many (discrete) stellar components one wishes to consider in the model. For present purposes we will consider the large-scale Galaxy to be rotationally symmetric and north-south symmetric, away from the central bulge, but we discuss explicitly the asymmetry of the central bulge below. The gross features of the Galaxy with which we are left include the thin disk, the thick disk, the (subdwarf) halo, the central ($r \lesssim 3$kpc from the Galactic centre) bulge, and a very central ($r \lesssim 1$kpc from the Galactic centre) structure. This last component may or may not be the same as the central bulge as defined above, and there may or may not be continuity between some or all of these components. This question is important for studies of Galactic evolution (*cf.* Gilmore *et al.* 1989) but not for our immediate overview purposes. Any continuum can be modelled at some level as a sum of discrete functions, and the amount and quality of extant data is such that a model with several discrete components has more than enough degrees of freedom to describe available observations.

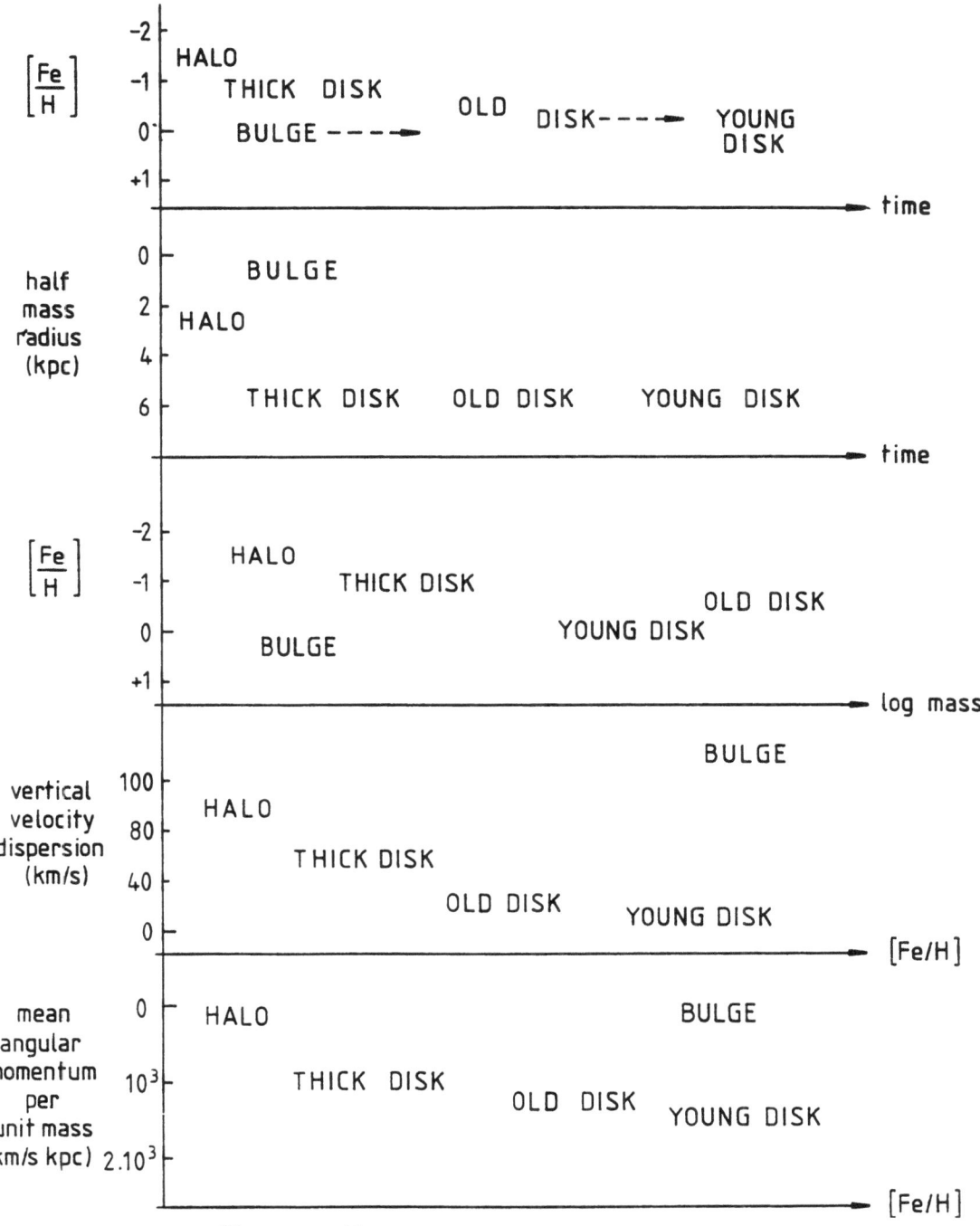

Figure 1: *Alternative views of the Galaxy*

These components can be represented in many parts of parameter space. Four such representations are shown in Figure 1, as a taste of the complexity of the relationship between the parts which make up the whole.

Given the number of (discrete) components one desires to include, it now remains to specify the general properties of each component.

2.1 The Central Bulge :– $r \lesssim 1\text{kpc}$:

The existence of this component is deduced in 2 ways – in a model-dependent way from the inner rotation curve, and directly from optical and IRAS counts of late M giants and from near IR integrated-light observations. The rotation curve modelling is somewhat complicated by the increasingly plausible evidence for bar-driven non-circular motions in the gas (*a priori* plausible, since most galaxies do not show a maximum in the inner rotation curve like that in the Galaxy), while the star count data are somewhat complicated since the stars counted are in a short-lived and poorly-understood evolutionary state. Thus one cannot reliably deconvolve a density gradient from an abundance and/or an age gradient. Most available data are adequately represented by a spatial density profile $\nu(r) \propto r^{-1.8}$ for $r \lesssim 1\text{kpc}$.

2.2 The Main Bulge :– $1 \lesssim r \lesssim 3\text{kpc}$:

The annulus between $\sim 10°$ and $\sim 30°$ from the Galactic centre is one of the least understood and yet one of the most significant regions in the Galaxy. It corresponds to the only non-disk regions of sufficiently high surface brightness in external galaxies that they can be studied, and yet has been almost entirely neglected in our Galaxy. Just sufficient star count data exist to show that the stellar distribution in this annulus is not consistent with models which do not include it as an extra component. Very preliminary indications suggest a scale length of $\sim 1\text{kpc}$ is appropriate for the density profile, but such data are complicated by the difficulty in reliable corrections for reddening. The available data are not adequate to define the structure parameters consistently, due to difficulties with photometry in crowded fields, the complexities of patchy reddening, and the need to obtain quite large amounts of data to define the density and colour distributions adequately. Until better data exist it is not possible to include a reliable description of the dominant stellar population within $\sim 30°$ of the Galactic centre in extant star count models. Near infrared data are crucial here.

2.3   The Subdwarf Halo:

There are several alternative density profiles which one might wish to consider to describe the stellar halo distribution. To a precision which is adequate for comparison with presently available data, the subdwarf system can be described by the density distribution which corresponds to an $r^{1/4}$ -law when seen in projection. This density profile is, for $r/r_e \gtrsim 0.2$ or $r \gtrsim 0.5$kpc (latitude $\gtrsim 5°$), given by (Young 1976):

$$\nu_{halo}(r) \propto \frac{\exp\left[-7.669(r/r_e)^{1/4}\right]}{(r/r_e)^{7/8}}. \tag{5}$$

In this equation r is the distance from the Galactic centre, which is related to the radial R and vertical z distances from the observer to a star by

$$r^2 = R_0^2 + R^2\cos^2 b - 2RR_0\cos b\cos\ell + z^2 \tag{6}$$

where

$$z = R\sin b \tag{7}$$

with R_0 the solar galactocentric distance (~ 7.8kpc), b and ℓ the Galactic coordinates of the line of sight, and the transformation

$$z \Rightarrow (c/a)z \tag{8}$$

allows for an oblate distribution with axis ratio c/a. The unknown quantities in equation 5) are the scale factor r_e, the radius which encloses 50% of the total halo luminosity, and the proportionality constant. r_e can be measured directly, and a value of $r_e \approx 2.7$kpc is widely quoted from surface photometry. The proportionality constant may be merged with the corresponding proportionality constant which arises in the normalisation of the luminosity function, since only the product of the luminosity and density functions is observable. This is determined by counting subdwarfs near the Sun.

An alternative density distribution, which is that derived directly from RR Lyrae and globular cluster studies, is a power law $\nu(r) \propto r^{-n}$, or allowing for a 'core radius' r_c and combining the normalisation constants into f_{halo}, the fraction of all stars near the sun which belong to the halo

$$\nu_{halo}(r) = f_{halo}\nu_{r=0}\frac{(r_c^n + R_0^n)}{(r_c^n + r^n)}. \tag{9}$$

A suitable value for r_c is ≈ 1kpc, though this value is neither well-determined nor critical. The normalisation constant, the fractional number of all stars in the solar neighbourhood which follow this density law and are described by this colour-magnitude relation, is $\approx 1/800$, with an uncertainty of at least 50% in this value.

The luminosity function appropriate for the field subdwarfs can be constrained in a variety of ways. Samples of field subdwarfs with good distances have been analysed, and show that, within the very large sampling noise, there is no significant difference in the luminosity function of subdwarfs with [Fe/H] $\lesssim -1$ and that of solar abundance disk stars for $5 \lesssim M_V \lesssim 15$. (This similarity imposes some constraints on star formation theories, showing that the low mass spectrum is almost independent of the metallicity through the range when the effect of metals on the high temperature cooling function changes from being negligible to being dominant.) Direct modelling of star count data reaches similar conclusions regarding the similarity of the shape of the luminosity functions of halo and disk stars. For evolved stars one may utilise available age data. Effectively all stars with [Fe/H] $\lesssim -0.8$ are as old as the globular clusters. Hence, all evolved halo stars have almost identical masses, and the luminosity function is determined entirely by stellar evolution, rather than by the stellar initial mass function. One may therefore use the observed luminosity functions of evolved stars in globular clusters with some confidence.

Some complexities remain, however. The most important of these is the treatment of the horizontal branch. The second parameter problem alluded to earlier means that considerable uncertainty is involved in treating horizontal branch stars, and that one must be guided entirely by observation. There is suggestive, though not conclusive, evidence that one would expect more blue horizontal branch stars than are seen, so that the second parameter problem afflicts subdwarfs near the sun. A further complication is photometric evidence that globular clusters show significant differences in their main sequence luminosity functions, with a possible systematic change in the slope of the luminosity function with metallicity. The significance of this is complicated by the fact that there are corresponding, and not well defined, systematic changes in the stellar mass-luminosity and absolute magnitude-colour relations with metallicity, due to the effect of changing opacity on the stellar atmosphere and interior. The significance of any resulting systematic changes in the stellar *mass* function remain to be studied in detail. Since the relationship (if any) of the field stars to globular clusters remains unknown this adds a further note of warning, and another interesting parameter which may be measured, to star count modelling.

2.4 The Thick Disk:

The vertical density profile of the thick disk is shown in figure 9 of Kuijken & Gilmore (1989), and is adequately represented for $1000 \lesssim z \lesssim 3000 \text{pc}$ by an exponential

$$\nu(z) = \nu_{z=0} \exp(-z/h_{z,td}). \tag{10}$$

The current best estimates for the normalisation constant and scale height are 4% and 1000pc, respectively. The normalisation constant $\nu_{z=0}$ is roughly an order of magnitude larger than that for the halo stars, though again with an uncertainty of about 50% in this value. Note that the normalisation in this case is *not* the fractional number of thick disk stars near the sun, but the numerical value required to model the stellar distribution a few kpc above the plane assuming the density profile of equation 10. The relationship of this numerical value to the actual number of thick disk stars in a volume near the sun is a steep function of the Galactic $\mathcal{K}_z(z)$ law (*cf.* Kuijken & Gilmore 1989) and the stellar velocity distribution function. We note in passing that a noticeably oblate $r^{1/4}$ density profile ($c/a \approx 1/4$) also provides a good description of the data.

The radial profile of the thick disk is still poorly known. The Basel star count surveys suggest that a radial exponential is an adequate description, so that the two-dimensional (rotational symmetry is assumed remember) density distribution becomes

$$\nu(r,z) \propto \exp(-z/h_{z,td} - r/h_{r,td}). \tag{11}$$

The radial scale length $h_{r,td}$, based on available star count modelling and by assumption from photometry of other disk galaxies is the same as the radial exponential scale length of the thin disk, or $\sim 4 \pm 1 \text{kpc}$. The numerical value of this scale length is discussed further in Gilmore *et al.* (1990).

The colour-magnitude relation $\Im$ for the thick disk is defined rather well. The mean metallicity of the thick disk is well determined (*cf.* Gilmore *et al.* 1990) to be like that of the metal–rich globular cluster system, while the age of at least the metal–poor part of the thick disk is also similar to the age of the globular cluster system. For more metal–rich thick disk stars the situation remains unclear, pending further observations. Thus one should adopt the colour-magnitude relation of a metal–rich globular as the most suitable choice, but beware of the possibility that it may be inadequate, in being too red, for stars near the turnoff. A similar situation applies for the choice of an appropriate luminosity function for the thick disk. The luminosity function of a metal–rich globular cluster is the most appropriate choice available, but the possibility remains that it will lack an adequate mass range, corresponding to an age range, near the turnoff.

137

2.5   The Thin Disk:

Models of the spatial distribution of stars in the thin disk suffer from two serious difficulties: the system is a complex mixture of ages, velocity dispersions, and metallicities; and there is a lot of observational information available. Subsidiary complications, particularly the effects of large and irregular obscuration, further confuse the issue. To provide a reliable description of the stellar distribution within ~ 200pc of the Galactic plane one would have to model the star formation history, the present distribution of ages, chemical abundances and velocities, and the reddening distribution, and would also have to know the $\mathcal{K}_z(z)$ force law near the plane to high precision. If that information were available there would be no point in building crude statistical models of the data. Conversely, when near infrared surveys become available, a wealth of fundamental studies of the structure and the evolution of the Galactic disk will be possible for the first time. By considering currently available optical data only, we restrict the discussion to distances $\gtrsim 200$pc from the plane. This effectively restricts the modelling to a description of the distribution of old disk stars with vertical velocity dispersion $\gtrsim 15$kms^{-1}. For this restricted case, the considerable volume of extant data shows an exponential to be an adequate description of the spatial density distribution, though a vertical distribution described by $\nu(z) \propto \text{sech}^2(z/2z_0)$, with z_0 the equivalent exponential scale height, is an equally satisfactory description.

The radial distribution of the old disk is invariably assumed to follow an exponential distribution, though the true distribution is almost entirely uncon-strained by observation. The strongest evidence in favour of a radial exponential is the argument that other disk galaxies are (more or less) exponentials. Some indirect dynamical arguments and some limited photometric data support a ra-dial scale length of $\sim 4 \pm 1$kpc for the old disk (*cf.* Gilmore *et al.* 1990). As the reader will have deduced from this uncertainty, the radial distribution is poorly constrained since it cannot be seen: interstellar extinction forms a very effective screen between the real old disk and our models, and *vice versa*. Once again, we await with interest the forthcoming near infrared surveys.

3   WHAT IS THE GALACTIC BULGE?

3.1   Surface Brightness Measurements

Since the Galaxy is not extragalactic, and occupies rather a large solid angle on the sky, the interpretation of surface photometry must proceed in a model dependent way. Surface photometry at low Galactic latitudes ($b \lesssim 20°$) is of course very severely affected by interstellar obscuration, which requires careful consideration. This is especially important for studies of the Galactic disk. At higher latitudes obscuration is less of a serious problem, though local obscuration near the Sun still requires careful treatment. One proceeds by adopting some model for the spatial luminosity distribution of the halo, calculating the line-of-sight integral of this distribution, subtracting from that model a model for the disk contribution, and then correcting for a model of the distribution of inter-stellar extinction. One then compares this resulting model to observations at a range of Galactic latitudes to determine the several parameters (central surface brightnesses, density profiles, characteristic scale lengths, axial ratios) of the halo and disk luminosity distributions. In practice the good data are limited to a narrow range in surface brightness, due to the dominance of the disk and extinction problems at low latitudes, and the faintness of the halo at high latitudes. Thus, as one might imagine, the resulting parameters are not determined to high precision.

The most careful attempt to carry out an analysis of this type is that by de Vaucoulers & Pence (1978). They showed that their surface brightness data over a range in surface brightness of 1.5 magnitudes between Galactic latitudes 15° and 30° were consistent with a luminosity distribution which follows the $r^{1/4}$ -law, with an assumed effective radius $r_e = 2.7$kpc. The c/a axial ratio of the halo was equally well fit by models with $c/a = 1$ and models with $c/a = 0.6$. Other combinations of r_e and c/a were not tried.

The difficulties imposed on models of this type by interstellar extinction may be mitigated using observations in the near infra-red ($\sim 2 - 2.5\mu$m), where the effects of reddening are substantially reduced. Suitable 12μm data from the IRAS satellite, and 2.4μm observations by Hiromoto *et al.* (1984) and Hayakawa *et al.* (1981) were analysed for this reason by Garwood & Jones (1987). They investigated a variety of $r^{1/4}$ models, and deduced an axial ratio $c/a = 0.25$. They also showed that an $r^{1/4}$ -law halo, with the normalisation in the central regions of the Galaxy fixed by the infrared data, was inconsistent with the observed space density of extreme Population II subdwarfs near the Sun.

Thus, the most likely interpretation of available surface photometric data is that the Galactic halo is consistent with, but does not require, an $r^{1/4}$ –law distribution with effective radius $r_e \sim 2.5$kpc and axial ratio $c/a \lesssim 1$. In the central few kpc of the Galaxy an additional, flatter distribution is seen, with a shorter spatial scale length. The relationship between this central component and the rest of the Galaxy remains problematic, and is discussed further below.

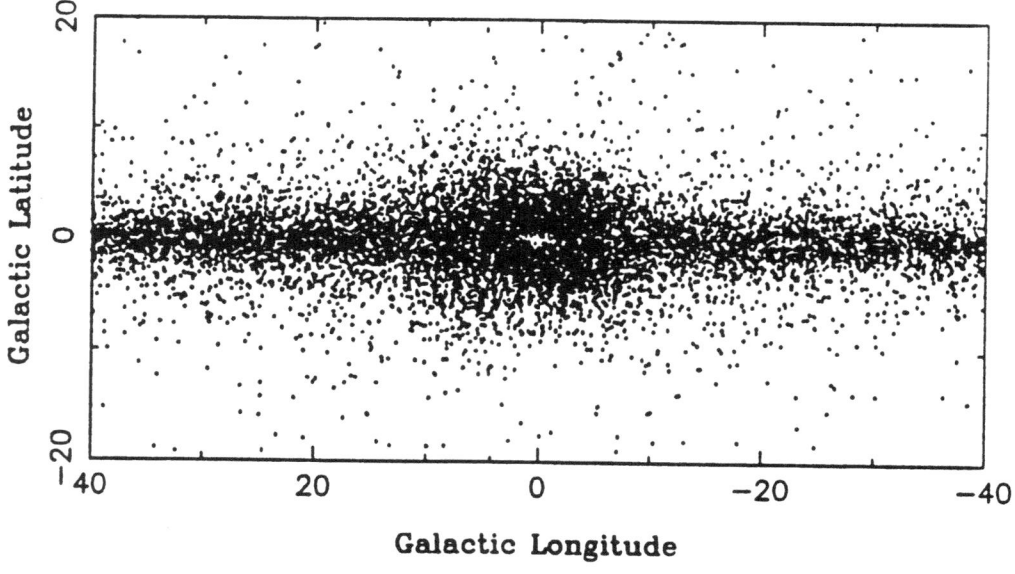

Figure 2: *The distribution of those IRAS sources with $0.4 \lesssim f_{12\mu m}/f_{25\mu m} \lesssim 1.6$ and $1\ Jy \lesssim f_{12\mu m} \lesssim 8\ Jy$, where $f_{x\mu m}$ is the IRAS flux at the stated wavelength. This colour and flux range have been chosen so as to maximise the contrast between the bulge and the disk (from Harmon & Gilmore 1988).*

3.2 IRAS Source Results

The large scale distribution of sources which are luminous at $12\mu m$ and $25\mu m$ was mapped by the IRAS satellite, and forms a striking disk–bulge picture which is complementary and almost completely orthogonal to optical stellar studies (Habing 1987). From the IRAS colour data one can select those sources which are luminous (dust-enshrouded) stars, and exclude those hotter stars which are simply nearby and highly reddened. The resulting spatial distribution is shown below.

The sources outlining this distribution can be shown to be a mixture of two types of late-type giants, with those in the bulge being intermediate-mass AGB stars ($1.2 \lesssim \mathcal{M}/\mathcal{M}_\odot \lesssim 3$, Harmon & Gilmore 1988) and those in the disk being a mixture of intermediate mass and low mass stars with dust shells. The parameters of the large-scale spatial distribution have been derived by Habing (1987) who shows that the extended distribution forms two disks. About 80% of the stars form a thin disk (scale height $\lesssim$ 200pc) with radial exponential scale length 4.5kpc, and with a cutoff at $\sim$ 9kpc, or about at the solar circle. The remaining 20% of stars form a thick disk, with scale height $\sim$ 2kpc, radial exponential scale length ~ 6kpc, and no evident cutoff. The details of these distributions however remain uncertain due to residual contamination of the IRAS samples with objects other than late type stars.

These disk parameters may be understood in terms of the two types of star detectable by IRAS. IRAS could see low mass stars with high optical depth dust shells, and higher mass young AGB stars with high mass-loss rates. For low mass stars there is a correlation between the optical depth of the dust shell in the late stages of evolution (the Mira variable stage) and metallicity, with those stars which have [Fe/H] $\gtrsim$ −1 having the highest optical depth shells (Harmon & Gilmore 1988). These same Mira variables have pulsation periods from ~ 150 days to ~ 200 days, and have been known for many years to outline a thick disk. In fact, the Vatican conference stellar population classification scheme used these variables to define the Intermediate Population II.

The thin disk IRAS sources are predominately higher mass AGB stars – Miras and OH/IR stars. These stars are young, and hence their distribution reflects that of the young disk and the molecular gas which corresponds to regions of current and recent star formation. The distribution of molecular clouds is now well-defined (Solomon & Rivolo 1987) and also shows a cutoff at about the solar circle, in excellent agreement with Habing's analysis of the IRAS stellar data.

The IRAS sources in the central few degrees of the Galaxy outline the central bulge, which is hidden from optical study by interstellar obscuration. The detectable outer edge of the IRAS bulge in Figure 2 in fact is near Baade's Window, where optical studies are first possible. IRAS and optical data are therefore nicely complementary. The spatial distribution of the central IRAS bulge ($4° \lesssim |b| \lesssim 10°$, where the lower latitude limit is set by satellite confusion) is well described by a roughly spherical exponential with scale height 375pc, corresponding to a half-light radius (~ 1.68 exponential scale lengths) of ~ 600pc. That is, the central bulge contains at least some young stars (those seen by IRAS, *cf.* Harmon & Gilmore 1988) and has a scale length a factor of about 5 smaller than that followed by more metal–poor halo stars. In the absence of adequate age data for the sources in the central few degrees of the Galaxy it is not yet possible to tell if the small scale length is due to a steep abundance gradient in an old stellar population, or to the presence of a 'discrete' young central component of the Galaxy.

3.3 Is the Bulge a Bar?

The suggestion that the bulge is bar shaped is not new. de Vaucoulers (1964) is one example, while W.B. Burton has consistently argued that the inner rotation curve is an *apparent* rotation curve, so that the inner maximum means either an inner mass concentration or, equally, an asymmetric mass distribution. Such suggestions remained difficult to test in detail until recent IR survey data became available.

The recent availability of infra-red survey data has allowed a detailed analysis of the spatial distribution of the central regions of the Galactic Bulge. Several authors have emphasised that the stars detected by IRAS are predominately long period variables, distributed in a very flattened concentrated inner bulge (Habing 1987, Harmon & Gilmore 1988, Nakada *et al.* 1991, Weinberg 1992). It was noticed early in these analyses (*e.g.* Harmon & Gilmore 1988 figure 3a) that the spatial distribution of IRAS sources is asymmetric, with the sources being systematically brighter at positive longitudes. Uncertainties in the IRAS photometric calibration were expected to have both this sign of effect and to be consistent with its amplitude, so that the most conservative conclusion had to be that calibration problems were the explanation.

Ground-based studies of the IRAS sources however (*e.g.* Whitelock 1992) showed the flux asymmetry to be real, thus providing direct evidence for bar structure. Detailed analyses of the spatial distribution of IRAS sources (Nakada *et al.* 1991; Weinberg 1991), and the $2.4\mu m$ surface brightness distribution (Blitz & Spergel 1991) both detect asymmetry in the luminosity distribution. Kinematic analyses of the galactic central gas (Binney *et al.* 1991) also provide dynamical consistency with a model in which the central regions of the Galaxy are barred. The resulting model is reasonably consistent between all these analyses, and suggests a bar pointing nearly along the line of sight to the Sun – an offset of only 16° fits the data best (Blitz *et al.* 1993). This value may seem implausibly small, but in fact it is just a selection effect – any larger value would have made the bar easier to see, and its existence would no longer be in dispute.

3.4 Is the 'Bulge' Part of the 'Halo'?

The luminosity distribution of the central few kpc of the Galaxy out of the plane is dominated by a component – the 'bulge' for present purposes – which has a half-light scale length of about 0.5kpc. This is particularly apparent in near infrared data, but is also evident in careful optical studies. The relationship of the extended stellar halo, represented by the subdwarfs near the Sun, to the bulge is quite unknown. It is important to remember that the bulge corresponds to the only non-disc contribution to the luminosity of external spiral galaxies which is of sufficiently high surface brightness to be studied.

Available data provide a somewhat confusing picture of the properties of the Galactic bulge. Analysis of those stars dominant at $12\mu m$ in the IRAS survey suggests they are intermediate age long period variables (Harmon & Gilmore 1988). The majority of the bulge population at low Galactic latitudes must be older than the Sun, and may be as old as the metal-rich globular clusters (Terndrup 1988). Chemical abundance data for a sample of K giants in Baade's Window shows them to be metal rich, with modal abundance perhaps twice solar (Rich 1988). The distribution of abundances for these stars is consistent with that expected from the simple model of chemical evolution with a closed box (no inflow or outflow), but with effective yield significantly higher than that derived from observations in the solar neighbourhood. Similar abundance data for planetary nebulae and RR Lyrae stars (Gratton *et al.* 1986) however provides a modal abundance of one-half solar, consistent with the same effective yield as is seen near the Sun. A point which is related to this is why the metal-rich stars in the bulge should be appropriate as templates in stellar population studies of giant elliptical galaxies. The absolute magnitude of the bulge is about -18 while big ellipticals are several magnitudes brighter. Given the well-established relations between absolute magnitude and colour and metallicity (or at least line strength), it is really very unlikely that the integrated spectrum of the bulge can look like that of a much bigger galaxy. It would be interesting to investigate this further, as it suggests that the very metal rich stars must provide a minor component of the integrated light of the Galactic bulge.

Kinematic data from a variety of studies which are currently underway are beginning to quantify the larger scale kinematics of the Bulge. The Bulge is rotating, with a roughly linear rotation curve rising to a peak of ~ 80km/s at ~ 2kpc from the centre. The velocity dispersion profile drops steeply down the minor axis, from a central value of ~ 120km/s to ~ 80km/s by a latitude of $10°$. By analogy with studies of the bulges of external spirals, one might have expected the Galactic bulge to be rotating, with both velocity dispersion and mean rotation about one-half of the asymptotic amplitude of the HI rotation curve, or about 100 kms^{-1}. Straightforward application of the radial Jean's equation to the kinematics of stars near the Sun also predicts a value for the velocity dispersion of both the disk and the halo stellar populations of $\sim 125 kms^{-1}$ near the Galactic centre. Systematic rotation will reduce this value, while triaxiality of the central bulge potential may either increase or decrease it substantially, depending on the geometry of the situation.

There are several fundamental properties of the bulge which could be determined from straightforward observations. These include determination of the abundance distribution for stars of sufficiently low luminosity so that dredgeup has not affected their atmospheric abundances, thereby providing a distribution function of abundances which is representative of that at the time of stellar formation. A sufficiently large sample of stars must be observed to clarify the following points:

i) Are the very metal-rich stars a tail of a distribution which is represented by the abundance distribution seen in the old planetary nebulae and RR Lyrae stars, or *vice versa*;

ii) Where are the very metal-rich old disk (and thick disk) stars which are expected in significant numbers if there really is a radial abundance gradient in the disk;

iii) Where are the subdwarfs, like those near the Sun, whose density distribution must also peak in the centre of the Galaxy?

If these several groups of stars are intimately related one might expect a smooth distribution function to be found. If however they represent discrete phases of the evolution of the Galaxy a multi-modal distribution can be expected, like that found for the globular cluster system. Kinematic and spatial distribution data for these same classes of stars are necessary for a serious analysis, and could readily be obtained.

One of the most important properties of the bulge which is amenable to test is the age range of the metal-rich stars. While it is often assumed that these stars are as old as the globular clusters there is in fact no strong evidence in favour of this hypothesis. It would indeed be mildly surprising if that population of stars which is the youngest in chemical terms, in that the greatest number of generations of massive stars must have had time to evolve and explode before its formation, and which is young in dynamical terms, in that a substantial amount of dissipation of binding energy occurred before star formation, was at the same time among the oldest in a chronological sense.

There is a simple test of the hypothesis that the very metal-rich stars in the central bulge are indeed as old as the globular clusters, and that is to determine the ratios of oxygen (preferably) or the α elements (Mg,Ca,Si,Ti). Old subdwarfs near the Sun and giants in globular clusters show a systematic overabundance of oxygen and the α elements, with [O/Fe] and [α/Fe] $\sim +0.4$ for [Fe/H] $\lesssim -1$ [Figure 3]. The explanation of this behaviour is reviewed in Gilmore *et al.* (1989), and is simply that these elements are produced only in supernovae from

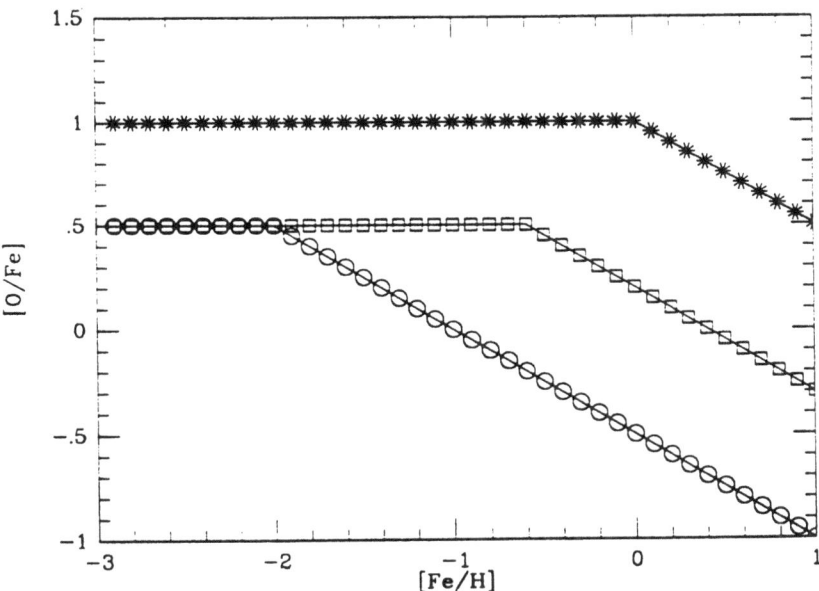

Figure 3: *The relationship between the oxygen to iron ratio and stellar metallicity. The top line shows the effect of an IMF biassed towards high mass stars, the circles and squares show a local IMF with slow and rapid star formation respectively. Location of the Galactic bulge on this plot will determine its IMF and formation rate.*

high-mass progenitors, and tend to be dominant in the very early stages of the evolution of a stellar system. After about 1Gyr (Type I) supernovae from lower-mass progenitors dominate the elemental production in the Galaxy, so that all element ratios tend towards the solar values. This age corresponds to a metallicity of about -1dex.

If the central bulge really is as old as the globular clusters, it must also have element ratios consistent with dominance of its chemical evolution by high-mass short-lived supernova progenitors. Thus, determination of the ratios [O/Fe] and [α/Fe] for the bulge stars with [Fe/H] $\approx +0.3$ is a unique test of the hypothesis that the bulge is as old as the globular clusters. If these ratios are significantly less than $+0.4$, then the bulge is younger than the clusters.

4 THE GALACTIC CENTRE

The Galactic central regions are buried behind very many magnitudes of optical extinction. Infra-red and radio astronomy provide effectively all the precise information we have available, though the improving spatial resolution of X-ray and γ−ray satellites is beginning to allow very high energy studies to contribute.

To put the spatial scales in context, an angular scale of 1 degree corresponds to $\sim 175\text{pc}$, 1 arcminute corresponds to $\sim 2.9\text{pc}$, and 1 arcsecond corresponds to $\sim 0.05\text{pc}$

The morphology revealed by infrared and radio mapping is extremely complex. For historical reasons the Galactic centre is often described in a mix of radio and infrared alphabet and number designations by constellation. The larger scale morphology is conveniently classified as being Sagittarius (Sgr) A, B, C, D, and E. Sgr E is a sum of many HII regions, Sgr D is a sum of supernova remnants (SNR) and regions of current star formation, Sgr C is a mix of very large filaments and a compact shell, Sgr B is an extremely dense set of HII regions and complex filaments, while Sgr A contains the central part of the Galaxy, and includes representatives of pretty well every known astrophysical object and a dramatically wide range of physical conditions.

A very important point to remember is the place of the Galactic centre in the larger scheme of things. Very complex mixtures of structures, as we see in SgrA and its neighbours, are not unusual. M83 has also been studied with reasonable spatial resolution (Gallais *et al.* 1991), and shows a very complex mix of nucleus, bridge, arc, bar, ... structure. A very large number of different physical structures with very different physical conditions contribute to the total output flux. Single component 'models' for the central regions of galaxies observed at low spatial resolution need to be considered in this context. Secondly, in spite of its great psychological and scientific interest, the central regions are relatively quiescent. The radio flux from the central few hundred parsec of our Galaxy is an order of magnitude less than that of the median flux from 'normal' spirals, while the central compact radio source SgrA* is 10,000 times fainter than those sub-parsec sized sources we can see in other normal spirals. It is helpful to remember this when considering explanations of the phenomena which invoke extreme objects.

The morphological properties of the Galactic centre are extremely complex, with structure visible on all measured scales. Detailed representations of various parts of the whole can be found in several articles in a recent conference proceedings (Morris 1989) and in many review articles (eg Blitz *et al.* 1993).

4.1   The Energy Balance

One of the more interesting specific questions one can ask about the central regions is the energy balance. The most important terms are summarised in Table 4 below.

TABLE 4

ENERGY BALANCE IN THE CENTRAL REGIONS

ENERGY		Parameters	Energy Density ergs cm^{-3}
Gas Pressure	$E_p \sim nkT$	$n \sim 10^5$, $T \sim 7000K$	$E_p \sim 10^{-7}$
Magnetic Pressure	$E_B \sim B^2/8\pi$	$B > 10mG$	$E_B \gtrsim 4 \times 10^{-6}$
Gravitation	$E_G \sim GM\rho/r$	$r \lesssim 1pc$, $M \gtrsim 3 \times 10^6$	$E_G \sim 3 \times 10^{-5}$
Tidal Forces	$E_t \sim GM(\Delta r)^2 \rho/r^3$	$\Delta r \sim 0.1pc$	$E_G \sim 3 \times 10^{-7}$

The importance of each contribution to the pressure varies from place to place. Gravity is clearly dominant in the central few pc or so, but not at very much larger distances from the centre. In the radio filaments the very high polarization shows the gas dynamics to be dominated by the poloidal magnetic field. Cloud motions are MHD waves. In the central few hundred pc the gas pressure is kept high by a pervasive very hot ($T \sim 10^8 K$) ISM. The resulting pressures are several hundred times greater than pressures in molecular gas elsewhere in the Galaxy.

4.2 Star Formation in the Centre?

4.2.1 Should One Expect Current Star Formation?

Very high pressures might be expected to encourage very high rates of star formation. In fact, one of the major topics of current activity is the star formation rate and recent history in the central regions. While magnetic pressure within clouds is certainly relevant here, another severe constraint arises from the tidal forces.

For stars to form their progenitor gas cloud must be stable. The limit of stability is just the Roche density, which is of interest in the very central regions,

$$\rho \geq M_{centre}/r^3 \sim 4 \times 10^{10} \frac{M_{BH}}{10^6 M_\odot} \left(\frac{0.1pc}{r}\right)^3 cm^{-3} \qquad (12)$$

for tidal stability near a putative central point mass of mass M_{BH}, and more

147

generally

$$\rho \geq 3 \times 10^7 \left(\frac{r}{1pc}\right)^{-2} \left(\frac{\sigma}{80km/s}\right)^2 \text{cm}^{-3} \tag{13}$$

Most observed GMC in the central regions have densities of order 10^5cm^{-3}, suggesting no star formation in the very central regions. The additional support against collapse provided by magnetic pressure also has to be considered. However, higher densities can certainly be provided in transient collisions and shocks, so that in principle some star formation in the very central regions might occur. One would not expect an extremely high rate, however.

4.2.2 Is There Current Star Formation?

The observational evidence for current star formation a little out of the central regions is now convincing. Far IR studies have shown that OB stars ionize the filaments of molecular clouds to provide the large scale thermal filaments. Spectra of individual hot supergiants are available (eg Allen *et al.* 1990) while near IR energy distributions (eg Moneti *et al.* 1992) show many sources to be young massive stars still embedded in dust. Nearer to the Galactic centre sources such as IRS7 are now well established to be massive supergiants, with a derived age of order 60 million years in that case.

Very near the Galactic centre a significant number of high-excitation He emission stars have been detected recently. The IRS16 'source' has now been resolved into a cluster of a large number of stellar sources, of which at least 6 are very hot stars with He emission (Eckart *et al.* 1992). That is, the infrared source IRS16 associated (at low spatial resolution) with the Galactic centre, is a cluster of extremely hot, and quite likely young, stars. The advantages of the infrared for observing even hot sources behind extinction are again evident, as is the case for continuing star formation in the Galactic centre.

4.3 The Great Annihilator

There are several distinctly unusual objects in the central regions of the Galaxy. Among the most unusual is the Great Annihilator. This a luminous and variable source of the 511keV electron-positron annihilation line. Perhaps 5% of the time something creates $\sim 10^{10}$ tonnes of positrons. And not much else. The positrons in their turn annihilate rapidly and efficiently with electrons, producing a pair of 511keV photons for each annihilation. This source is highly variable, and so far as is known is unique in the Galaxy. With the poor spatial resolution until recently available with γ—ray detectors, it could be located only to within some degrees of the Galactic centre. Effectively all that could be reliably deduced was that the source must be small to be visible, having size of order 1000km. Thus it could not be a very massive black hole.

The almost invariable assumption made that this source was associated with the Galactic centre compact radio source SgrA* is a salutary warning! The logic was impeccable – a unique object is found poorly located near a unique place. Surely the thing and the place must be the same! Wrong. Another example of the same logic, also wrong, is Sco X-1, the brightest Galactic X-ray source. This source is nearly perfectly aligned inside a double radio source, prompting the obvious assumption that the radio 'lobes' have been ejected from Sco X-1. With the benefit of hindsight, and more useful of proper motion data, it is now clear that perverse chance aligned a unique X-ray object on the line of sight to an unrelated background source. Similarly with the Great annihilator, which is now known to lie some 1 degree from the centre.

Mirabel *et al.* (1992) detected the variable source as part of a VLA program monitoring the Galactic central regions for transient sources, and were able to derive a precise position. So what is the Great Annihilator? Most likely a neutron star (or perhaps a stellar mass black hole, though that is not required) which happens to lie near the surface of a very dense molecular cloud. The high gas densities in the cloud provide the necessary high particle densities to allow efficient annihilation in a small volume, and so let us see this remarkable source. It is interesting to note that a similar positron source far from a dense cloud could not be seen, as its positron flux would disperse before annihilation, leading only to a very low surface brightness background. Such things may in fact be relatively common in the Galaxy. The Great Annihilator is visible more because of the presence of a very dense molecular cloud than because it is a source of positrons. Why then is it near the Galactic Centre? Since most of the high density molecular gas is there. Similar selection biases may be relevant to 'unique object in a unique place' arguments, for example for the radio source Sgr A*.

The Great Annihilator was to some extent predicted before its discovery, as the following lines illustrate:
As one great furnace flamed
Yet from those flames no light
But rather darkness visible.   (John Milton, Paradise Lost).

4.4 Sgr A*

The unresolved radio source SgrA* has not yet been certainly identified at wavelengths shorter than the high frequency radio, though some recent mm observations may be relevant. (If these are indeed of SgrA* then a very dense dust component is required.) A possible detection at $2\mu m$ has been suggested, but again the flux observed is low, and the likelihood of chance projection from a star in the very dense central stellar cluster is quite high. An upper limit on the X-ray flux in the 2–30keV range is $F_X \lesssim 4 \times 10^{35}$erg/s. The overall luminosity is thus very low, being some 10000 times less luminous than compact radio sources of the type studied in other 'normal' galaxies. While SgrA* is unique in our Galaxy, it isn't very luminous.

Radio data show a size distribution of $\theta_{1/2} = 0".0015\lambda^{2.05}$, consistent with either a source with an r^{-1} internal density profile, or to a point source broadened by interstellar scattering. The radio brightness temperature, $T_B \gtrsim 10^9$K excludes a thermal source. The final direct piece of information is a limit on its transverse velocity, from radio proper motion measurements, of $V_t \lesssim 40$km/s.

It has long been suggested that the centre of the Galaxy is a 'natural' place to find a very massive black hole, if such a thing were to exist, and so speculation that SgrA* may be it has been continued. Such an explanation has difficulty with the low source luminosity of course: massive black holes were after all introduced into astronomy to explain superluminous sources, not dramatically under-luminous ones. But nonetheless, it would be nice if our Galaxy did have such a thing for us to study, and SgrA* is the nearest we have to a candidate. Hence the interest. But is it very massive? There are arguments both for and against.

4.4.1 Why SgrA* cannot be massive:

1: Tidal limit on star formation: Recent high spatial resolution spectroscopic imaging studies of the Galactic central IRS16 stellar cluster show HeI line emission, consistent with the cluster containing a significant number of very hot stars. The simplest explanation of these is that they are young and massive. For stars to form their progenitor gas cloud must be stable. The limit of stability is just the Roche density (see equation 12), which is of interest in the very central regions.

Such extremely high gas densities are not readily available, suggesting that the presence of young stars is a suggestive argument against the presence of a massive compact central object.

This argument is suggestive rather than conclusive, as there are two counter arguments. Collisions between clouds or supernova shocks may induce transient extremely high density regions, in which star formation might occur. Alternatively, the stars may not be young at all. The stellar densities in the very central regions are sufficiently high that stellar collisions should be common. For the very distended atmospheres of red giants, the probability of a collision approaches unity. Such a collision would be an efficient way to strip off the outer atmosphere, exposing a hot, but low mass core.

2: Black holes should be bright. SgrA* is faint. Given the direct observational evidence for copious quantities of matter in the Galactic central regions, one must hypothesize that the current rate of accretion onto a central object is abnormally low. SgrA* will be a more luminous source for our descendants.

3: The dynamical centre of mass must correspond to the luminous centre of mass. SgrA* is significantly offset from the centre of the IRS16 stellar cluster. The dynamical timescale for orbits round a central massive object is

$$t_d \sim 1000 \times \left(\frac{M_{BH}}{10^6 M_\odot} \right)^{-1/2} \left(\frac{r}{1.5pc} \right)^{1.5} \text{yr.} \qquad (14)$$

Phase mixing occurs on a few dynamical timescales, or a timescale less than the lifetime of the observed stars. Thus, IRS16 should be distributed symmetrically about the centre of the relevant potential, which is therefore not at SgrA*.

It is important to remember that all these are plausibility arguments, and none is individually robust. It is true however that *a posteriori* the distribution of stars in the central regions of the Galaxy is certainly *not* what one would have expected for a gravitational system dominated by a central point mass.

4.4.2 Why SgrA* can be massive:

1: The most general argument in favour of SgrA* being a unique object is that it is a unique (in the Galaxy) radio source at a unique location. Is it plausible that such a peculiar object should be found just an arcsecond or so away from the precise Galactic centre, without it in fact being the exact Galactic centre? Well, yes! The examples noted above of Sco X–1 and the Great Annihilator illustrate the weakness of this argument.

2: The argument which is potentially the easiest to make precise is the observed peculiar motion of SgrA* relative to the dynamical centre. If the peculiar motion is vanishingly small, then SgrA* must be at the centre. Gravity doesn't allow any other choices. The present limits on the transverse motion of SgrA* from radio proper motion data provide a limit of ~ 40km/s. This is low enough to be moderately unusual, but not so low as to be interesting as yet. When this limit is improved by another 2 orders of magnitude or so it will be interesting either way.

3: Direct determination of the mass distribution in the central regions from dynamical studies is the only conclusive experiment. Only gravity can measure mass. Careful dynamical studies will determine both the location of the centre and the central density of the mass distribution in the next few years. The methodology is the application of the Collisionless Boltzmann Equation (CBE), in the form of the velocity moments. Converted into observables it can be written

$$M(r) = \frac{r\sigma^2(r)}{G} \left[\frac{\mathrm{d}lnn_*(r)}{\mathrm{d}lnr} - \frac{\mathrm{d}\sigma^2(r)}{\mathrm{d}lnr} + \frac{V_{rot}^2(r)}{\sigma^2(r)} - 2\beta \right] \qquad (15)$$

In this equation the terms in the square bracket on the rigth hand side have the following physical meanings:

i) $\mathrm{d}lnn_*(r)/\mathrm{d}lnr$ is the radial stellar density gradient. For a power law density distribution $n_*(r) \propto r^{-\alpha}$ this gradient is just the power law index α. Observationally., from near-IR surface brightness mapping, the numerical value is $\alpha \sim -1.8$.

ii) $\mathrm{d}\sigma^2(r)/\mathrm{d}lnr$ is the radial profile of the velocity dispersion, and is physically the pressure gradient contributing to the support of the stellar system against gravity. Determination of this is discussed further below.

iii) $V_{rot}^2(r)/\sigma^2(r)$ is the angular momentum contribution to support of the density profile against gravity. Observationally the contribution provided by this term in the very central regions is small, though it is dominant farther out from the centre in the disk.

iv) 2β is a correction term to allow for anisotropy in the stellar velocity distribution. It is possible that the transverse velocity dispersion, and hence that part of the contribution to the pressure support of the system, is different to that observed down our line of sight. There may be a radial bias in the orbits in the very centre, or a tangential bias. The amplitude of this term remains unknown, pending adequate proper motion data for the stars in IRS16.

Determination of the enclosed mass at some radius using this equation requires a mixture of data determining the mean rotation velocity of material in closed circular orbits, the random velocity dispersion and the surface brightness distribution, to define the density profile. One uses different tracers at different radii. At large distances HI data are adequate, though the triaxiality of the bulge noted earlier must be treated explicitly for distance more than a few tens of parsecs from the centre. Closer to the centre OH/IR stars are ideal. Within a few parsecs of the centre a mix of stellar velocities, for the HeI emitting sources, the velocity dispersion of the unresolved $2\mu m$ light, representing the bulk of the K giants in the central cluster, the neutral and the ionized gas are all utilised. Each of these tracers has strengths and weaknesses, with specific reservations raised in every case. For example, is the gas in the very centre really on closed ballistic orbits, or are there significant MHD effects? Are the hot stars in IRS16 really a relaxed central cluster? If so, why is the median velocity so different from zero? And so on. However, for distances farther than about 1 pc from the centre all determinations are in adequate agreement, and are consistent with dominance of the potential by the central stellar cluster. Within the central 1pc there is some evidence – about two standard deviations is probably a realistic level of significance at present – that there is an excess unidentified and unresolved 'point' mass, with a mass of perhaps $10^6 M_\odot$.

A point worth especial note is the density profile of the central stellar cluster. The density profile derived from the observed $2\mu m$ surface brightness distribution is

$$\rho(r) \propto r^{-1.8\pm0.1} \qquad 1pc \lesssim r \lesssim 200pc \qquad (16)$$

This is sufficiently similar to the density profile $\rho(r) \propto r^{-7/4}$ expected for a dynamically relaxed stellar cluster bound to a central black hole to have excited comment. However, the cluster is certainly not in this state, its own mass exceeding that of any central component by some orders of magnitude. The similarity is illusory.

The stellar density profile is also similar to the density profile $\rho(r) \propto r^{-2.23}$ for a cluster after core collapse, with no central black hole, and to the density profile $\rho(r) \propto r^{-2}$ for an isothermal sphere. This range of possible similarities precludes direct use of the stellar density profile as an argument in favour of, or against, and specific physical model.

It remains the case that there is some evidence in favour of a central compact object associated with the centre of the Galaxy, some evidence against, and several plausibility arguments which suggest that SgrA* is not quite at the dynamical center of the Galaxy. If that is true, what might it be? Perhaps the Great Annihilator provides another clue here, as a young neutron star moving through very dense molecular gas might have many of the observed properties.

In this case, the proper motion data should soon see a signal, and the Galactic centre would be relegated merely to the place with the highest concentration of stars, gas, energetic events and mass in the Galaxy. Perhaps that is sufficient for it to retain some interest.

5 LOW MASS STARS AND ALL THAT

One of the great scientific challenges for the future where the near infrared will be crucially important is the study of very low mass stars, and substellar objects – "brown dwarfs". To put these studies in context we will consider two aspects of the problem. First, dynamical methods which determine how much mass can be in very low mass objects in the immediate neighbourhood of the Sun. Second, direct determinations of the luminosity function of very low mass luminous stars, to provide a reliable base from which to extrapolate the number of subluminous 'stars' which might exist. Prior to that, we provide a brief overview of the objects themselves, to emphasise their intrinsic interest in addition to their interest as a possible location of 'dark' matter.

5.1 Brown Dwarfs

Brown dwarfs are not brown. They would perhaps be better named as dark red midgets. The basic definition is however a self-gravitating object which is of such low mass that it never reaches a stable equilibrium nuclear burning state. This corresponds to objects of mass less than about 8% of the mass of the Sun, or about 80 times the mass of Jupiter. Such objects rapidly cool to surface temperatures below 1000K, and are of small surface area. Thus, they are both extremely low luminosity sources in the visible, and are of intrinsically very low luminosity. Their material is predominately hydrogen in a metallic state, they are always fully convective, and have dense yet thin atmospheres. The atmospheres contain only $10^{-11} M_\odot$ of mass, yet control the appearance of the brown dwarf.

The detailed internal and atmospheric physics of brown dwarfs is a complex and fascinating subject in its own right. The interested reader is recommended to read an excellent review by Burrows & Liebert (1993). The whole of this physics may be summarised, for present purposes in considering the detectability of such object, in two simple relations:

$$T_e \approx 1550K \left(\frac{10^9 yr}{t}\right)^{0.324} \left(\frac{M}{0.05 M_\odot}\right)^{0.827} \left(\frac{\kappa_R}{10^{-2}}\right)^{0.088} \qquad (17)$$

$$L_{BOL} \approx 4 \times 10^{-5} L_\odot \left(\frac{10^9 yr}{t}\right)^{1.297} \left(\frac{M}{0.05 M_\odot}\right)^{2.641} \left(\frac{\kappa_R}{10^{-2}}\right)^{0.35} \qquad (18)$$

These show well the low temperatures and luminosities relevant for brown dwarf studies, and the consequent need for sensitive infrared searches.

We now consider the dynamical evidence relevant to calculation of the number of brown dwarfs which might exist near the Sun.

5.2 Mass Densities Near The Sun

A description of the amount and spatial distribution of mass near the Sun is an essential prerequisite for any study of the dynamical evolution of the solar neighbourhood. Knowledge of the amount of baryonic matter which has ever been available to the interstellar medium is clearly necessary for chemical evolutionary models, while the current spatial distribution of old stars and the fraction of matter currently in the ISM are also clearly of importance for both chemical and dynamical processes. A common factor linking all these distributions is the Galactic vertical force law, $K_z(z)$. It is this force law which provides a measure of the total amount of matter near the Sun, which measures the present–day gas fraction, which allows conversion of local kinematic data to weighted contributions to a column through the Galaxy, and which identifies or precludes local dark matter as a significant source or sink term in chemical evolution.

The distribution of mass in the Galactic disk is characterized by two numbers, its local *volume* density ρ_o and its total *surface* density $\Sigma(\infty)$. The local *volume* mass density — *i.e.* the amount of mass per unit volume near the Sun, which for practical purposes is the same as the volume mass density at the Galactic plane has units of $\mathcal{M}_\odot pc^{-3}$. The best available determination of the local mass density in *identified* material is $\sim 0.1 \mathcal{M}_\odot pc^{-3}$, with a very-poorly defined uncertainty of perhaps as much as 25% in this value. Comparison of this value with that determined from dynamical analyses is required to test for the existence of dark matter associated with the Galactic disk.

The integral surface mass density has units of $\mathcal{M}_\odot pc^{-2}$, and is the total amount of disk mass in a column perpendicular to the Galactic plane. It is this which is required for the interpretation of rotation curves and the large-scale distribution of mass in galaxies. Recent determinations of this surface mass density lead to values of about $50\mathcal{M}_\odot\mathrm{pc}^{-2}$ for that fraction of the mass associated with the Galactic disk. Thus the Galactic disk provides only about 50% of the total Galactic potential at the solar galactocentric distance. The nature of the mass which generates the other half of the potential remains unknown.

5.2.1   Measurement of the Galactic Potential

All determinations of the mass distribution in the Galactic disk (volume or surface densities) require a solution of the collisionless Boltzmann equation. In view of the inconvenience of general solutions of this equation derived from real data, in practise one utilises its vertical velocity moment, the vertical Jeans' equation:

$$\mathcal{K}_z = \frac{1}{\nu}\frac{\partial}{\partial z}(\nu\sigma_{zz}^2) + \frac{1}{R\nu}\frac{\partial}{\partial R}(R\nu\sigma_{Rz}^2) \tag{19}$$

where $\nu(R,z)$ is the space density of the stars, and $\vec{\sigma}_{ij}(R,z) = \langle v_i v_j\rangle - \langle v_i\rangle\langle v_j\rangle$ their velocity dispersion tensor.

The first term on the right hand side of equation 19 is dominant, and contains a logarithmic derivative of the stellar space density $\nu(z)$, and a derivative of the vertical velocity dispersion, σ_{zz}. Since the dominant (by mass) old disk stellar population in the solar neighbourhood is tolerably well described by an isothermal stellar population, the part of this term containing the derivative of the space density dominates the determination of $\mathcal{K}_z(z)$ near the Sun. This point is not often appreciated adequately, but means that one should determine stellar density profiles with even greater care than that required for the velocity dispersions.

The second term in the Jeans' equation (19) describes the tilt of the stellar velocity ellipsoid away from the local cylindrical–polar coordinate system in which velocity dispersions are measured. One therefore needs the R-gradients of σ_{Rz} and of ν. There are no general analytical solutions for this term, as it depends on the unknown "third integral" of the motion. Estimates of its importance may be derived by numerical integration of orbits in potentials which are thought to be realistic approximations to that of the Galaxy, and are described in detail by Kuijken & Gilmore (1989a,b).

Given a measurement of the gravitational field $\vec{\mathcal{K}}(R,z)$ in an axisymmetric galaxy, the total density ρ of gravitating matter follows from Poisson's equation:

$$\nabla\cdot\vec{\mathcal{K}} = -4\pi G\rho. \tag{20}$$

In the case of a disk galaxy we can express the R-gradient in $\nabla \cdot \vec{\mathcal{K}}$ in terms of the observed circular velocity at the Sun, v_c, or in terms of the Oort constants of Galactic rotation A and B

$$\rho = -\frac{1}{4\pi G} \left\{ \frac{\partial \mathcal{K}_z}{\partial z} + 2\left(A^2 - B^2\right)\right\}. \tag{21}$$

For a disk galaxy with an approximately flat rotation curve the second term is small within a few kpc of the disk plane (Kuijken & Gilmore 1989a provide a more exact calculation; for an exactly flat rotation curve $A^2 - B^2 \equiv 0$ at $z = 0$), so we can integrate in z to obtain the total column density $\Sigma(z)$ between heights $-z$ and z relative to the disk plane $z = 0$:

$$\Sigma(z) = \int_{-|z|}^{|z|} \rho(z)dz = \frac{|\mathcal{K}_z|}{2\pi G} - \frac{(A^2 - B^2)}{\pi G}|z|. \tag{22}$$

The physical interpretation of a determination of the Galactic $\mathcal{K}_z(z)$ force law can now be seen. In effect, one measures the pressure–gravity balance of the *collisionless* stellar 'fluid'. The hydrodynamic analogy following from the description of the collisionless Boltzmann equation as the equation of stellar hydrodynamics is particularly appropriate here. The dominant first term on the right hand side of equation 19 contains a logarithmic derivative of the stellar spatial density $\nu(z)$, and the stellar velocity dispersion σ_{zz}. The spatial density term plays the role of a scale height, the velocity dispersion is analogous to a temperature, and the product $\nu\sigma_{zz}$ is a pressure.

5.2.2   Determination of the disk surface mass density

Because high-energy stars are present at all heights above the Galactic plane, measurements of the potential very close to the plane still require knowledge of the high-energy tail of the distribution function. Therefore either the tail of the velocity distribution at low z, or the density *and* potential at high z, are required to measure the potential at low z, and hence to deduce the local volume density of matter ρ_o. The phase-space distribution function we discuss below, however, depends on the density only at points farther from the plane than the height at which data are being analysed. It is possible to capitalise on this insensitivity to the detailed shape of the potential (equivalently, the detailed mass distribution) near the plane to derive the potential at large distances from the plane from high-z data alone. Since a measurement of $\mathcal{K}_z$ at any height relates directly to the total surface density integrated to that height, this is extremely useful, allowing us to obtain meaningful determinations of the surface mass density of the Galactic disk from high-z data alone.

157

In most $\mathcal{K}_z$-studies, the density $\nu(z)$ is known to better precision than the velocity distribution. Instead of fixing the parameters of the latter, and then using these to model the density gradient, it is therefore preferable to work in the other direction, and predict the velocity distribution of a tracer in different model potentials, given its density. These velocity distribution models can then be compared with the observed velocity data using maximum likelihood techniques.

Given a distribution function $f_z(E_z)$ and a potential $\psi(z)$, we can calculate the density $\nu(z)$, which is just a moment of f_z:

$$\nu(z) = \int_{-\infty}^{\infty} f_z(z, v_z) dv_z = 2 \int_{\psi(z)}^{\infty} \frac{f_z(E_z)}{\sqrt{2\left(E_z - \psi(z)\right)}} dE_z. \tag{23}$$

Reparameterizing the z-height in terms of the potential ψ, we have

$$\nu(\psi) = 2 \int_{\psi}^{\infty} \frac{f_z(E_z)}{\sqrt{2(E_z - \psi)}} dE_z. \tag{24}$$

This equation is an Abel transform, which has the well-known inversion

$$f_z(E_z) = \frac{1}{\pi} \int_{E_z}^{\infty} \frac{-d\nu/d\psi}{\sqrt{2(\psi - E_z)}} d\psi, \tag{25}$$

so that there is a unique relation between $\nu(\psi)$ and $f_z(E_z)$. Because of this equivalence of $\nu(\psi)$ and $f_z(E_z)$, there is a triangular mathematical relationship between the three functions $\psi(z)$, $\nu(z)$ and $f_z(E_z)$: any one of them can be deduced from the other two. Abel inversions are somewhat unstable, but not as unstable as taking a direct derivative of the data.

It is important to note that equation 25 shows that $f_z(E_z)$ depends on the density only at points where the potential exceeds E_z, *i.e.* beyond the point $z = \psi^{-1}(E_z)$. It is this property which allows the derivation of $\mathcal{K}_z(z)$ at large z independently of the poorly known distribution of mass near the plane.

An analysis technique based on equation 25 has been devised by Kuijken & Gilmore (1989a,b) for the determination of $\mathcal{K}_z(z)$, and $\Sigma(z)$. The essential feature of that analysis is that one avoids the assumption of isothermality, by instead postulating a range of potentials $\psi(z)$, and for each of them calculating $f_z(z, v_z)$ from $\nu(z)$. The range of model distribution functions was then compared to an observed distribution function of velocity–distance data using maximum likelihood, and used to select the best-fitting model potential.

Kuijken & Gilmore (1989a,b) determine a dynamical total surface mass density $\Sigma_\infty = 46 \pm 9 \mathcal{M}_\odot \mathrm{pc}^{-2}$. These same authors integrate the local observed

158

volume mass density in stars through their derived potential, and add the directly observed mass in the inter-stellar medium, to derive an observed integral surface mass density of $\Sigma_{\infty,obs} = 48 \pm 8 \mathcal{M}_\odot \mathrm{pc}^{-2}$. There is thus evidence that there is no significant unidentified mass associated with the Galactic disk.

It is perhaps useful to emphasise just what it was that the Kuijken & Gilmore analysis measured. Their data provide a robust measurement of the *TOTAL* potential below an effective distance from the plane of about 1.1kpc. This value is equivalent to $70 \pm 9 \mathcal{M}_\odot \mathrm{pc}^{-2}$. A more model-dependent conclusion is to use the Galactic rotation curve to deconvolve this *total* potential into disk and halo components. It is this analysis which leads to a *disk* surface mass density of $\sim 50 \mathcal{M}_\odot \mathrm{pc}^{-2}$.

5.2.3   The local volume mass density

Local determinations of the *volume* mass density have until recently suggested the existence of a large amount of dark matter in the disk. The sensitivity of determinations of the local volume mass density ρ_o to uncertain data lies in the modelling of the stellar velocity distribution near the Galactic plane, and in the determination of the stellar density distribution with distance from this plane. Both F dwarf and K giant tracer samples have been analysed to determine ρ_o, with both having produced a result of $\rho_o \sim 0.20 \mathcal{M}_\odot \mathrm{pc}^{-3}$, where the identified mass provides $\rho_{o,obs} = 0.10 \mathcal{M}_\odot \mathrm{pc}^{-3}$ (Bahcall 1984).

More recently however, Bahcall *et al.* (1992) have shown that the F-star samples analysed previously were affected by a systematic calibration error, while a new data set is available for the K giant sample. Bahcall *et al.* analysed the new data set, and determined a difference between the observed and the dynamical mass densities at the $\sim 1.8\sigma$ level, which is not statistically significant by normal criteria. The most recent analysis of that same new data set (Fuchs & Wielen 1993) also shows no significant discrepancy between the dynamical and the identified mass in the Solar Neighbourhood.

Thus, there is now good dynamical data showing that there is no significant unidentified mass within ~ 1kpc of the Sun. Nonetheless, the 'identified' mass budget still includes a significant *NUMBER* of brown dwarfs, even if their total mass is not dynamically very significant. We now consider the number of such stars which are awaiting observation.

6 HOW MANY LOW MASS STARS ARE THERE?

There are two almost independent determinations of the stellar luminosity function (LF) for low mass stars. One is based on detailed analysis – trigonometric parallax, kinematics, high angular resolution imaging – of the complete sample of stars very near the Sun. The best studied data set is restricted to the sky north of $\delta = -20°$ and within 5.2 pc. The second LF is determined from photometric data for stars at larger distances – 50 pc to 100 pc being possible – with distances derived by photometric parallax.

Each of these LF's has advantages and disadvantages. The 5.2 pc LF is based on high precision data for single stars, since most binaries will have been resolved or detected. However, the sample is small, so that the statistical precision of the LF is poor. The more distant sample has very many stars, so that random errors are unimportant. In this sample however close binaries remain unresolved, so that LF is that of stellar systems, not of single stars. Additionally, it is essential that the photometric parallax method – essentially application of the absolute magnitude-colour relation in reverse – is corrected from its local calibration to be appropriate to the field sample of relevance. As well as this calibration, the various contributions to the scatter in the absolute magnitude-colour (CM) relation must be understood, so that appropriate Malmquist corrections can be applied.

The important parameters in this experiment are the distributions of ages, chemical abundances, binaries, and measuring errors, each of which can change systematically from local to distant samples, and each of which should be modelled. With this diversity a Monte Carlo approach is most effective.

6.1 Modelling the Luminosity Function

Those variables which need to be considered in the modelling include a mixture of things one must calibrate ("noise") and things one wants to learn ("signal").

6.1.1 Pre main-sequence evolution

Low-mass stars evolve onto the main sequence over a considerable period. During this time predictions of the luminosity are complicated by asymmetric dust shells and disks, rapid mass loss, chromospheric activity and reddening from residual molecular cloud material. Thus quantification of this effect is difficult.

Fortunately, for present purposes, most stars at high latitudes are old enough to be free of such complexities. When modelling luminosity functions in young open clusters, however, the problems can be considerable.

6.1.2 Main-sequence stellar evolution

Stars evolve up and across the main-sequence during their lives. Evolutionary tracks are available for intermediate and high mass stars, and for abundances up to about solar. For lower masses and higher abundances some extrapolation is needed. Analogous to this of course one must adopt an age distribution. For high latitude field stars a uniform distribution is appropriate.

6.1.3 Chemical Abundances

Chemical abundances systematically affect the location of the main sequence. The correction for this is straightforward in principle. Complexities arise in quantifying the abundance effect at low luminosities, and in knowing the relevant abundance distribution ("metallicity gradient") to adopt. For low luminosities, where very few subdwarfs with good parallaxes are known for an empirical calibration, extrapolation of atmospheric models is required. For abundance ranges of relevance, some observational data are available. The most reliable method is however to investigate the allowed range of possibilities (Kuijken and Gilmore 1989), both in the mean and range.

6.1.4 Parallax Errors

Parallax errors in both the nearby LF sample and in the sample defining the colour-luminosity relation (if these differ) must be considered. Both the random and the resulting systematic bias need consideration, following Lutz & Kelker (1973).

6.1.5 Binary Stars

The number of unresolved binaries is a critical parameter which varies from sample to sample. The number will be least in the 5.2 pc sample, as it is both close (enhancing spatial resolution) and well studied. In the sample used to define the CM relation it will be greater, though many wide binaries will have been found and perhaps rejected from the sample, during parallax studies. For the field LF sample the spatial resolution of the mostly photographic surveys exceeds several arcsecs, so that the full binary properties of the LF are relevant. It is of course possible that there is a correlation between primary and secondary properties in the binary mass function, or a correlation between the incidence of binarism and primary mass. All these possibilities must be included in the range of models.

6.1.6 Galactic Structure

When comparing the space density of stars per unit volume of samples in the Galactic Plane and far from the Plane it is clearly necessary to allow for the systematic change in space density with distance, the density law. In fact, this exercise is in large part designed to determine this.

The effect of all these contributions to the error budget is to confuse a reliable comparison of the nearby single star luminosity function with the distant stellar system luminosity function. The apparent complexity of this process however illustrates an important general point. It is usually more reliable to know the allowed range of values of a function or a parameter than it is to know the "best" available determination.

In this case calculation proceeded using a range of parameters in a Monte Carlo process, as described by Kroupa *et al.* (1990, 1991, 1993). The first requirement is to check that reasonable ranges of parameter space are considered, by reproducing a well understood case. Here the most appropriate example is the "cosmic scatter" of $\sigma \sim 0.^m3$ about a mean colour-magnitude relation for well-studied trigonometric parallax stars. With that successfully achieved one can begin to do astrophysics.

The specific null hypothesis of relevance to be tested is: Is there a single stellar mass function, convolved with a realistic stellar binary distribution and a Galactic disk scale height, which can be converted through a consistent mass-luminosity relation to reproduce both the local single star luminosity function and the more distant stellar system luminosity function?

Following calibration of the modelling technique as described below, it is next necessary, prior to answering the question, to consider the mass-luminosity calibration.

162

6.2   Conversion of Luminosity to Mass

In the section above we have discussed conversion of apparent colour of a stellar system to the space density of single stars as a function of their intrinsic luminosity, the luminosity function. A more fundamental function is the space density of single stars as a function of their mass, the present day mass function (PDMF). For the low mass stars of interest here, where mass loss during that part of their main sequence lifetimes which has occurred to date is small, the mass function is a good approximation to the initial mass function (IMF) generated by the star formation process. For higher mass stars, where evolutionary corrections are large, derivation of the IMF from the PDMF is a complex exercise. [An excellent discussion of this more general problem is provided by Scalo (1986).]

The stellar luminosity function $\Phi(M_v)$ is directly related to the stellar mass function $\zeta(\mathcal{M})$ as

$$\Phi(M_v) = \frac{dN}{dM_v} = -\frac{dN}{d\mathcal{M}} \cdot \frac{d\mathcal{M}}{dM_v} = -\zeta \frac{d\mathcal{M}}{dM_v}. \tag{26}$$

The important point here is that the *gradient* of the mass-luminosity relation is the function which relates the LF to the IMF.

The existence of structure in the mass-luminosity relation thus generates structure in the LF, even for a smooth underlying IMF. This is illustrated in Figure 4, which shows both the LF for single stars within 5.2 pc of the sun (Figure 4a) and the observational constraints on the mass-luminosity relation.

The most obvious features of the LF are a flat section near $M_v \sim +7$, and a maximum near $M_v \sim +12$. Were the mass-luminosity relation to have a nearly constant gradient then these features would be reproduced in the IMF, and would represent properties of the physics of star formation. Unfortunately, they are artifacts of molecular physics, in that the mass luminosity relation has structure at these luminosities. This is illustrated in Figure 4c, which shows the *gradient* of the mass-luminosity relation. The important feature of Figure 4c is that features in the mass-luminosity relation correspond to those in the luminosity function. This simple correspondence means that the stellar mass function lacks structure. The structure in the mass-luminosity relation is understood in terms of changes in the opacity sources, and consequent changes in the equation of state, of stellar atmospheres. Further discussion is provided by Kroupa *et al.* (1990 ,1991, 1993).

163

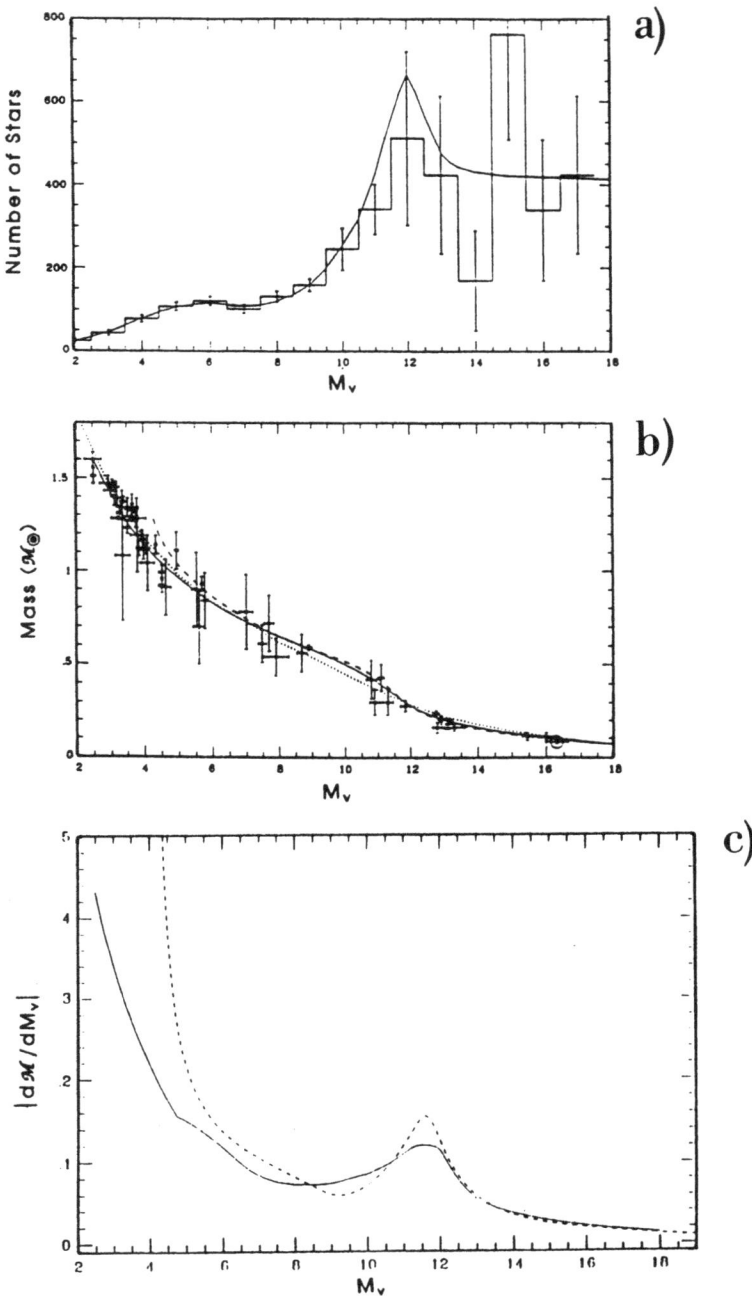

Figure 4: *The top panel shows the observed local luminosity function, the centre panel the mass-luminosity relation, with data from Popper (1980), and the bottom panel the absolute value of the slope, dM/dM$_v$, of this relation. Note how the shape of this function mimics that of the stellar luminosity function.*

For modelling purposes, the relevant feature of the result shown in Figure 4c is that a *range* of mass functions and a *range* of mass-luminosity relations must be allowed for in the Monte Carlo process.

6.3   Monte Carlo Modelling - An Example

Given the preliminaries above we are now ready to begin. One wishes to include all the observational and physical effects discussed. In practice, one proceeds by Monte Carlo modelling. This rather formidable name means simply including random sampling into computer modelling. Statistical comparison of the range of results allowed by adjusting each parameter within its formal error range with the observational data then provides a statistically optimal value for each parameter, and an "error bar" about that value.

When a large number of parameters is present it is important to calibrate the method on a well studied case. In this example, the scatter in the H-R diagram for well studied trigonometric parallax stars is well reproduced. The many tests and checks are illustrated in the papers of Kroupa *et al.*, and need not be repeated here.

Rather, we summarise the results in Figure 5. Figure 5 shows the nearby star luminosity function, as in Figure 4, together with two determinations of the field star system luminosity function. Also shown is the resulting stellar mass function, converted into the observational plane, which is consistent with all available data on the absolute number of single stars near the Sun and their distribution with luminosity and colour, and with the apparent number of stellar systems to ~ 100 pc from the Sun, and their distribution with colour and apparent magnitude. The lower panel shows the stellar IMF which is consistent with all these data.

The astute observer will note that this mass function shows brown dwarfs (or at least, stars at the lowest mass to burn hydrogen, and by implication 'stars' of somewhat lower mass) to be the most common objects by number. Remember that earlier dynamical studies show that brown dwarfs are not important contributors to the local mass density. They are however out there, waiting to be seen.

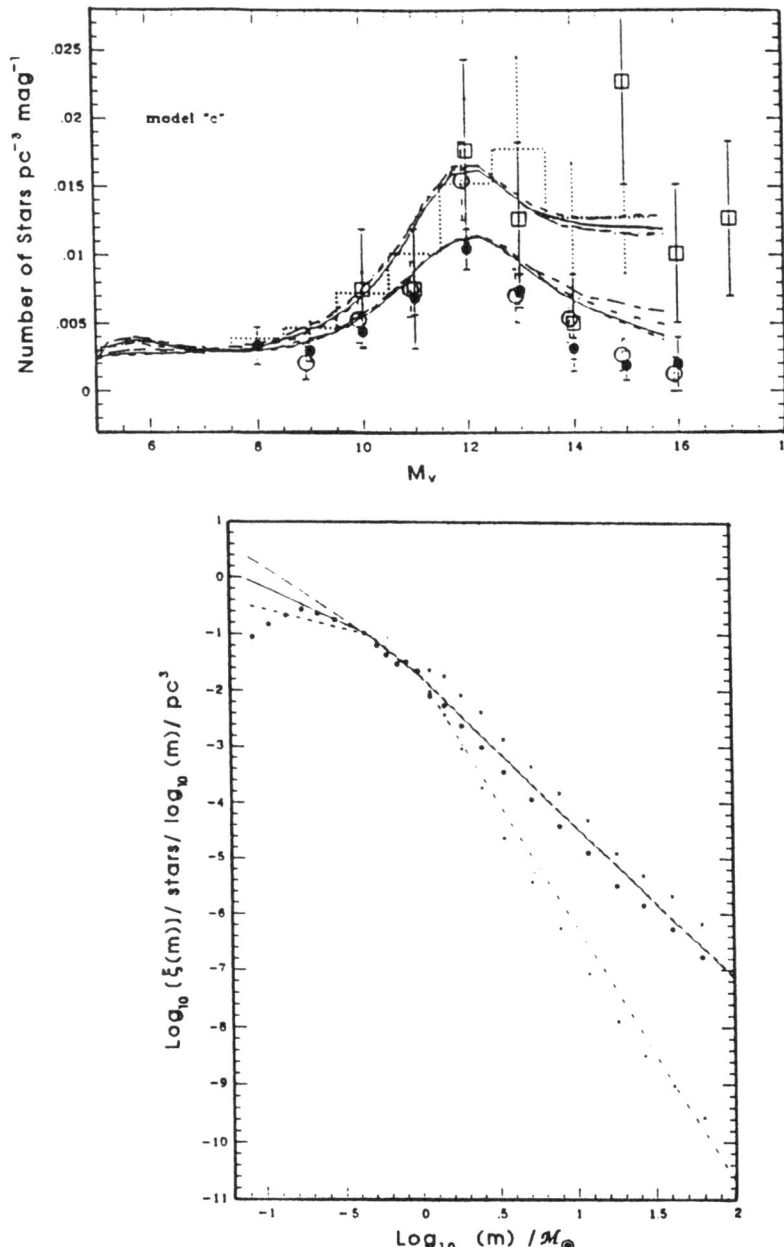

Figure 5: *Top panel: The single star luminosity function from Figure 4a (open squares) and two photographic system luminosity functions (Reid and Gilmore 1982, open circles; Stobie et al.1989, solid circles). The solid curve is the single stellar IMF which is most consistent with all observational data, converted into the observational plane. Bottom Panel: The stellar initial mass function (solid line) and present day mass function (dotted line) which are most consistent with available data.*

6.4   Is it worth it?

Following this extensive and detailed modelling, one wonders naturally if this is all necessary. At one level of course it is only when all known effects are included in models that one can be confident of the answers. Thus an explicit demonstration that the Malmquist corrections are well-understood and applied correctly is necessary so that more complex derived quantities can be derived. Examples include the stellar IMF discussed above, and the surface mass density of the Galactic disk derived by Kuijken and Gilmore (1989).

As well as this, such modelling sometimes provides useful new predictions, or explanations of results which were not part of the original study. An example here is the apparent very steep decrease in the scale height of very late M stars reported by Hawkins (1988). He reports a decrease in disk apparent scale height from $\sim$ 200 pc for stars with $V-I \sim 2$ to only $\sim$ 75 pc for stars with $V-I \sim 3.5$.

This dramatic decrease is difficult to understand as a normal effect of the age-scale height relation for field stars, as it would require that almost all stars with $M_v \gtrsim +12$ have been formed in the last $\sim$ 1 Gyr. This result is rather an artefact of a bias towards equal masses in binary stars as one approaches the minimum mass for hydrogen burning.

If all binary stars are formed so that the two masses are chosen independently from the same IMF, then most secondaries will come from the most common stars, the late M dwarfs. If such a star has a massive primary companion it will have only a small effect on both the luminosity and colour of the system, and will be modelled as described earlier. A low mass primary however can have only a nearby equal mass *luminous* secondary. This directly gives a very large increase in the system luminosity, which will tend to a factor of two change at the lowest masses. [Lower mass companions of course have zero luminosity.]

This change provides a very considerable correction to the colour-magnitude relation, and leads to a considerable change in the apparent scale height of the distribution of stars with decreasing luminosity.

Interestingly, one could invert the process, and deduce the mass ratio distribution of the lowest mass stars from careful derivation of the scale height change. This remains to be attempted.

167

7   CONCLUSIONS

The Milky Way Galaxy is our nearest example of one of those large spirals which dominate the luminosity density of the Universe. Because we live inside it near the Plane, our view is in large part obscured by inter-stellar material. Infra-red astronomy allows a unique view through that obscuration, thereby considerably increasing both the detail and the reliability of our knowledge of the Milky Way. Some of the most obscured regions are the most intrinsically interesting, with the Galactic Centre being the most obvious example, so that here the infrared is essential for any progress.

Intrinsically low temperature objects are also best studied in the near infra-red. The lowest mass stars and brown dwarfs are of particular interest here, and again we await progress in infra-red observations to advance these subjects.

Science is built up with facts, as a house is built up with stones. But a collection of facts is no more a science than a pile of stones is a house. (*Poincare*) In that spirit these lectures have emphasised the models and the understanding of the Milky Way which have been built up from many observations in many wavebands. The greatest advances happening now are happening in the infra-red, as new array technology begins to be exploited. We can confidently expect a similar school to this in a few years time to provide a more detailed, more complete, and no doubt in some ways different, understanding of our Galaxy.

You and I are fortunate to be working in astrophysics at a time of such rapid technological change, and I look forward to the many important results you will produce in the next few years. To reverse the spirit but not the words of the old Chinese curse, *may you live in interesting times.*

8   REFERENCES

Allen, D., Hyland, A., & Hillier, D. 1990 *Mon. Not. Roy. astr. Soc.* **244** 706

Arnold, R.A. & Gilmore, G. 1992 *Mon. Not. Roy. astr. Soc.* **257** 225

Bahcall, J.N., 1984 *Astrophys. J.* **287**, 926

Bahcall, J.N., Flynn, C., & Gould, A. 1992 *Astrophys. J.* **389**, p 234

Binney, J., Gerhard, O., Stark, A., Bally., & Uchida, K., 1991 *Mon. Not. Roy.*

astr. Soc. **252** 210

Blitz, L. & Spergel, D., 1991 *Astrophys. J.* **379** 631

Blitz, L., Binney, J., Lo, K.Y., Bally, J., & Ho, T.P., 1993 *Nature* **361** 417

Burrows, A., & Liebert, J., 1993 *Reviews of Modern Physics* in press

Casertano, S., & van Albada, T.S., 1990 In *Baryonic Dark Matter* eds. D. Lynden Bell & G. Gilmore (Reidel: Dordrecht) p 159

Copeland, H., Jensen, J. O. & Jørgensen, H. E., 1970. *Astron. Astrophys.,* **5**, 12

Eckart, A., Genzel, R., Krabbe, A., Hofman, R., van der Werf, P., & Drapatz, S., 1992 *Nature* **355** 526

Fich, M., & Tremaine, S. 1991 *Ann Rev Astron Astrophys* **29** 409

Fuchs, B., & Wielen, R., 1993 In *Back to the Galaxy* (3rd Maryland Astrophysics Conference, College Park, Oct 1992, in press

Gallais, P., Rouan, D., Lacombe, F., Tiphene, D., & Vaughlin, I., 1991 *Astron. Astrophys.* **243** 309

Garwood, R. & Jones, T.J. 1987, *Publ. Astr. Soc. Pacific,* **99**, 453.

Gilmore, G., Wyse, R.F.G. & Kuijken, K. 1989, *Ann. Rev. Astron. Astrophys.,* **27**, 555.

Gilmore, G., King, I.R., & van der Kruit P.C. 1990. *The Milky Way as a Galaxy* (University Science Books: Berkeley)

Gilmore, G., Reid, N. & Hewett, P., 1985. *Mon. Not. R. astr. Soc.,* **213**, 257

Gratton, R.G., Tornambe, A., & Ortolani, S. 1986, *Astron. Astrophys.,* **169**, 111.

Habing, H.J. 1987. In *The Galaxy,* eds G. Gilmore & B. Carswell (Reidel : Dordrecht) p173.

Harmon, R.T. & Gilmore, G. 1988, *Mon. Not. Roy. astr. Soc.,* **235**, 1025.

Hawkins, M. R. S. 1988. *Mon. Not. R. astr. Soc.,* **234**, 533.

Hayakawa, S., Matsumoto, T., Murakami, H., Uyama, K., Thomas, J., & Yamagami, T. 1981, *Astron. Astrophys.*, **100**, 116.

Hiromoto, N., Maihara, T., Mizutani, K., Takami, H., Shibai, H. & Okuda, H. 1984, *Astron. Astrophys.*, **139**, 309.

Kroupa, P., Tout, C., & Gilmore, G. 1990, *Mon. Not. R. astr. Soc.*, **244** 76

Kroupa, P., Tout, C., & Gilmore, G. 1991, *Mon. Not. R. astr. Soc.*, **251** 293

Kroupa, P., Tout, C., & Gilmore, G. 1993, *Mon. Not. R. astr. Soc.*, **in press**

Kuijken, K. & Gilmore, G., 1989a, *Mon. Not. Roy. astr. Soc.* **239** 571

Kuijken, K. & Gilmore, G., 1989b, *Mon. Not. Roy. astr. Soc.* **239** 605

Kuijken, K. & Gilmore, G., 1989c, *Mon. Not. Roy. astr. Soc.* **239** 651

Lutz, T., & Kelker, D., 1973, *Publ. Astron Soc Pacifc* **85** 573

Moneti, A., Glass, I., & Moorwood, A., 1992, *Mon. Not. Roy. astr. Soc.* in press

Morris, M. (editor), 1989, *The Center of the Galaxy* (Kluwer: Dordrecht)

Mirabel, I.F., Rodriguez, L., Cordier, B., Paul, J., & Lebrun, F., 1992, *Nature* **358** 215

Nakada, Y., Deguchi, S., Hashimoto, O., Izumiura, H., Onaka, T., Sekiguchi, K., & Yamamura, I., 1991, *Nature* **353** 140

Popper, D.M., 1980, *Ann. Rev. Astron. Astrophys.* **18**, 115

Reid, I. N. & Gilmore, G., 1982, *Mon. Not. R. astr. Soc.*, **201**, 73.

Rich, M., 1988, *Astron.J.*, **95**, 828.

Scalo, J. M., 1986, *Fundamentals of Cosmic Physics*, **11**, 1.

Solomon, P.M., & Rivolo, A.R., 1987, In *The Galaxy* eds G. Gilmore & R. Carswell (Reidel: Dordrecht) p 105

Stobie, R. S., Ishida, K. & Peacock, J.A., 1989, *Mon. Not. R. astr. Soc.*, **238**, 709.

Terndrup, D. 1988, *Astron. J.*, **96**, 884.

de Vaucoulers, G. & Pence, W.D. 1978, *Astron. J.*, **83**, 1163.

de Vaucoulers, G., 1964, in *The Galaxy and the Magellanic Clouds* eds F.Kerr & A. Rodgers (Aust Acad. Sci) p195

Wielen, R., Jahreiss, H., & Kruger, R, 1983, *The Nearby Stars and the Stellar Luminosity Function* eds davis Philip & Upgren, (Davis Press) p163.

Weinberg, M., 1991, *Astrophys. J.* **384** 81

Weinberg, M., 1992, *Astrophys. J.* **392** L67

Whitelock, P.A., 1992, in *Variable Stars and Galaxies* Astron. Soc. Pacific Conf. Ser. vol. 30 in press

Young, P.J. 1976, *Astron. J.*, **81**, 807.

Galaxies in the Infrared

Charles M. Telesco

NASA-Marshall Space Flight Center
Space Science Laboratory, ES63
Huntsville, Alabama, U.S.A.

1 EXTRAGALACTIC INFRARED EMISSION: AN OVERVIEW

The IR spectral region is a rich hunting ground in extragalactic astrophysics. Galaxies glow with the IR emission from stars of all ages and from gas and dust that has been energized by these stars, by supernovae, and probably by nuclear black holes. It has even been proposed that some high IR luminosities correspond to the dissipation of the kinetic energies of galaxies in collision. IR radiation is not only ubiquitous, but it is relatively impervious to interstellar extinction, and so we can peer into the cores of galactic nuclei that are completely obscured visually. The IR domain extends from 1 μm to 300 μm and has been sub-divided into the near-IR (1-5 μm), the mid-IR (5-30 μm), and the far-IR (30-300 μm) regions. Sub-millimeter emission extends from 300 μm to 1mm.

The schematic spectral energy distribution of a "starburst" galaxy that is undergoing an active episode of star formation in its center is shown in Fig.1. I also indicate a "cirrus" spectrum, emitted by benignly heated dust in a more quiescent galaxy, and two types of energy distributions that arise in at least some Seyferts and other active galactic nuclei (AGNs): non-thermal power-law emission and emission from hot dust heated by the exotic nuclear object. It is possible for starburst, cirrus, and AGN emission to occur in a single galaxy.

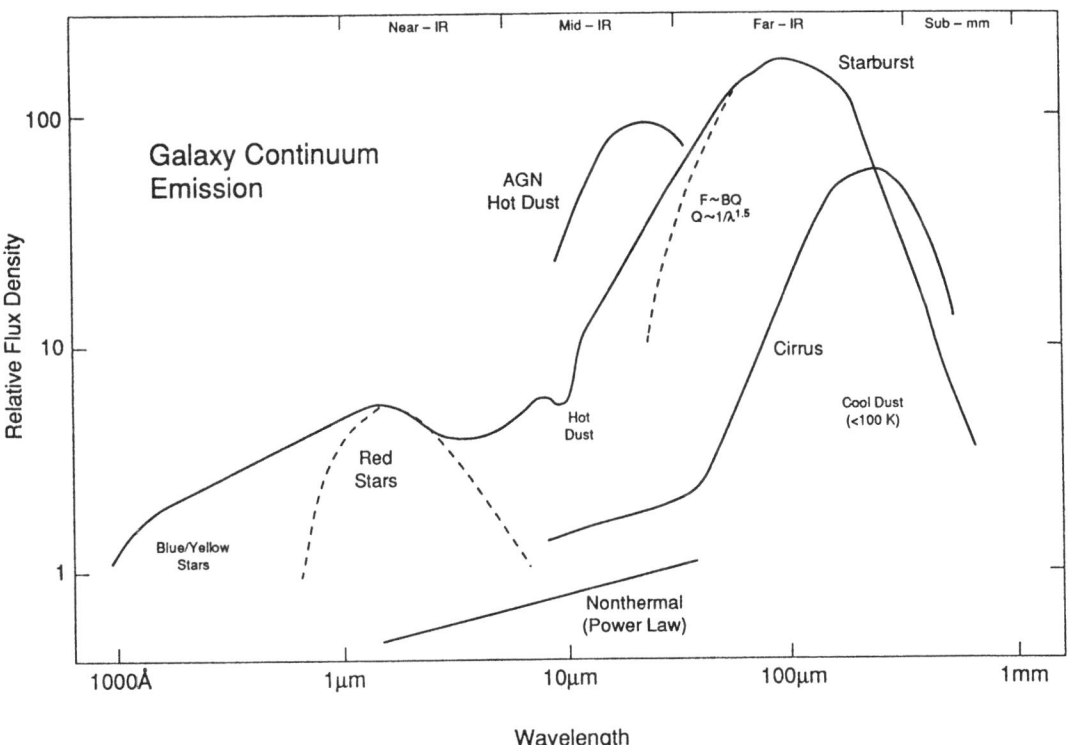

Figure 1: *Schematic overview of spectral energy distribution in a galaxy*

The near-IR continuum emission in **starburst galaxies**, especially at 1-3 μm, is usually dominated by photospheric radiation from red giant and red supergiant stars, with smaller detectable contributions often arising from ionized gas (bremsstrahlung), very hot dust, and blue stars. Somewhere between 3 μm and 10 μm, depending on the strength of recent star formation, the continuum energy distribution begins to rise steeply with increasing wavelength as the reradiated emission from hot dust becomes more important. The mid-IR-emitting dust has been heated to several hundred degrees by the young massive stars. The energy distribution reaches a maximum in the 80-100 μm region, falling off rapidly with wavelength thereafter. The dust particles emitting this far-IR radiation have characteristic temperatures of 30-50 K. Much cooler dust should be observable at sub-millimeter wavelengths, but even relatively large masses of very cool dust can be difficult to see because the luminosity is dependent on temperature.

In **quiescent galaxies**, in addition to photospheric near-IR emission from red stars, one observes throughout the spectrum emission from dust particles heated by a redder (compared to that from massive stars) interstellar radiation field characteristic of an older stellar population; this is often referred to as "cirrus" emission because of its filamentary appearance reminiscent of cirrus clouds. There is also the possibility that dust shells around evolved stars make a contribution to the mid-IR emission in those galaxies. Nonthermal emission may be observed in some **active galactic nuclei** (AGNs), but in at least one Seyfert galaxy, the active nucleus has an energy distribution peaking in the mid-IR, indicative of hot dust probably heated by the non-stellar object.

Certain infrared spectral features are apparent even with very low spectral resolution. The CO absorption band near 2 μm originates in the photospheres of red giants and red supergiants, and a broad absorption feature at 8-14 μm is due to intervening silicate dust. With somewhat higher spectral resolution, one often sees throughout the 3-12 μm region a group of five relatively narrow emission features attributed to very small dust grains or large molecules. The whole infrared region is full of emission lines from atomic and molecular gas which has been photoionized, shocked, or otherwise collisionally excited.

2 KEY INFRARED DIAGNOSTICS

2.1 Infrared Emission Lines from Gas

Infrared **fine structure transitions** (also called forbidden transitions) occur between levels with excitation temperatures of a few hundred degrees, and so the atoms and ions are easily excited collisionally in many environments of astrophysical interest. The populations of the levels depend on the collision rates and on the radiative transition probabilities from the relatively long-lived excited states (the transitions are "forbidden" to electric dipole radiation), so that the ratios of various line strengths indicate gas densities and temperatures (see, e.g., Stacey 1989; Spinoglio & Malkan 1992). When the gas temperature is much higher than the excitation temperature, the line strengths (particularly for two or more lines originating from different levels of the same ion) are diagnostic primarily of the gas densities, whereas when the excitation and gas temperatures are comparable, the line strengths depend on both temperature and density. In addition, the energy distribution of the exciting source can be deduced by determining which ionization states of the gaseous components are present. I emphasize that most IR fine structure lines are unaffected by interstellar extinction and are incisive probes of even the most heavily obscured regions.

The number of fine-structure lines which one expects to see throughout the IR is very large, and the potential for characterizing the interstellar medium and the sources which power it in other galaxies is enormous. Spinoglio & Malkan (1992) give the ionization potentials and critical densities appropriate for numerous IR fine structure lines. (The critical density is that density at which the collisional de-excitation and radiative de-excitation are equal; for densities below the critical density, collisional de-excitation is negligible, and radiative transitions dominate.) Not all of these lines have been observed in our own galaxy, let alone in other galaxies. The full development of the field awaits such facilities as ISO, SOFIA, and SIRTF.

The strengths of **recombination lines** is determined primarily by the rate at which gas is ionized. In a starburst, the photoionization rate is a useful direct measure of the number of massive stars. The near-IR recombination lines of hydrogen at 4.05 μm (Brackett α) and 2.17 μm (Brackett γ) are strong, easily observable from the ground, and usually suffer only modestly from the effects of interstellar extinction. Since A(2.2 μm)$\approx$0.1A$_V$, their ratio is a useful measure of the extinction to regions with very high visual obscuration, for which the Balmer decrement is virtually useless as a measure of the extinction. Other IR recombination lines which have special relevance to extragalactic observations are the H I λ1.28 μm (Paschen β) and the He I λ2.06 μm lines.

The fine structure and recombination lines discussed so far originate in neutral and ionized atoms. However, at least one astrophysically significant molecule, H_2, has been detected in other galaxies at IR wavelengths (although many submillimeter molecular lines have also been detected). Numerous H_2 lines due to transitions between **vibration-rotation** states are emitted in the 2-4 μm region in Galactic star-formation regions (e.g., Beckwith et al. 1983), with several of these, including the v = 2-1 S(1) line at 2.248 μm and the v = 1-0 S(1) line at 2.122 μm, being strong enough to observe in other galaxies. The H_2 is thought to be excited either collisionally in shocks or by absorption of 912-1108 A photons followed by fluorescent de-excitation. Depending on which lines are observed, the observations can be used to determine values of the interstellar extinction, shock velocities, and UV energy densities.

2.2 Emission from Dust

Almost invariably, review talks and journal articles about the IR emission from galaxies begin with some form of the statement "Much of the luminosity of galaxies is emitted in the infrared." In fact, this comment always refers to the IR continuum radiation emitted by dust, particularly at far-IR wavelengths. One usually finds that the IR energy distribution is a broad bump, with the maximum flux density occurring somewhere between 60 and 200 μm. I show in Fig. 2 the IR energy distributions for various Galactic sources that are first-order templates for the interpretation of the extragalactic IR emission. These are typical of the sources used by others for this purpose. For most IR sources, the main body of the broad emission bump closely resembles a blackbody emission spectrum that is modified by a wavelength-dependent emission efficiency of the form λ^n, where the value of n is usually taken to be in the range -1 to -2. IR emission in excess of this blackbody on the short-wavelength side of the bump is usually present and indicates that there is dust at temperatures higher than that characterizing most of the bump.

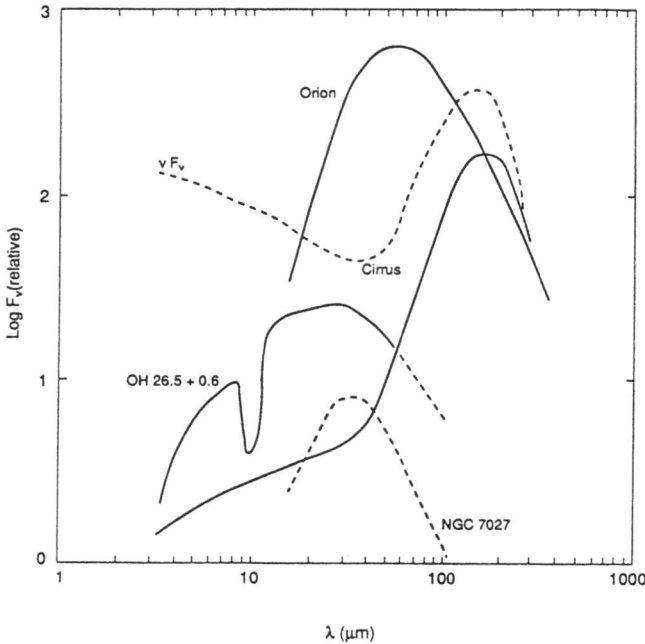

Figure 2: *IR energy distributions for Galactic template sources.*

The region of the Orion nebula is the nearest (450 pc) site of **active star formation** (see review in Genzel & Stutzki 1989). In Fig. 2, I show the energy distribution for the 1'-diameter region, corresponding to 0.13 pc, centered on the IR cluster (Werner et al. 1976). This cluster of young, massive stars emits $\sim 10^5$ $L_\odot$, of which 90% emerges at $\lambda > 30$ μm. It is buried in a dense, molecular cloud subject to complex radiative transport effects which include the absorption by outlying dust in the molecular cloud of near-IR and mid-IR radiation emitted by dust near the inner edges of the cluster cavity, resulting in the steep decline of flux at shorter wavelengths. This IR energy distribution peaking at 60-100 μm is characteristic of compact HII regions containing OB stars in dense molecular clouds where nearly all of the dust is cooler than 100 K (e.g., Churchwell, Wolfire, & Wood 1990). OB stars which have dispersed their shroud of molecular material can still heat the remaining cloud from the outside, with the Trapezium stars and their visible companions in Orion being a notable example.

A unique discovery of the IRAS survey was that the Galaxy is laced with a vast filamentary structure that resembles cirrus clouds. This **cirrus emission** is most evident at 60-100 μm, but is also clearly present at 12-25 μm (Fig. 2). About 80% of the cirrus is emitted by dust mixed with the extensive distribution of low density HI, with the remainder mixed with diffuse HII (Boulanger & Perault 1988, BP88). The interstellar radiation field (ISRF) heats the cirrus-emitting dust. In the solar neighborhood, the ISRF mean intensity $4\pi J$ = Uc = 2.17×10^{-2} ergs cm^{-2} s^{-1}, and the energy density is U = 0.45 eV cm^{-3} (Mathis et al. 1983). Only ~7% of the ISRF luminosity is UV radiation at $\lambda < 0.25$ μm (Mathis et al. 1983), but the efficient absorption by dust of the UV accounts for about 50% of the cirrus emission (BP88). This UV is emitted by OB stars, and so stars younger than 10^8 yr account for about half of the cirrus emission in the solar neighborhood, with the remaining half of the cirrus luminosity being powered by older stars. Therefore, the cirrus emission in the solar neighborhood, which is often used as a template for IR emission in other galaxies, reflects to some extent relatively recent star formation, a fact that must be taken into account in such comparisons, especially with "quiescent" galaxies.

The cirrus energy distribution shown in Fig. 2 shows a break near 40 μm. About 40% of the IRAS cirrus flux is emitted at $\lambda < 40$ μm, and BP88 estimate that ~25% of the cirrus emission between 5 μm and 1 mm emerges there. The dust grains emitting most of the radiation at $\lambda > 40$ μm are in equilibrium with the ISRF. However, the dust grains emitting the shorter-wavelength radiation are at temperatures of several hundred degrees even though they are generally too far from heating sources to achieve these high temperatures if they are large "standard-size" grains (~0.1 μm) in radiative equilibrium with the ISRF. It is now believed that this radiation is emitted by very small grains (size ~0.001 μm) which, because of their small heat capacity, are transiently heated to high temperatures upon absorption of a single photon (Andriesse 1978; Sellgren 1984). Based on a group of emission features in the 3-12 μm region observed in Galactic reflection nebulae and other sources, polycyclic aromatic hydrocarbon molecules (PAHs), or some variation thereof, are thought to constitute one component of this population of small grains (Leger & Puget 1984). A large fraction of the cirrus emission in the IRAS 12 μm band may arise in these bands (BP88), but one or more additional components of small grains are needed to account for the 25 μm and at least part of the 60 μm emission (Draine & Anderson 1985). Absorption of UV photons appears to be important in heating the small grains and PAHs (e.g., Puget & Leger 1989), but recent studies (Sellgren, Luan, & Werner 1990) have shown that visual photons may also play a role.

When intermediate mass stars (2-8 $M_\odot$) ascend the asymptotic giant branch, they rapidly lose mass and may become completely obscured by dust formed in the ejecta. These red giant and supergiant stars heat their **dust shells** to several hundred degrees, and the shells may contribute detectably to the integrated IR continuum emission in a quiescent galaxy. I show in Fig. 2 the IR energy distribution for the OH/IR dust-shell star OH26.5+0.6 (Werner et al. 1980; Olnan et al. 1984). [The OH maser emission results from pumping of the molecule by IR photons (e.g., Werner et al. 1980) and is not relevant to the current discussion.] The IR energy distribution is significant throughout the mid-IR and the shorter far-IR wavelengths.

Finally, I show in Fig. 2 the IR energy distribution for the planetary nebula NGC 7027. I have included this object because it is the most extreme example I can find of a Galactic H II region powered by a hot star. The central star has a temperature of ~1.7×10^5 K (see references in Telesco & Harper 1977), and, although there is a massive neutral shell surrounding the H II region (Bieging, Wilner, & Thronson 1991), nearly all of the IR emission appears to originate from within the H II region (Telesco & Harper 1977). NGC 7027 may be useful as a template for an **ionized gas ball** energized by a relatively hard UV radiation field; the average energy for a Lyman continuum photon in NGC 7027 is ~50 eV (Telesco & Harper 1977), compared to ~18 eV for an O8.5 star (Telesco, Decher, & Joy 1989).

2.3 Emission from Stars

Photospheric emission from stars is most evident in galaxies at wavelengths shorter than ~5 μm. Red stars are especially bright in the near-IR. In Fig. 3a (adapted from Doyon, Joseph, & Wright 1990) I show the loci of JHK colors for giant and dwarf stars. Red supergiant stars have near-IR colors that are barely distinguishable from red giants. The vectors in the figure indicate how the colors are modified by the presence of blue stars (A0), bremsstrahlung emission (H II), emission from hot dust, and extinction (A_V). Late-type stars have a photospheric CO absorption band at $\lambda > 2.3$ μm. The depth of the CO band increases with decreasing stellar effective temperature and with increasing

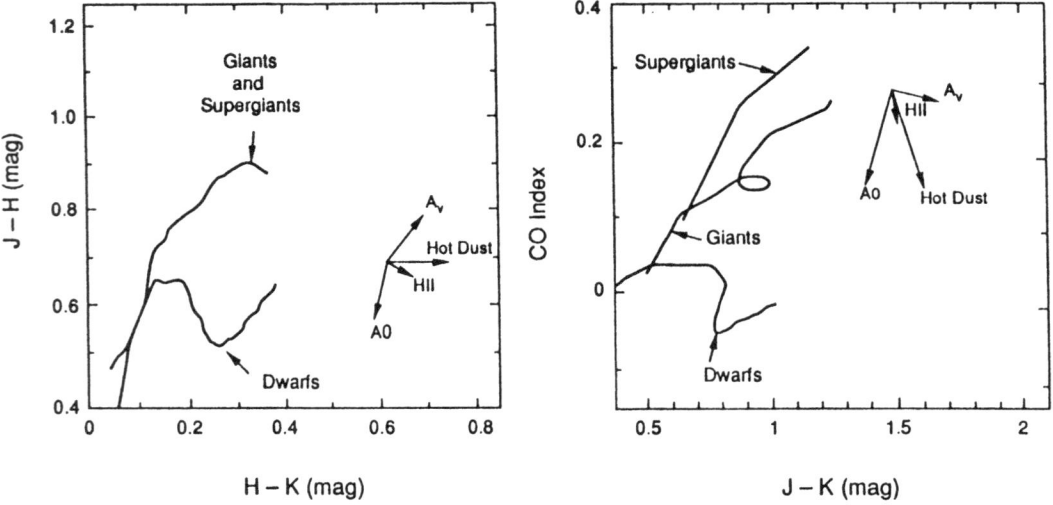

Figure 3: *Near-IR colors and the CO index for stars.*

luminosity. As shown in Fig. 3b (also adapted from Doyon, Joseph, & Wright 1990), the depth of the CO feature, as represented by a CO index, can be used to distinguish between red giants and red supergiants. Very late-type stars also have an H_2O absorption feature near 1.9 μm, the strength of which increases with lower stellar effective temperature (Aaronson, Frogel, & Persson 1978).

180

3 QUIESCENT GALAXIES

It is easier to define a quiescent galaxy by what it is not than by what it is. In the present discussion, a quiescent galaxy is one in which current and recent star formation and/or an AGN provide an insignificant fraction of the bolometric luminosity. In a real sense, quiescent galaxies are the starting points for discussions of starburst and AGN galaxies, since they represent the physical framework within which these other activities occur.

Aaronson, Frogel, Persson, their colleagues, and other investigators have shown that the 1-3 μm radiation from elliptical galaxies and the bulges of early-type spirals are dominated by photospheric emission from red giant stars (e.g., Frogel et al. 1978; Aaronson, Frogel, & Persson 1978). They did this by comparing the galaxian JHK colors and strengths of the 2.3 μm CO and 1.9 μm H_2O absorption bands to those of individual stars and star clusters in the Galaxy. These properties best match those of giants in the range K5 to M5. However, the JHK colors do not match those of a *single* giant spectral type. Instead, they originate in a composite population of red stars, so that one is seeing progressively redder giants as one shifts from shorter (J) to longer (K) wavelengths.

There is evidence that the near-IR colors of these benign stellar populations are redder closer to the nucleus and with increasing galaxian luminosity. These color effects are partly attributable to differences in metallicity (see, e.g., Rieke & Lebofsky 1979) and to an increased contribution from younger stars as one observes at larger radii (Aaronson 1977). Scans in the UBVRIJHKL photometric bands by Sandage, Becklin, & Neugebauer (1969) extending ±60" (200 pc) from the nucleus in M31 show an unambiguous nuclear reddening at the shortest wavelengths compared to the IR region. B, I, and K scans extending to more than ±20' (4 kpc) in M31 (Hiromoto et al. 1983) show a very smooth symmetric structure at K, whereas the bumps and troughs attributable to extinction and blue stars are seen at B and I; beyond ±5' (1 kpc), an exponential disk dominates the bulge population in K light, whereas such a disk is increasingly difficult to define as one considers shorter wavelengths. The point I want to make here is that even apparently quiescent galaxies are inhomogeneous and have something interesting to tell us, and near-IR observations provide unique information about the distributions of stellar populations and obscuring dust. Indeed, Block & Wainscoat (1991) demonstrate that a broad reassessment of the classification of galaxies based on near-IR imaging is worthwhile.

Ellipticals and lenticulars are generally regarded as the most quiescent among galaxies. As a class, they have relatively small amounts of interstellar matter and therefore little fuel to feed either star formation or an AGN. However, they are not devoid of circumstellar and interstellar gas and dust, as IR observations indicate. Jura et al. (1987) showed that about half of the bright ($B_T < 11$) Shapley-Ames ellipticals were detected by IRAS at 100 μm. Considering the low level of activity apparent in these galaxies, it is reasonable to suppose that this 100 μm radiation is the elliptical-galaxy counterpart of far-IR cirrus emission from cool dust observed in the Galaxy (Section 2.2) and quiescent spirals (see below). With a ~6"-

diameter beam centered on the nuclei, Impey et al. (1986) detected about 30% of their optically selected sample of 65 ellipticals at 10 μm. For those ellipticals detected at 10 μm, an average of ~40% of the 10 μm flux is not photospheric emission, although in most cases the JHKL colors are like those of red-giant photospheres. By comparing small-aperture groundbased and large-aperture IRAS data, Knapp et al. (1992) infer that the 12 μm IRAS emission from optically bright ellipticals is extended like the giant stars; the ratios of these two sets of data, after application of a variety of corrections, are consistent with the ratio expected for an "$r^{1/4}$-law" brightness distribution considered typical for ellipticals and the bulges of spirals. Based on a comparison with the 2.2 μm emission, the extended 12 μm flux is generally found to be higher than expected for red-giant photospheres. It is also higher than expected for the hot componant of cirrus emission if the ratio of 12 μm and 100μm cirrus fluxes for ellipticals is the same as that for normal spirals (Helou et al. 1989). Knapp et al. (1992) conclude that about one-third of the 12 μm light from quiescent E galaxies is due to circumstellar shells associated with mass loss from red-giant stars. The energy distribution for these shell stars peaks in the mid-IR (Fig. 2). It is germane that Frogel & Whitford (1987) point out in their study of red giants in the Galactic bulge that long-period variables (like OH/IR stars), which have the most prominent dust shells among red giants, probably contribute only ~20% of the mass loss there; ~83% if the mass loss comes from the far more numerous M5-M7 giants with only modest mass loss.

Knapp et al. (1992) have estimated the mass-loss rates (dust + gas) from evolved stars in ellipticals based on the 12 μm excesses and the assumption that the relationship between the mid-IR luminosity and the mass-loss rate is like that observed for ~1 $M_\odot$ stars now evolving through the giant stage in the Galaxy. Another way of making this estimate is to assume that the outflow is driven by radiation pressure (e.g., Jura et al. 1987). Since the photon momentum is hν/c, where c is the speed of light, the radiative momentum transport per second through a shell is L/c. If a fraction f of this luminosity is absorbed by the dust ($f \times L = L_{IR}$), then the force imparted to the dust is $f \times L/c = d(Mv)/dt = v \times dM/dt$, where dM/dt is the mass-loss rate, and v is taken to be the terminal velocity (~15 km s^{-1}). The two methods give comparable values: the total mass-loss rate from a typical E galaxy is 0.1 $M_\odot$ yr^{-1}, in reasonable agreement with values derived from optical studies (Faber & Gallagher 1976). In 10^{10} yr, mass loss from evolved stars should therefore contribute ~10^9 $M_\odot$ of gas to the interstellar medium, a value that is about an order of magnitude higher than actually observed in ellipticals (Gallagher et al. 1975). This discrepancy has led to proposals in which gas is removed from the galaxies by, for example, hot winds (Faber & Gallagher 1976).

Among the quiescent spiral galaxies, M31, usually classified as type Sb I-II, is probably the most thoroughly studied at IR wavelengths. Because it is only 690 kpc away and subtends nearly a degree, detailed maps of M31 have been obtained despite the low spatial resolution of IRAS (Walterbos & Schwering 1987). The structure of M31 is generally the same in all four IRAS bands: the IR emission is complex and extends across the whole visible disk, with the dominant IR features being a "ring" at a radius of about 10 kpc and a bright source coincident with the nucleus. Whereas the nuclear IR emission arises in a region with no evidence of "activity," the IR ring is coincident with OB stars, H II regions, dark lanes, and atomic and molecular gas. Interestingly, all giant H II regions are close to bright IR sources, but not all discrete IR sources are associated with H II regions.

Walterbos & Schwering show that the integrated IR energy distribution of M31 resembles that of Galactic cirrus in which cold dust and hot, transiently heated small grains in diffuse atomic clouds are heated by the general interstellar radiation field. Most of the emission has a low 60-100 μm color temperature (for a λ^{-2} emissivity) of only 20-23 K like that of the cool cirrus component. Less than 10% of the total IR luminosity of 2.6×10^9 $L_\odot$ is due to heating by OB stars associated with regions of recent ($<10^7$ yr) star formation. This latter result is surprising in view of the coincidence of the IR ring with the ring of population I constituents, but most of the emission in the ring in fact must not be intimately associated with the star-formation regions. In addition, much of the emission actually comes from "inter-arm" regions outside the ring.

The mid-IR emission in the disk of M31 is distributed like the cool cirrus component, consistent with its being from transiently heated small grains like those associated with Galactic cirrus. However, the origin of the mid-IR radiation from the nucleus is less clear. Extrapolating from the nuclear near-IR flux, Soifer et al. (1986) show that about half of the 12 μm emission from the center of M31 is photospheric emission from giant stars (the energy distribution for the central 4' is shown in Fig. 4). They propose that the rest of the 12 μm and 25 μm radiation, which has a color temperature of ~370 K, is emitted by dust in circumstellar shells, as described above for ellipticals. Indeed, the ratio of excess mid-IR flux and total stellar mass in the nucleus of M31 (a measure of the relative number of shell stars in the population) is comparable to that in the Galactic bulge where the mid-IR is attributed to dust shells (Soifer et al. 1986). In addition, the mass-loss rate per shell star in M31 is estimated to be ~$2 \times 10^{-6} M_\odot$ yr^{-1}, consistent with mass-loss rates for Galactic shell stars (see Soifer et al. 1986). However, Walterbos & Schwering (1987) argue that the uncertainties in the analysis by Soifer et al. allow sufficient room for much of the excess nuclear mid-IR flux in M31 to be emitted by hot cirrus dust. Taking the ratio F(12 μm) = 0.042 × F(100 μm) to be characteristic of cirrus (Helou et al. 1989; Knapp et al. 1992), the expected 12 μm cirrus flux from the central 4' of M31 is 1.6 Jy. Based on the near-IR emission, the 12 μm photopheric flux is 2.5 Jy (Soifer et al. 1986), so that 0.8 Jy, or only 16% of the total 12 μm flux from the M31 bulge, may originate in circumstellar shells. However, the issue remains unresolved since we cannot be sure that the cirrus emission in an Sb bulge, where blue stars are rare, is like that in spiral disks, where blue stars are common. Additional uncertainty arises from the fact that UV photons, which are probably sparse in the nucleus of M31, are apparently not *required* for transient heating of small grains (Sellgren, Luan, & Werner 1990).

Two additional examples help illustrate properties of quiescent galaxies. Using 12-200 μm maps of the edge-on Sb galaxy NGC 4565, Engargiola & Harper (1992) showed that the IR luminosity accounts for 16% of the bolometric luminosity. About half of the total L_{IR} of $1.1 \times 10^{10} L_\odot$ is from warm dust associated with recent star formation, nearly all of which (80%) is in the disk and associated with features that are probably edge-on spiral arms. The central few kiloparsecs emit only a small fraction (14%) of the IR luminosity, with that being almost entirely cirrus. M33, of type Sc II-III, may be typical of late-type quiescent galaxies. Rice et al. (1990) find that the luminosity of 1.2×10^9 L$_\odot$ emerging between 5 μm and 1 mm accounts for only about 20% of the luminosity emitted at λ > 0.1 μm. Consistent with its late Hubble type, which is partly defined by the prominence of the spiral arms, many discrete IR sources coincide with spiral-arm H II regions; individual H II regions have IR luminosities in the range $2\text{-}70 \times 10^6$ L$_\odot$, comparable to the most luminous Galactic star-forming regions. The discrete H II regions account for about 60% of the IR luminosity from M33, with the remainder being cirrus-like.

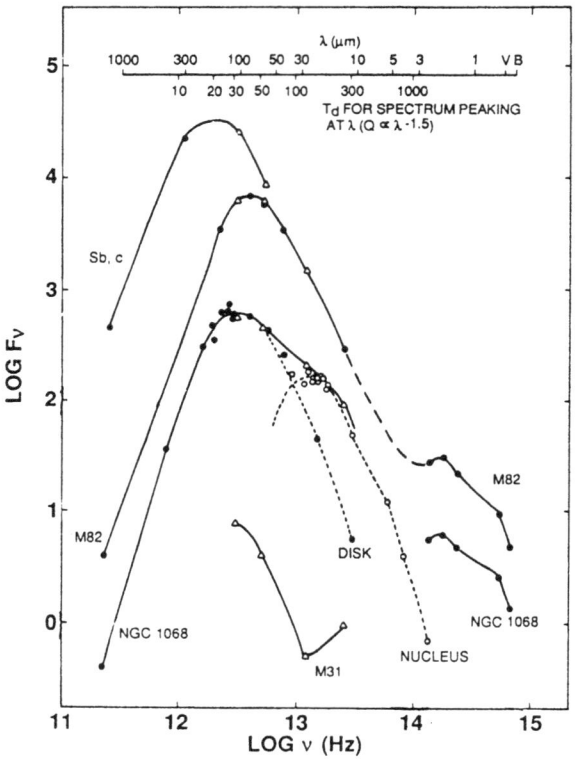

Figure 4: *IR energy distributions for galaxies discussed in the text.*

The impression one gets from studies such as those described above is that most of the IR luminosity from quiescent disk galaxies is emitted in the disk. The available data suggest that recent star formation may account for a small (M31) or large (M33) fraction of the IR luminosity, probably depending on the Hubble type of the galaxy. This conclusion is consistent with limited survey data. For example, Chini et al. (1986; see also Cox & Mezger

1987; Cox, Krugel, & Mezger 1986) separated the IR energy distributions for the central regions of 18 Sb and Sc galaxies into cirrus and starburst components. The average energy distribution is shown in Fig. 4. They estimate that typically 40% of the IR radiation from the central 7 kpc of these galaxies and from the disk of the Milky Way (total $L_{IR} \approx 2 \times 10^{10}$ $L_\odot$) is emitted by warm dust in regions of star formation. The remainder is cirrus emission.

4 STARBURST GALAXIES

Weedman et al. (1981) coined the term "starburst" to indicate qualitatively that a galaxy is experiencing notably strong or intense star formation. Although many astronomers dislike the term, I find it useful for distinguishing this type of galaxy from both quiescent ones and those, usually termed "active galaxies," in which non-stellar nuclear objects may generate much of the luminosity. Here we consider those galaxies in which a significant amount of the large IR luminosity is generated by recently formed stars or processes associated with these stars.

4.1 IR Observations of Starburst Galaxies

4.1.1 Energy Distributions

The IRAS survey obtained the largest body of IR photometry of galaxies. Most of these data are in four broadbands centered at 12, 25, 60, and 100 μm. Helou (1986) showed that the IRAS colors of normal (i.e., non-AGN) galaxies are reasonably well defined. I show in Fig. 5 Helou's two-color plot for galaxies unresolved by IRAS and with high-quality fluxes. The galaxy colors lie in a swath stretching from the lower right to the upper left in the diagram. The straightforward interpretation suggested by Helou for this locus of colors is that it represents a mixing line, with cirrus emission dominating the colors to the lower right, starburst emission dominating in the upper left, and a mixture of these two determining a galaxy's specific colors between the two extremes. By referring to the IR energy distributions of cirrus and star forming regions shown in Fig. 2, it is easy to see the basis for this trend. The cirrus dust emitting the *far-IR* radiation is very cool (low 60-to-100 μm ratio) because the intensity of the ISRF is low, and the *mid-IR*-emitting cirrus dust is hot (high 12-to-25 μm ratio) because those grains are very small and transiently heated. The higher radiative energy densities in starbursts heat up the larger grains, thereby increasing their 60-to-100 μm color temperature. The flux from those grains increases dramatically with temperature (σT^4), and the energy distribution shifts to shorter wavelengths. Emission from these larger grains, which are cooler than the small transiently heated ones, begin to dominate the mid-IR, thus resulting in a lower 12-to-25 μm color temperature. The distribution of galaxy colors in the two-color plot can also be reasonably well understood as the locus generated as the energy density that heats the dust is increased from the cirrus values (lower right) to the starburst values (upper left), as shown by models (Desert, presented in Helou 1986).

The IR energy distributions also permit some discrimination between starbursts and AGNs. De Grijp, Miley, & Lub (1987) present 25-60-100 μm two-color plots for normal galaxies, HII/starburst galaxies, and AGNs, which include type 1 and 2 Seyferts and optically selected quasars. Although there is significant overlap among the samples, it is evident that the centroids of the colors differ for the various groups. The more active galaxies have bluer 60-to-25 μm colors and, less marked, bluer 100-to-60 μm colors, corresponding to hotter dust. The conclusion is that a component of hot dust is present in active galaxies that is not present in starburst and other normal galaxies. In one well studied example, which we will discuss in detail later in these lectures, this hotter dust clearly resides very near the AGN and is probably heated by a non-stellar object.

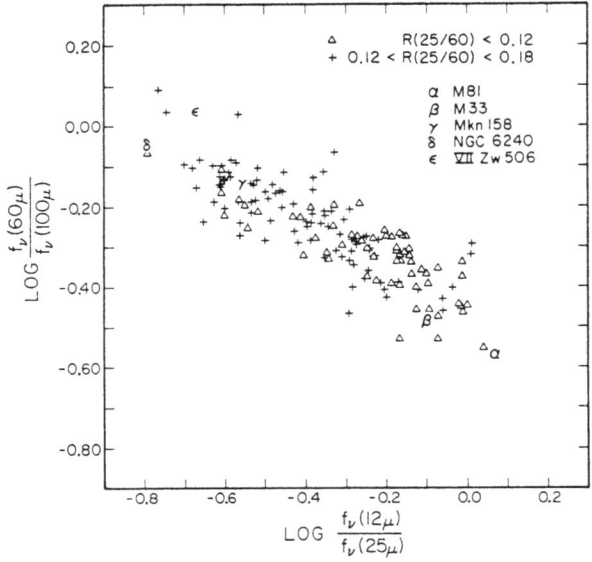

Figure 5: *IRAS two-color diagram for mostly normal (non-AGN) galaxies.*

We see that the broadband IR energy distribution of a galaxy can permit an assessment of the type of source which powers the IR emission, and in particular to distinguish starbursts from other types of activity that can generate the IR luminosities. The reliability of this assessment depends on where an object's colors fall in the two-color plots, since there are some regions where different classes overlap more. Rowan-Robinson and his colleagues (e.g., Rowan-Robinson & Crawford 1989) have made a strong effort to separate the contributions from starbursts, cirrus, and AGNs to the IR emission from a large sample of IRAS galaxies; they assume canonical distributions for each class, and they show that generally these separations are straightforward and plausible. Of course, rarer objects, almost by definition, may not be easily classified. For example, the energy distribution of the extremely luminous galaxy IRAS 10214+4724 resembles that of a starburst (Clements et al. 1992), but the environment and conditions that lead to such activity may be so unusual that the template energy distributions are inapplicable. It turns out that the broadband IR energy distribution for M82, for which $L_{IR} = 3 \times 10^{10} L_{\odot}$ (Telesco & Harper 1980), is typical for starburst galaxies of modest luminosity (see comparison in Roche et al. 1991). M82's IR energy distribution is shown in Fig. 4. Note how M82's energy distribution resembles closely that of the Orion IR cluster (Fig. 2).

Combining the broadband far-IR data with mid-IR spectroscopy which they normalized using the IRAS 12 μm flux, Roche et al. (1991, and references therein) produced the composite and average starburst galaxy energy distributions shown in Fig. 6. In addition to the broad far-IR peak, the average energy distribution shows a depression near 10 μm. Part of this depression, the so-called silicate feature, is attributable to absorption by cool silicate dust, but it is also evident that mid-IR emission features, discussed below, tend to enhance the depth of the feature, especially the unidentified or PAH features at 7.7, 8.7, and 11.3 μm. (The 12.8 μm line is from neon.) As Roche and colleagues have shown, one must correct for these emission features in order to accurately determine the true amount of absorption by silicates. As an interesting aside, consider the IRAS Low Resolution Spectra spanning the 8-23 μm region of the starburst galaxies NGC 253 and M82 (Cohen & Volk 1989). The two spectra look very much alike between 8 μm and 13 μm, which is the part of the spectrum accessible from the ground. However, by using the LRS data to correct for the emission features, we see that NGC 253 still seems to have a reasonably strong silicate absorption, but, surprisingly, M82 may *not* have silicate absorption. This is puzzling because most of the mid-IR emission in the LRS beam should come from the starbursts in both galaxies, and smaller-beam spectra (e.g., Gillett et al. 1975) of those starburst regions are nearly identical for the two galaxies, with a continuum apparently detected *between* the 11.3 μm feature and the Ne line. This must be telling us something important about the differences in the spatial variations in the dust properties in the two galaxies, as first noted by LeVan and Price (1987).

Near-IR continuum emission in excess of that expected from the photospheres of red giant stars is evident in starburst galaxies. In his near-IR survey of nearby spirals (recession velocities $\leq$ 3000 km s^{-1}) with IRAS far-IR luminosities greater than 3×10^9 L$_\odot$, Devereux (1989) found that the mean 2.2 μm luminosity of the galaxies with L(10 μm) $\geq 6 \times 10^8$ L$_\odot$ is a factor of three greater than those with lower 10 μm luminosities and a factor of four greater than those "normal" spirals with low far-IR luminosities (Devereux, Becklin, & Scoville 1987). In addition, the 2.2 μm emission is greater, and the H-K color is redder, in those spirals with more 10 μm. Most of these galaxies have H II-region-like optical spectra, and so the IR emission is powered by starbursts. Doyon (1991) showed that merging galaxies thought to be undergoing starbursts are more luminous per galaxy at 2 μm by factors of three to six than the average Virgo spiral (M$_K$ $\approx$ -20.7, Devereux et al. 1987). Hot dust, bremsstrahlung, and red supergiants associated with the starbursts must account for this excess emission. Near-IR spectroscopy by Doyon (1991) on a small sample of starburst and, possibly, AGN galaxies with IR excesses shows that many of these have strong CO features. For example, his spectrum of the ultraluminous galaxy NGC 1614, presented in Fig. 7, shows a deep CO absorption. For NGC 1614 and several other galaxies, Doyon determines a spectroscopic index of 0.30±0.02. Elliptical galaxies and spiral bulges, where red giants dominate the near-IR light, have a CO index of ~0.2 (Fig. 3), so that red supergiants appear to emit much of the excess near-IR light in some starburst galaxies.

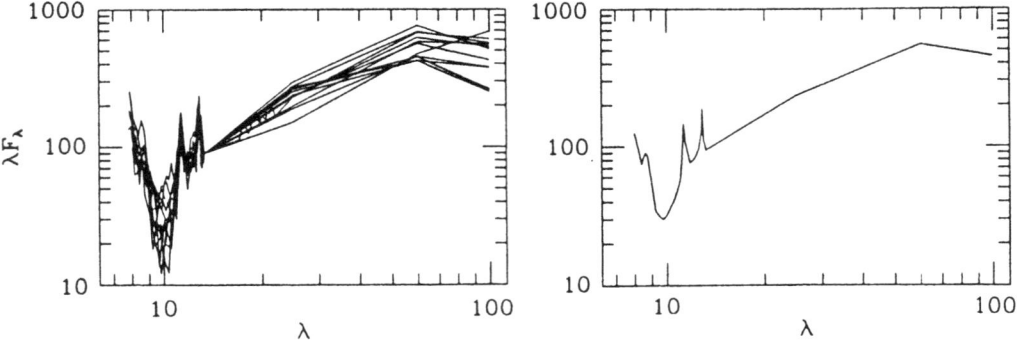

Figure 6: *Composite (left) and average (right) starburst IR energy distributions.*

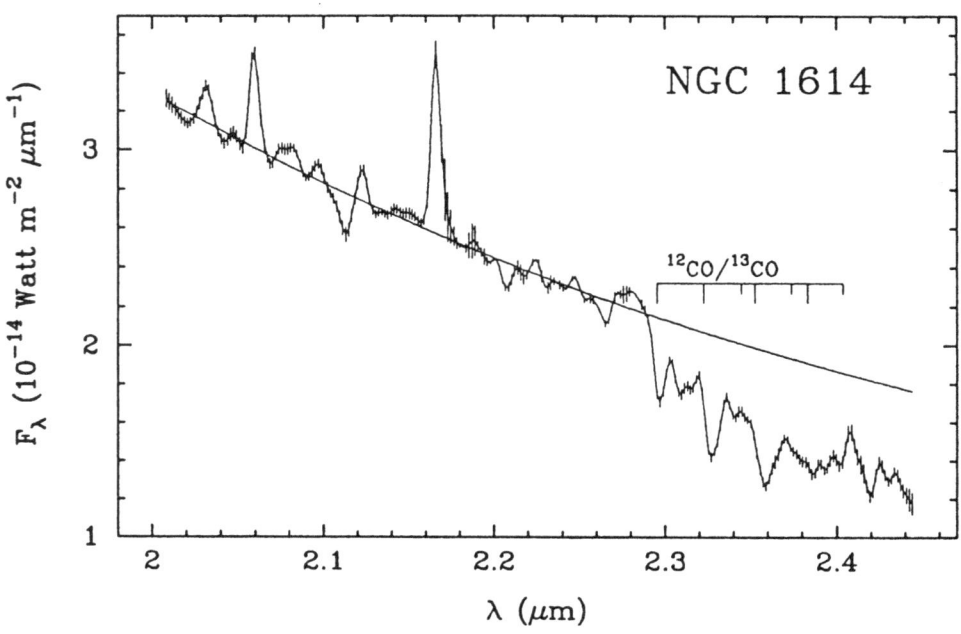

Figure 7: *Near-IR spectrum of NGC 1614 showing CO absorption feature. Note also the bright Brγ and He I emission lines.*

4.1.2 Spatial Distributions

The spatial distribution of the IR emission holds important clues to the nature and genesis of the sources that power it. From a purely practical standpoint, broadband mapping and imaging is always easier than spectroscopy in the same wavelength region, and so images of a starburst guide more detailed observations and provide the context within which they are interpreted. One fundamental task of imaging is to determine how the IR luminosity is distributed. The IR energy distribution of starbursts peaks at 80-100 μm, with typically 80% of the IR luminosity being emitted at $\lambda > 30$ μm. Because the atmosphere is opaque in the 30-300 μm region, all of the far-IR maps of galaxies have been obtained with the Kuiper Airborne Observatory (KAO) and IRAS. Both of these have relatively small telescopes, so that the angular resolution of the maps is typically 1' and no better than several tens of arcsec. Thus, we cannot see details smaller than ~200 pc in the far-IR even for the closest starbursts (e.g., M82 at 3 Mpc).

Mid-IR imaging has the enormous advantage that it can be done with arcsec angular resolution at large groundbased telescopes. The mid-IR has therefore been used to trace the detailed spatial distribution of the far-IR luminosity. The spatial correlation between the mid-IR and the far-IR breaks down on scales smaller than a few parsecs in starbursts, as judged by maps of Galactic star forming regions, but in extragalactic starbursts the correlation seems to hold at the highest spatial resolution of >10 pc. Because starburst galaxy IR energy distributions are similar, we can probably estimate the far-IR luminosity from the mid-IR luminosity at a particular position in a starburst to within a factor of two (see Telesco, Dressel, & Wolstencroft 1993). This capability permits us to use the mid-IR maps to infer how the numbers of young stars, and therefore the recent star formation rates, depend on location. Comparison of mid-IR scans (Telesco, Decher, & Joy 1989) and far-IR scans (Joy, Lester, & Harvey 1987) of the M82 starburst provide direct evidence that the mid-IR is a good tracer of the far-IR emission; although there are differences in the wings of the scans, the overall similarity is impressive. Telesco et al. (1991) also show that there is no discernible trend, or wavelength-dependent morphology, between 10 μm and 30 μm in M82. The mid-IR does appear to be a good tracer of the far-IR luminosity emitted by warm dust heated by the young stars. However, in general the mid-IR is probably not a good tracer of the cooler dust evident in the submillimeter region (Telesco et al. 1991).

About thirty starburst and AGN galaxies have been mapped in the mid-IR, with most of this work appearing only recently (Telesco, Dressel, & Wolstencroft 1992). In nearly all cases, this groundbased mapping has been of the central regions of galaxies for which noteworthy activity had been found previously at other wavelengths or suggested by IRAS. Nearly all of the maps have been made at 10 μm with angular resolutions of 4"-6" either by raster-scanning single detectors (e.g., Telesco & Gatley 1984; Becklin et al. 1980) or with the NASA-MSFC 20-bolometer array which has 4"-square pixels (e.g., Telesco et al. 1987, 1988, 1989, 1991). Maps with ~1"-resolution made with the new generation of integrated arrays are just beginning to appear (e.g., Keto et al. 1992; Telesco & Gezari 1992; Pina et al. 1993).

Even with arcsec resolution, many of the central starburst regions in IR-luminous galaxies (e.g., NGC 6946, see below) are unresolved, but many others show complex extended emission. Consider the SABc galaxy NGC 5236 (M83), which is 4.7 Mpc away. As shown in Fig. 8, the 10 μm emission is distributed across the central ~20" (500 pc) and has an irregular morphology with two peaks. At optical wavelengths, a dark lane cuts diagonally across this region, presenting an optical morphology that is very different from that in the mid-IR (see Fig. 4 in Telesco 1988). Significantly, the centroid of the starburst is displaced ~ 170 pc from the nuclear stellar mass peak (the true nucleus); although the starburst encompasses the nucleus, the nucleus is not the preferred site of recent star formation even though it represents the bottom of the gravitational potential well. We will explore possible reasons for this in Section 4.3.

A starburst and an AGN can coexist is the same galaxy. Mid-IR imaging can define their relative roles in generating the IR luminosity, because the AGN is expected to be more compact than the starburst region. For example, the 10 μm map (Fig. 9) of the type 2 Seyfert galaxy NGC 1365 (Telesco, Dressel, & Wolstencroft 1993) shows that, while the brightest mid-IR peak coincides with the Seyfert nucleus, a zone of HII regions located about 600 pc away is also luminous in the IR and may actually dominate the total IR luminosity because the ratio of far-IR emission and mid-IR emission may be much higher for the starburst than for the AGN. The difference in starburst and AGN IR energy distributions and the utility of mid-IR mapping to delineate the relative roles of the two phenomena is well illustrated by the nearby (18 Mpc) type 2 Seyfert NGC 1068 which has an IR luminosity of 3.2×10^{11} L$_\odot$ (Telesco, Harper, & Loewenstein 1976). Assuming that the IR radiation is from dust grains, Telesco et al. (1976) showed that the far-IR luminosity from NGC 1068, which is ~50% of the total IR luminosity, must be emitted by a source that is more extended than the Seyfert nucleus: the minimum solid angle Ω for the source at temperature T emitting flux F is that of a blackbody with intensity B(T), such that F = ΩB. Subsequent mid-IR and far-IR observations confirmed this prediction, showing that the Seyfert nucleus dominates the emission at $\lambda < 30$ μm and a kiloparsec-size starburst dominates the far-IR luminosity (Telesco et al. 1984; Telesco & Decher 1988). Even though the Seyfert nucleus in NGC 1068 emits most of the mid-IR radiation, the extended disk has low-surface-brightness mid-IR emission that is bright enough to image.

What the mid-IR mapping of starburst galaxies reveals is that the IR distributions exhibit a wide range of scale sizes, from less than 60 pc to more than ~3 kpc, with 300-400 pc being typical. These sources have a variety of morphologies. Some are elongated and asymmetric, while others are symmetric and double lobed. IR bright AGNs sometimes have prominent extranuclear starbursts. The higher resolution images being made with the new generation of mid-IR arrays is beginning to reveal the detailed structure of these starbursts. In a 12.4 μm image of M82 (Telesco & Gezari 1992) made with an angular resolution of ~1", corresponding to 16 pc, several bright emission knots are apparent. If we assume that the IR energy distributions for the individual knots are identical to that of the starburst as a whole, the 5-300 μm luminosity of the brightest knots is $1-2 \times 10^8$ L$_\odot$, which can be supplied by 10^4-10^5 massive stars. This mass range is comparable to that of Galactic globular clusters, but it is much larger than rich open clusters. Thus, young "super star clusters" may be embedded in the knots.

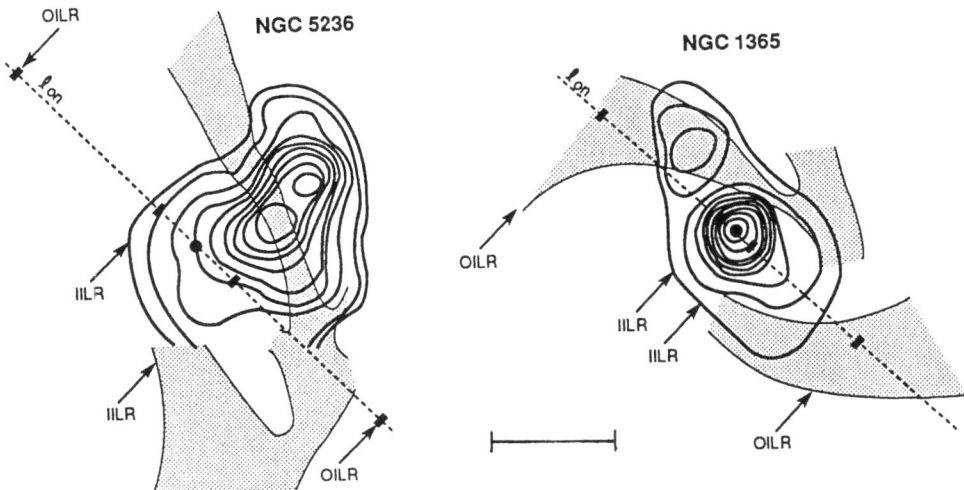

Figure 8 (left) and Figure 9 (right): *Maps of NGC 5236 (M83) and NGC 1365 overlaid with sketches of visually dark lanes. Also indicated for each galaxy is the line of nodes (l.o.n.) and the positions where it intersects the <u>outer</u> inner Lindblad resonance (OILR) and the <u>inner</u> inner Lindblad resonance (IILR). Filled circles indicate the positions of the nuclei, and the bar at bottom corresponds to 10".*

Two additional comments are in order about the mid-IR (and by inference the far-IR) spatial distribution. First of all, there is an excellent spatial correlation between the mid-IR and the nonthermal-radio distributions in starburst regions (e.g., Ho et al. 1989; see Fig. 4 in Telesco 1988). This relationship, which is at least part of the explanation for the *global* radio-IR correlation among galaxies (Condon, Anderson, & Helou 1991), presumably results from the coexistence of the young stars, which power L_{IR}, and their supernova offspring, which generate the nonthermal radio radiation by the synchrotron process. Starburst models, of the type discussed below, support this conjecture (e.g., Rieke et al. 1980).

Secondly, we have known for many years that there is a global correlation between the 2.6 mm CO line luminosity and the IR luminosity (Telesco & Harper 1980; Rickard & Harvey 1984; Young et al. 1986), but only recently, with the increased availability of mid-IR maps of starburst galaxies, have we begun to acquire insight into the detailed spatial relationship between these two components. In contrast to the relationship between the nonthermal-radio and IR distributions, which are nearly always well-correlated spatially, the distributions of the dense interstellar gas (evident in CO maps or as dark lanes in visual images) and the IR luminosities show significant differences. For example, the dark lanes intersect the circumnuclear starburst peaks in NGC 1365 and NGC 5236 but are much more extended than the starbursts (Figs. 8 and 9). In galaxies for which interferometric CO maps are available, such as IC 342, Maffei 2, NGC 253, NGC 6946 (Section 4.3), NGC 1068, and NGC 1097,

191

the IR peaks do generally coincide with gas peaks, but in most of these cases the gas is much more extended than the starburst in the sense that $L_{IR}/L(CO)$ is much lower away from the starburst (Telesco, Dressel, & Wolstencroft 1993). As discussed in Section 4.3, these differences between the distributions of the starbursts and dense interstellar matter can be understood as being a consequence of the effects that bars have on interstellar gas and the evolutionary state of the starburst.

We have focussed on the mid-IR and far-IR properties of central starburst regions, but it is worth noting that these starbursts, spanning the central few hundred or even few thousand parsecs, do not dominate the IR luminosity in most galaxies usually regarded as starburst galaxies. Comparing the IRAS disk-integrated and PSC 60 μm fluxes (Rice et al. 1988) for NGC 5236 (M83), for example, one finds that about 60% of the flux is emitted from regions outside the central ~1' (1.4 kpc). Telesco et al. (1993) find that central starbursts account typically for about half of the mid-IR emission in their sample of starburst galaxies. Of course, the starburst regions are usually small, so that the IR surface brightness is much higher there than for the average disk region; the typical starburst in the Telesco et al. (1993) sample is ~1% the angular size of the IRAS mid-IR beam, so that the starbursts are about one-hundred times brighter than the average surrounding disk. In his detailed study of the starburst galaxy NGC 6946, Engargiola (1991) shows that only 15% of the total IR luminosity of 8×10^{10} $L_{\odot}$ originates in the central 3.5 kpc-diameter region; the nuclear starburst (Fig. 18) is actually only a few hundred parsecs in size. Most (70%) of the IR luminosity in NGC 6946 is generated by recent star formation in the disk, with a marked concentration to the spiral arms (Engargiola 1991). However, there is strong evidence that the central concentration of IR luminosity in starburst galaxies increases with increasing luminosity (e.g., Wang & Helou 1992).

As we have intimated, knowing the IR morphology provides the basis for understanding the origin of starburst activity because we can determine the relationship of the current star formation with the interstellar gas from which the stars form, and we can examine how the patterns of star formation are shaped by the global galactic dynamics. *Near-IR* mapping plays an especially important role in the latter pursuit. The near-IR radiation in quiescent galaxies is dominated by the light of red giant stars. These giant stars are a small fraction of the stellar mass, which is mainly in the low-mass main sequence stars, but the giant stars trace out the distribution of that stellar mass. Therefore, near-IR images of a quiescent galaxy tell us how the dominant stellar mass is distributed. This is also the case in starburst galaxies, although one also often sees emission from red supergiants (which have the same IR colors as red giants), blue stars, hot dust, and ionized gas. The fact that the near-IR is much less affected by extinction than is visible radiation makes near-IR imaging especially valuable in studying the mass distributions in extremely dusty starburst galaxies.

Near-IR imaging usually permits one to locate the stellar nuclear mass peak. Knowing how a starburst is distributed with respect to the nucleus is fundamental to the exploration of how the starburst originated. In addition, in strongly interacting systems, identifying the nuclei amid the chaos can tell us about the degree of merging, including, of course, whether there are indeed two nuclei (see, e.g., Bushouse & Werner 1990; Carico et al. 1990). For example, the ultraluminous galaxy Arp 220 has a very complex optical morphology (see, e.g., Fig. 10 in Sanders et al. 1988) which suggests that its high luminosity is somehow caused by a collision between galaxies. A 2.2 μm image by Graham et al. 1990 reveals that Arp 220 does indeed have two nuclei separated by ~1" (350 pc), which confirms the impression, based on the optical appearance, of a violent merger.

In the presence of bright star formation and heavy extinction, the near-IR morphology can provide unique information about a galaxy's type. Maps at 2.2 μm of the well-known luminous system Arp 299, consisting of the galaxies IC 694 and NGC 3690, show that IC 694 has a large nuclear bulge, a region following an "$r^{1/4}$ law," extending out to ~4 kpc from the nucleus (Telesco, Decher, & Gatley 1985). Thus, IC 694 was probably of Hubble type Sb or earlier and, during the merger, may have received most of its current supply of gas from NGC 3690, which appears to have an exponential disk indicative of a later, more gas-rich, Hubble type (Telesco et al. 1985; see also Wynn-Williams et al. 1991). Mapping at 2.2 μm of the archetypal starburst/AGN galaxy NGC 1068 (Thronson et al. 1989) and the starburst galaxy M82 (Telesco et al. 1991) have revealed that these are barred galaxies, a conclusion that significantly affects our ideas about how the starbursts in those galaxies are generated.

Multi-color near-IR observations of starburst galaxies permit a much greater exploration of the starburst properties. The JHK two-color plot for positions in the unusual blue galaxy NGC 3310 (Telesco & Gatley 1984) show that the nucleus has normal red-giant colors, but many of the off-nuclear points in the starburst are much bluer, being displaced along the "A0" vector; at some positions, as much as 20% of the 2.2 μm light appears to be emitted by blue stars. In addition, ~30% of the 2.2 μm flux from a giant HII region is due to bremsstrahlung, consistent with the radio continuum flux observed there. The effects of heavy extinction and emission from hot dust can also be assessed by using these multicolor plots (e.g., Telesco et al. 1985), although it should be clear that the colors do not usually uniquely define the process. Although most of the near-IR work so far has been in the JHK bands, the L(3.5 μm) and M(4.8 μm) near-IR bands are potentially very useful even though the higher thermal background at these wavelengths makes their use more of a challenge. I show in Fig. 10 scans in the K, L, and U bands through the central starburst in M51 (Telesco, Decher, & Gatley 1985). The L band flux is a superposition of the light from old stars, evident most clearly at K, and structure due to the starburst which is evident at U and even more so at 10 μm (Fig. 1 in Telesco et al. 1985). A large L excess at 30" east of the nucleus is probably emission from dust, which has a L-10 μm color temperature of ~440 K. These data are crude by today's standards, but I think they help make the point that imaging at JHKLM with the new generation of arrays will permit detailed and useful studies of starbursts. This type of work is only beginning (see, e.g., Hyland et al. 1992).

4.1.3 Ionization Rates

Knowing the ionization rate N_i is fundamental to characterizing the energy sources in an IR-luminous galaxy. We have a fairly good idea of the relationship between the bolometric and Lyman-photon luminosities for normal stars (e.g., Panagia 1973), so that a necessary (but not sufficient) condition that a starburst power the IR luminosity might be that the ratio L_{IR}/N_i

inferred from observations be "star-like." However, for this ratio to be *exactly* star-like, *all* of the luminosity from the young stars must be absorbed ultimately by dust, and *all* of the photons more energetic than 13.6 eV must ionize hydrogen, with *all* of the recombination radiation then being absorbed by the dust. It would be very surprising indeed if these conditions were met in a starburst region; just look at the complexity of the Orion region, where only 20% of the total luminosity emerges in the IR (BP88), and a significant fraction of the Lyman luminosity may be absorbed by the dust (e.g., Aannestad 1989).

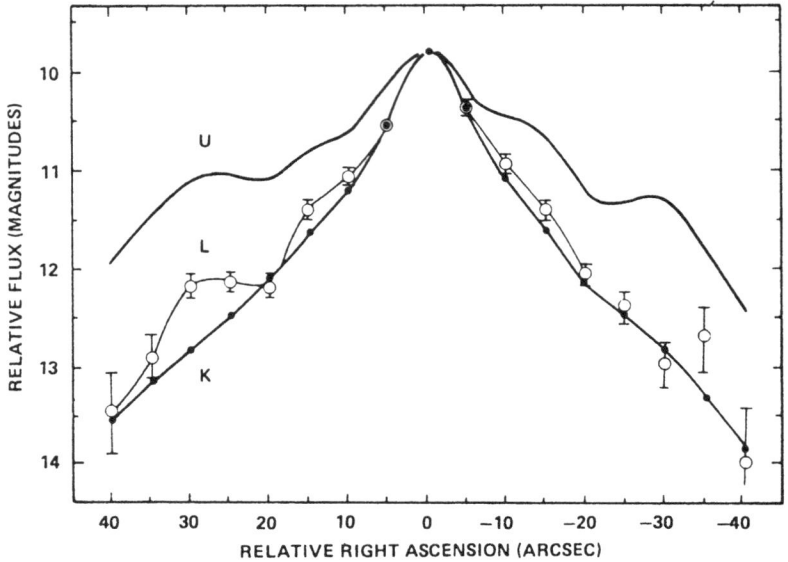

Figure 10: *K, L, and U scans through the center of M51.*

Because of their relative (but not absolute) immunity to extinction, the near-IR recombination lines of hydrogen are well suited for determining the ionization rate in galactic nuclei. I show in Fig. 11 the relation between the 10.8 μm and Brγ luminosities for the central regions of a sample of IR-luminous galaxies studied by Telesco, Dressel, & Wolstencroft (1993). The axes are also labeled by the values of N_i, derived from the line strength, and by the total IR luminosity, inferred by assuming that the IR energy distribution of each galaxy is similar to that of M82. The observed relation for Galactic radio-selected HII regions (Jennings 1975) and the expected relation for massive main sequence stars (Panagia 1973), assuming $L_{Bol} = L_{IR}$, are also shown. *We see that for this sample the IR emission is correlated with the line emission, as would be expected if a young stellar population heats the dust and ionizes the gas* (see also DePoy 1987, as presented in Telesco 1988). The luminosities and ionization rates are like those expected for ensembles of massive main sequence stars. Other similar IR studies of individual galaxies have reached the same conclusion (e.g., Kawara et al. 1987; Beck, Beckwith, & Gatley 1984), as have optical studies based on Hα and IRAS fluxes (Devereux & Young 1990; Leech et al. 1989).

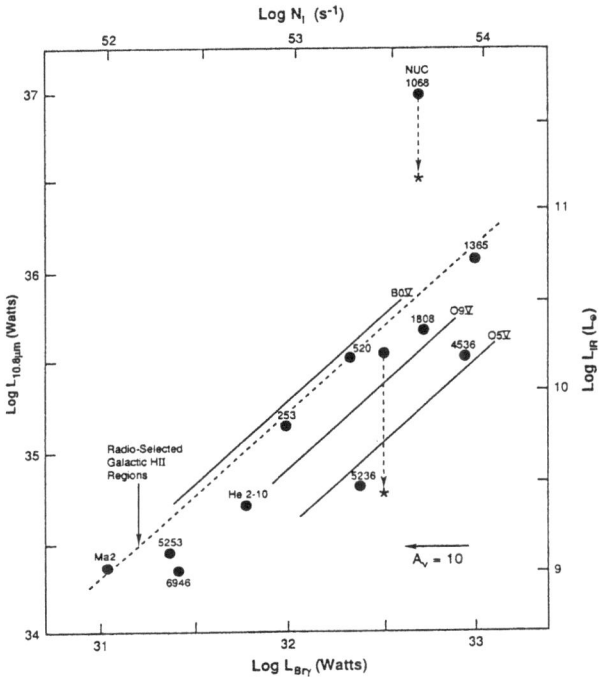

Figure 11: *Relationship between mid-IR and Brγ emission for several starburst and AGN galaxies. Also indicated are corresponding ionization rates and IR luminosities.*

Assuming that young stars do indeed power the IR continuum and line emission, there are several possible causes for the scatter about a linear relation, in addition to observational and reduction error: 1. there is some extinction to the Brγ line; 2. the M82 IR energy distribution may not be appropriate for all galaxies; 3. the source/gas geometries differ among the sample, resulting in different amounts of stellar radiation being absorbed by the dust and gas; and 4. the stellar spectral types characterizing each starburst may differ among the sample. Relevant to point 2 is the fact that some of the sources may not be powered by stars, which is certainly true for the Seyfert *nucleus* (central few arcsec) of NGC 1068 and probably also for the Seyfert NGC 5506; however, after using the actual IR energy distributions for those galaxies to derive their L_{IR} (star symbols), they are seen to deviate significantly from the stellar relations. DePoy (1987) finds that L_{IR}/N_i is higher than the starburst value for several Seyferts and Seyfert candidates, but it has not been confirmed (e.g., Kawara et al. 1987) that AGNs systematically have anomalous values of this ratio. The point I want to emphasize here is that starburst galaxies do generally follow the expected relationship, which provides a useful but not ironclad constraint on the nature of sources which generate large IR luminosities.

4.1.4 Excitation

Moderate-to-high resolution spectroscopy is the most reliable method of determining the excitation conditions in galaxies. The ionization states and the relative strengths of the emission lines can tell us about the spectral distribution of the radiation field that excites and ionizes the gas. The UV continua of the massive stars in starburst galaxies do not extend to frequencies as high as do the power-law continua in AGNs, and so gas in starbursts does not exhibit the high-ionization states observed in AGNs. Generally, astronomers have used optical emission lines to characterize the excitation properties of galaxies; the diagrams of optical line ratios developed by Baldwin et al. (1981) and Veilleux & Osterbrock (1987) have proven to be particularly valuable (see, e.g., Meadows et al. 1990). However, because they are so vulnerable to extinction, these optical lines may not tell the whole story. Spinoglio & Malkan (1992, hereafter SM92) used a photoionization code to determine the IR line ratios expected for a variety of ionizing energy distributions applicable to starbursts and AGNs. They point out that a starburst should not emit fine-structure lines from ions requiring $h\nu > 70$ eV for their formation. In the SM92 starburst models, which consider stars with effective temperatures $\leq 35,000$ K, Ne^{++} (ionization potential I.P. of Ne^+ is 41 eV) is the highest ionization species present, whereas fine-structure lines from Mg^{+4} [I.P.$(Mg^{+3}) = 109$ eV] and Ne^{+5} [I.P.$(Ne^{+4}) = 126$ eV] are expected to be strong in some Seyfert nuclei. The detection of the [S VI] 1.96 μm line [I.P.$(S^{+4}) = 73$ eV] in the Seyfert nucleus of NGC 1068 is consistent with that line being an AGN diagnostic (Oliva & Moorwood 1990).

Most of the large number of IR fine-structure lines have not been observed in other galaxies because the extragalactic lines are too faint to be detected with current instrumentation and facilities, a picture that will dramatically improve with the launch of ISO. However, some IR fine-structure lines are accessible from the ground. So far, the three most useful of these are the [Ne II] 12.8 μm, [Ar III] 9.0 μm, and [S IV] 10.5 μm lines. The ionization potentials of Ne^o, Ar^+, and S^{++} are 22 eV, 28 eV, and 35 eV, respectively. With this spread in values, even the detection of only one of these lines can tell us something about the shape of the exciting radiation field. In an extensive low-resolution ($\lambda/\Delta\lambda\approx 40$) spectroscopic survey of galaxies at 8-13 μm, Roche and colleagues (Roche et al. 1991, hereafter R91, and references therein) have shown that starburst galaxies emit almost exclusively the [Ne II] line, which indicates that the exciting stars are cooler than 35,000 K, or comparable in excitation to the Orion nebula (Lester, Dinerstein, & Rank 1979). Two interesting exceptions to this trend are the starburst galaxies NGC 5253 and II Zw 40 which show strong [S IV] lines, with the other two lines being undetected (R91). These appear to contain very young nuclear starbursts (e.g., Gonzalez-Riestra, Rego, & Zamorano 1987) containing stars as hot as ~50,000 K (e.g., Aitken et al. 1982). These two galaxies also exhibit unusually warm IRAS colors, consistent with the bluer photons and/or higher energy densities for the radiation field that heats the dust.

The detection of all three of these mid-IR lines and the determination of the line ratios should provide a powerful means to distinguish Seyfert, LINER, and starburst galaxies. Models by SM92 predict line ratios [Ar III] 9.0 μm::[SIV] 10.5 μm::[Ne II] 12.8 μm for Seyfert, LINER, and starburst galaxies of 1::3.7::1.3, 1::0.14::2.1, and 1::0.037::20, respectively. Currently available data (R91) do not permit one to distinguish between a LINER and a starburst based solely on the detection of the [Ne II] line; to date, this identification has been based primarily on optical spectroscopy (see below). Furthermore, detection only of the [S IV] line does not *ensure* that the source is a Seyfert, as demonstrated by its detection in the starburst NGC 5253 and the Seyfert NGC 5506. In addition, R91 point out that, contrary to expectation, the [S IV] line was not detected in their sample of Seyfert nuclei. Possibly (R91) the gas densities are much higher than considered in the models (which would result in collisional rather than radiative de-excitation of the fine-structure levels), or the line-to-continuum ratios require much higher spectral resolution that used so far (SM92).

In Section 4.2, we will look at the broader implications of the many far-IR fine structure lines observed in M82, the only galaxy that has been reasonably well observed in these lines. Here I would like to comment on what the far-IR line emission tells us about the exciting sources. Using a 48" beam in their cooled grating spectrometer on the KAO, Duffy et al. (1987) detected [O III] 51.8 μm, 88.4 μm, and [N III] 57.3 μm. When O^{++} is present in an H II region, the [O III] lines are the dominant cooling lines (e.g., Rubin 1985). The dependence on effective temperature of the ratio of the number of O^+-ionizing photons (I.P. = 35.1 eV) to the number of He^o-ionizing photons (I.P. = 24.6 eV) as a function of stellar effective temperature (Rubin 1985) is very strong. Using the ratios of the [O III] 52 μm and [N III] 57 μm lines to the H53α line, Puxley et al. (1989) and Bernloehr (1992) determined that the stellar effective temperature for the H II regions in M82 is 35,500 K, comparable to that for an O8 star. About 5×10^5 of these O8 stars, each emitting a bolometric luminosity of 5.4×10^4 $L_\odot$, are needed to produce the ionizing continuum. The expected total luminosity from this stellar population is therefore 2.7×10^{10} $L_\odot$, which is remarkably close to the detected IR luminosity of 3×10^{10} $L_\odot$ (Telesco & Harper 1980). Of course, not all of the stars are of spectral type O8, but since the most massive stars dominate the ionization and luminosity (see below), this spectral type is probably reasonably characteristic of the starburst in M82. The consistency of the line and continuum luminosities is impressive.

Doyon et al. (1992) are exploiting the ratio of the He I λ2.06 μm and H I Brγ line fluxes to determine the excitation in starbursts. The ionization potential of He^o is nearly twice that of H, and the line ratio depends steeply on stellar effective temperature up to ~38,000 K, about that of a 40 $M_\odot$ star. An important advantage of this technique is that the ratio is only weakly affected by extinction since the lines differ little in wavelength. The value of the line ratio is directly related to the volume of the H and He Stromgren spheres and "saturates" when they are equal. Most of the sample of eight starburst galaxies considered by Doyon et al. (1992) have excitation indicative of ~30 $M_\odot$ stars, but a few seem to have more massive stars.

Certain solid-state or macro-molecular IR emission features also have significant potential as diagnostics of the nature of the exciting sources. Near-IR and mid-IR studies (Moorwood 1986; R91, and references therein) have shown that starburst galaxies, like many Galactic sources, emit in features at 3.3, 7.7, 8.7, and 11.3 μm. As mentioned in Section 2.2, this group of features, which also includes one at 6.2 μm that is not observable from the ground, are thought to be produced through emission by large molecules upon absorption by each molecule of a single UV photon (Leger & Puget 1984). In one explanation, the UV-pumped molecules emit in these features during transitions between vibration states of the C-H bonds (Allamandola, Tielens, & Barker 1985). These features are never observed in "pure" AGNs. In the few cases in which they are observed in AGN galaxies (e.g., NGC 1365 and NGC 7469), there is strong evidence that they are emitted by a circumnuclear starburst. These features are usually accompanied by the [NeII] 12.8 μm line, which is also not detected in AGNs; the AGNs tend to have featureless mid-IR spectra. Consistent with our discussion in Section 4.1.1, Roche et al. (1991) find, in the IRAS two-color plot using 12, 25, and 60 μm fluxes, a clear separation between the galaxies with featureless 8-13 μm spectra and those dominated by UIR bands; the galaxies with mid-IR emission features have starburst IRAS colors. In addition, objects with silicate absorption features have IRAS colors intermediate between the starburst and AGN galaxies and tend to be Seyfert 2 galaxies, and so the silicate feature may be intimately linked with the AGNs themselves (Roche et al. 1991). Thus, it appears that Roche and his colleagues have established the mid-IR PAH features as an important hallmark of starbursts and a potentially powerful means by which to distinguish them from pure AGNs. Moorwood (1986) has done the same for the 3.3 μm PAH feature. Roche et al. (1991) suggest that the UV radiation field in AGNs is hard enough to destroy the PAHs.

4.2 Conditions in Starbursts

4.2.1 The Stellar Environment

In Section 4.1 we touched on the excitation and ionization properties of starburst galaxies because the types of IR observations discussed there are available for numerous galaxies. Currently, the broad range of IR spectral data from which we can infer the *detailed* conditions in starbursts are available only for M82, which we now consider. In the following discussion, I will draw principally from the work of SM92, Wolfire, Tielens, & Hollenbach (1990), and Stacey et al. (1991). The intensities of IR fine structure lines depend on the shape and luminosity of the ionizing continuum and the chemical composition and geometry of the gas. Clearly, some of the lines are useful for probing the ionized gas, while others are emitted from the neutral gas. In Fig. 12 I show a line diagnostic diagram for ionized gas in M82, adapted from SM92. As indicated, the two axes give predicted line ratios when the UV continua are emitted by blue stars in starbursts and by non-thermal (power-law) objects in Seyfert nuclei. The horizontal lines are loci of constant gas density n, and the more vertical lines are loci of constant ionization parameter U; for these lines, the [S III] ratio determines the gas density, and the [N III]/[Ne II] ratio determines U. The ionization parameter is defined as the ratio of the number density of H-ionizing (Lyman) photons to total H number density (H^+ + H^0 + $2H_2$) on the inner face of the illuminated cloud. For a given UV spectral shape, the ionization level is determined by U, which can be expressed as $U = Q(H)/4\pi r^2 nc$, where c is the speed of light. The M82 line ratios imply that the ionized gas is best characterized by log n = 2 and log U = -2.5, which implies that $Q = 1.2 \times 10^{11} r^2$. The [O III] 52 μm/H53α line ratio indicates that the average excitation conditions are those of an O8V star, for which log Q = 48.5. *The characteristic distance r of the excited gas from the illuminating star is thus ~ 1.6 pc.* As it turns out, this value of r equals the value derived for half the average separation between adjacent O8V stars that are required to produce the total ionizing and IR luminosities from M82, if we assume that the emitting volume is a cylinder 300 pc in diameter and 100 pc tall.

Additional understanding of the conditions in starbursts requires that we consider so-called photodissociation regions (PDRs). A PDR exists at the boundary of a molecular cloud and an H II region (see,e.g., Tielens & Hollenbach 1985; Stacey et al. 1991; Wolfire et al. 1990) where UV radiation with hv < 13.6 eV photodissociates H_2 and CO molecules and photoionizes heavy elements. In regions of relatively high UV flux (see below), the thickness of the PDR is determined by the absorption of the UV photons by dust. The gas in these regions is heated mainly by photoelectric ejection of electrons from dust grains, and the gas cooling is dominated by [O I] 63 μm and [C II] 158 μm emission. Stacey and colleagues (Stacey et al. 1991, and references therein) have made extensive observations of the [C II] 158 μm line in Galactic and extragalactic sources. The close association of this line with molecular clouds is demonstrated by the excellent linear correlation between the [C II] and CO line intensities. In addition, the [C II] line intensity is correlated with the IR intensity, which should be a measure of the UV intensity absorbed by the dust. However, the correlation is not linear, which implies that the more *intense* sources are not simply composed of greater numbers of the same IR objects contained in the less intense sources. The interpretation of this non-linearity (Tielens & Hollenbach 1985) is that deeper penetration of the molecular cloud by larger UV fluxes is impeded by the concurrent increase in the optical depth of dust. The models of PDRs constructed by Tielens and Hollenbach (1985) and the observed [C II] line strength and IR luminosity imply that the PDRs in M82 have gas densities of 10^4-10^5 cm^{-3} and UV energy densities of ~500 eV cm^{-3}, which is about ten thousand times larger than the UV energy density in the solar neighborhood. Based on a detailed analysis of the [C II], [O I], and [Si II] IR fine structure lines, Wolfire et al. (1990) infer comparable values for the UV energy density (200-300 eV cm^{-3}) and gas densities (5×10^4 cm^{-3}), as well as PDR gas temperatures near 400 K. The mass of gas in the PDR regions in M82 may be as much as 50% of the total gas mass, making PDR regions very significant energetically for this and other starburst regions (Stacey et al. 1991).

It is reasonable to suppose that dust grains are destroyed in the high energy densities in starburst regions. By comparing the IR color variations across the starburst in M82 to those in Galactic H II regions, Telesco, Decher, & Joy (1989) and Telesco, Dressel, & Wolstencroft (1993) showed that this is likely to be the case in M82 and other starbursts. Boulanger et al. (1988) determined the IRAS colors as a function of the UV energy density for the California Nebula, which is powered by an O7.5 III star, and concluded that ~80% of the 12 μm-emitting grains, which are thought to be small transiently heated particles, are destroyed in regions with UV energy densities greater than 8 eV cm^{-3}. Telesco, Decher, & Joy (1989) found that the color variations and energy densities across the M82 starburst region are similar to those in the California Nebula. They concluded that small dust grains are destroyed throughout the main body of the starburst but are abundant in its periphery, so that, in effect, the starburst is surrounded by a halo of mid-IR-emitting small grains. The dust which emits the mid-IR radiation deep in the starburst must be larger grains, which have a long lifetime against destruction, as well as very small grains that reside far enough inside molecular clouds that they are protected from the high energy densities.

In addition to the high energy densities, the nonthermal radio emission observed in starburst regions betokens a population of supernova remnants that have the potential to violently disrupt the interstellar medium, and there is strong evidence that this is indeed happening. Chevalier & Clegg (1985) and others (e.g., Heckman, Armus, & Miley 1990) have shown that massive bipolar outflows of gas from starburst galaxies are driven by supernovae which, in a starburst like M82, appear at a rate of one every few years. Shocked gas often results from these violent motions, and the near-IR emission lines of H_2 may be a useful, if still ambiguous, probe of this activity. In some starburst galaxies where these lines are observed, they can be attributed completely or partly to fluorescence (e.g., Doyon 1991), and the lines are then a measure of the UV radiation intensity. In others, the H_2 appears to be shock-excited. In our galaxy, shocks moving at 10-50 km s^{-1} excite H_2 to 1000-3000 K in many types of sources (e.g., Draine, Roberge, & Dalgarno 1983).

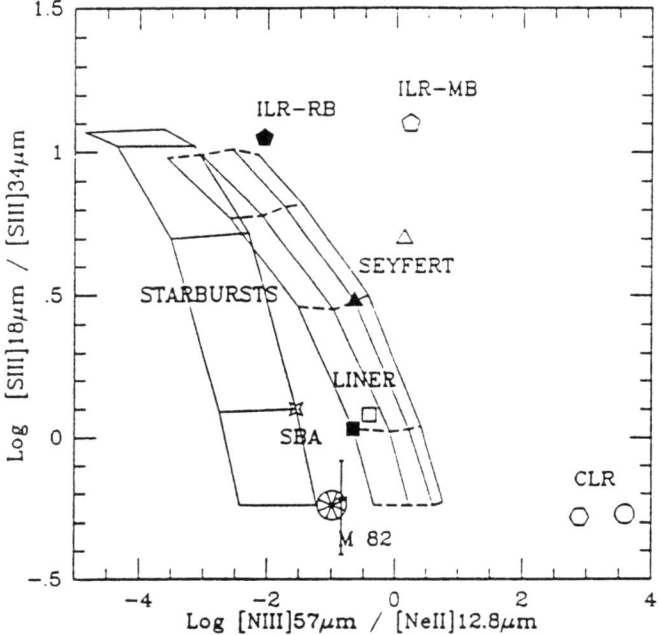

Figure 12: *Diagnostic diagram for far-IR fine-structure lines, from SM92. Positions for various emission-region models (SM92) are indicated. Starburst-model grid lines correspond to densities log n = 2, 3, 4, 5, and 6 (bottom to top) and ionization parameter log U = -3.5 and -2.5 (left to right). Observed and model ratios are indicated for M82.*

Extended emission from shocked gas has been observed in M82. Lester et al. (1990) found that the 1-0 S(1) line of H_2 is distributed across the starburst in M82, with the 2-1 S(1)/1-0 S(0) line ratio in the outer regions being indicative of shocks rather than fluorescence. The starburst region itself is best traced by the mid-IR, Brγ, and nonthermal-radio spatial distributions, which are similar to each other, but which differ from the distribution of shocked H_2. The ratio of the H_2 and Brγ line fluxes peaks strongly at the large-scale CO lobes, thought to be an edge-on molecular ring surrounding the starburst, rather than in the starburst region. Were the H_2 shocked in supernova remnants within the starburst, the S(1) line should be more nearly distributed like the starburst tracers. The relative enhancement of the H_2 emission near the CO lobes suggests that the hydrogen is shocked at the large-scale interface between the dominant starburst region and the surrounding molecular ring as the ring confines the violent gas motions generated in the starburst. The ring is then like a funnel, collimating these motions into the enormous bipolar outflow. The apparently global character of the H_2 excitation is in contrast with that producing the [Fe II] 1.64 μm line which is more nearly distributed like the nonthermal radio continuum, consistent with its being generated in individual supernova shocks throughout the starburst (Lester et al. 1990; Greenhouse et al. 1991).

4.2.2 Stellar Populations

Here we consider how IR observations have been used to draw sometimes controversial conclusions about the stellar populations, particularly the initial mass function, in starbursts. Models are often required in order to make detailed inferences about the stellar properties, and I will therefore sketch out the most salient features of simple starburst models. The first part of this discussion follows closely that in Telesco (1985).

The present-day, or current, mass function of a stellar population is simply the number of stars now in each mass interval, and the *initial* mass function (IMF) is the number of stars that actually form in each mass interval. (See Scalo 1986 for a comprehensive review of the IMF.) Except for a very young starburst, these two will differ because of the evolution of more massive stars away from the main sequence and possibly because the form of the IMF itself is time dependent. Let the IMF, the number of stars formed per year at time t in the mass interval m to m + dm, be represented by $\psi(m,t)\, dm = n(m)\, b(t)\, dm$. A star of mass m has main sequence bolometric luminosity l(m), Lyman photon production rate $n_i(m)$, main-sequence lifetime $t_{ms}(m)$, and post-main-sequence lifetime $t_{pms}(m)$. For a given IMF, one can determine the integrated properties of a stellar population by expressing these functions as piecewise continuous power-law functions of mass and integrating over the appropriate expressions (e.g., Telesco & Gatley 1984; Wright et al. 1988). For illustrative purposes I will consider this approach here. Keep in mind, however, that one may use the detailed evolutionary tracks for stars in each mass interval, following them from birth to death (among recent work see Doyon 1991; Doyon, Puxley, & Joseph 1992; Bernloehr 1992).

For the solar-neighborhood IMF, the total mass M(m), total luminosity L(m), and total Lyman photon rate $N_i(m)$ for all the main-sequence stars of mass m in a starburst are distributed with mass approximately as shown in Fig. 13a. The high-mass stars dominate both the luminosity and the ionization, whereas the low-mass stars dominate the mass. It is this inequitable division of power that makes it difficult to find out directly anything about the low-mass stars in a starburst. The mass-to-luminosity ratio (in units of $M_\odot/L_\odot$) of an unevolved stellar population is less than 0.01, whereas an old stellar population has M/L>3. Therefore, a low value of M/L is a sign of youth, indicating the possible presence of a starburst. In a galactic nucleus, the old pre-existing stellar population, which has M/L≈6, usually dominates the mass, whereas, a reasonably strong starburst occurring there can dominate the luminosity. As models like those considered below show (see Telesco 1985), in a strong starburst, such that 10% of the total mass becomes young stars in less than 10^7 yr, values of M/L<0.1 are expected for the composite population. Thus, M/L reflects both the youth of a burst and its strength. One finds that M/L $\approx$ 0.04 in the well-known starburst galaxies M82 and NGC 253, indicating that the luminosities from the starbursts overwhelmingly dominate those from the old stellar populations.

We consider a starburst that began T_o years ago and has proceeded at a constant rate: ψ (m,t) = n (m) $\equiv$ Cm$^\gamma$, where γ is the slope of the IMF, and C is a proportionality factor. The Salpeter value γ = -2.3 is often used to characterize the solar neighborhood, as have roughly similar relations presented in Miller & Scalo (1979) and Scalo (1986). If the IMF has a lower mass limit m_l and an upper mass limit m_u, the burst luminosity is:

$$L_{ms}(T_o) = \int_{m_l}^{m_o} n(m)\, l(m)\, T_o\, dm + \int_{m_o}^{m_u} n(m)\, l(m)\, t_{ms}(m)\, dm, \qquad (1)$$

where m_o is the mass of the main-sequence star for which $t_{ms} = T_o$. The first term represents the luminosity generated by stars with main-sequence lifetimes longer than the burst age (none is yet dieing), and the second term is the contribution from stars for which the deathrate equals the birthrate. We can also calculate the rate at which mass is converted into stars:

$$dM/dt \equiv M = \int_{m_l}^{m_u} m\, n(m)\, dm. \qquad (2)$$

Similarly, other model starburst properties, such as the total ionization rate, the post-main-sequence luminosity, and the supernova rate, can be calculated. The ratio of any two of these quantities is independent of the IMF proportionality factor C [recall: n(m) = Cm$^\gamma$]. Thus, for an assumed IMF shape and mass cutoffs, we can use the observed IR luminosity, for example, of a heavily obscured starburst to estimate the stellar mass-conversion rate, the ionization rate, etc.

The galaxy M82 is at the center of the controversy about starburst IMFs (Rieke et al. 1980 and references given below). We will apply our simple model to M82 to explore the origin of the controversial conclusions; we assume m_u = 60 M$_\odot$, which is probably too high (see Section 4.1.4 and below) but does not compromise the conclusions of this heuristic treatment. The primary observed quantities for the luminous starburst in M82 are the total luminosity L $\equiv$ L$_{ms}$ (comparable to L$_{IR}$), the ionization rate N$_i$ (from Brackett or radio recombination lines), the total dynamical mass (from IR emission lines or near-IR CO absorption band), and the near-IR luminosity and CO index emitted by red stars. We see in Fig. 13b that the ratio N$_i$/L is strongly dependent on burst age as long as the value of m_l is not too high; for lower values of m_l, the ratio continues to decline with burst age as the lower-mass stars accumulate on the main sequence. The ratio for M82 indicates a burst age $T_o \approx 2 \times 10^7$ yr, which implies the mass-conversion rates indicated in Fig. 13c. If m_l is as low as 0.1 M$_\odot$, comparable to that for the solar neighborhood, the star formation rate implied by the starburst luminosity is ~34 $M_\odot$ yr^{-1}. The mass of stars produced in the starburst would then be ~5 × 10^8 $M_\odot$, which equals the total dynamical mass of that region! Thus, if we feel more comfortable with a starburst mass comprising, for example, less than 20% of the total central mass, then $m_1 > 1$ $M_\odot$. This is the basis for the much-discussed conclusion that many starburst regions have IMFs deficient in low-mass stars; this conclusion has been drawn for galaxies other than M82 as well (e.g., Wright et al. 1988; Bernloehr 1992).

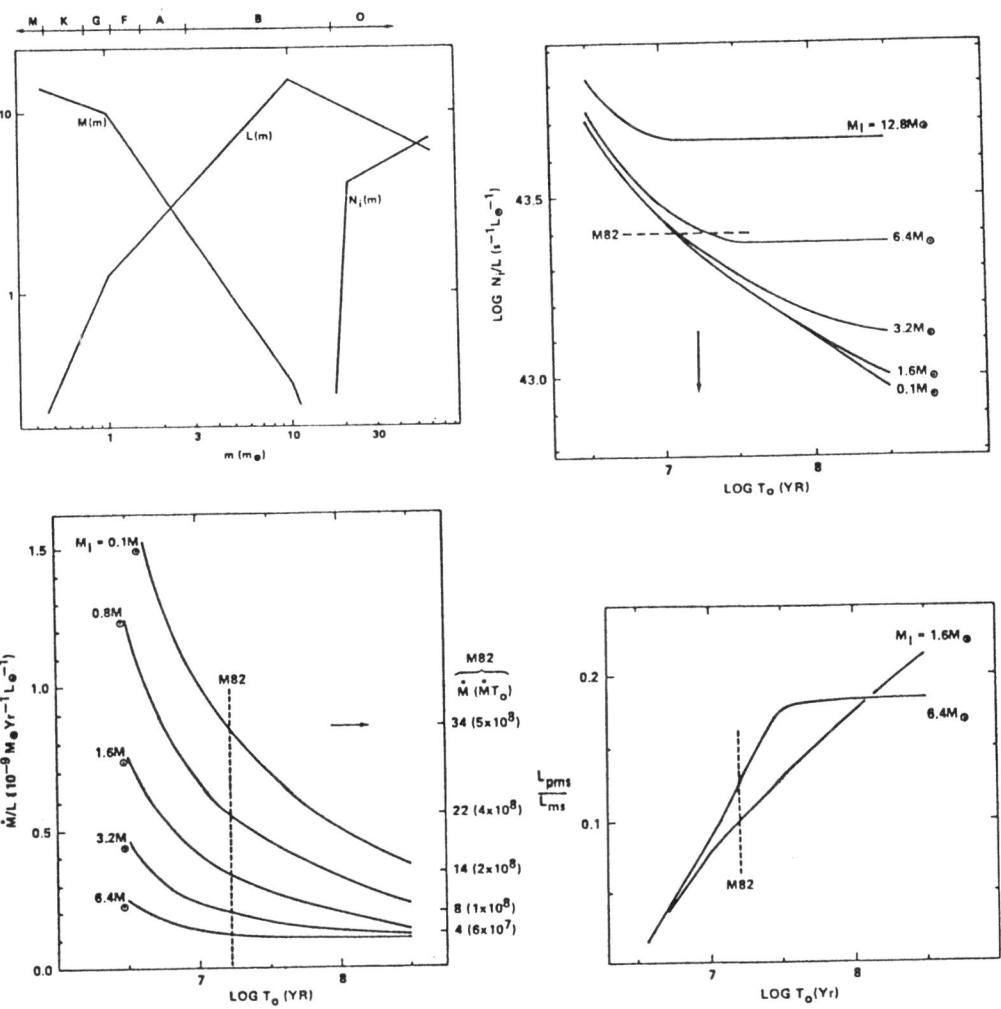

Figure 13: *For starburst models discussed in text: (a) upper left, distributions of mass, luminosity, and ionization rates as function of stellar mass on IMF; (b) upper right, evolution of ionization rate per unit luminosity for starburst with various lower-mass cutoffs and constant star formation rate; (c) lower left, evolution of star formation rate per unit luminosity; (d) lower right, evolution of the ratio of post-main-sequence and main-sequence luminosities.*

This model implies that the starburst post-main-sequence (pms) luminosity should be ~ 10% of the main-sequence luminosity (Fig. 13d). Considering the age of the starburst, these PMS stars must be red supergiants with a "characteristic" temperature of about 4500 K (see Harris 1976), and they can supply nearly all of the 2.2 μm (K-band) luminosity observed from the center of M82. In addition, the near-IR CO absorption feature should have the depth expected for red supergiants. More sophisticated starburst models also indicate that the K luminosity and CO feature in M82 should originate in supergiants (Bernloehr 1992), a prediction which is reasonably consistent with observations (Lester et al. 1990), although there are some interesting ambiguities. The stellar velocity dispersion of the central ~1" (16 pc) of M82 can be determined from the near-IR CO feature, as indicated in Gaffney, Lester, & Telesco (1993). The value of the dispersion implies that the stellar mass and M/L_K are comparable to that for the same size region in the Galactic Center and therefore that red *giants* probably dominate the near-IR emission at the very center of M82. Red supergiants must dominate the near-IR excess in the more extended starburst region, since M/L_K for the central 30" is an order of magnitude lower than for the central 1". This picture is consistent with the observation that the near-IR colors throughout the center are like those of reddened giants and supergiants (Telesco et al. 1991), which are virtually indistinguishable from their near-IR colors alone. However, a gradient is expected in the depth of the CO feature as one goes from the giant-dominated nucleus to the supergiant-dominated extended starburst, but no difference is observed (Lester et al. 1990), possibly due to the enrichment of metals in the bulge stars (Gaffney & Lester 1992).

The conclusion that starburst IMFs are deficient in low mass stars, particularly in M82 for which the most data are available, is subject to many uncertainties. First, the observed value of L_{IR} may not be the total starburst luminosity; Rieke (1988) suggests that roughly half of the starburst starlight escapes out of M82's plane (and perpendicular to our line of sight) without being absorbed by dust. Second, probably a significant fraction of the stellar ionization radiation is absorbed by dust; Wood & Churchwell (1989) estimate that 50% - 90% of these photons are absorbed by dust in Galactic compact H II regions. Third, the near-IR extinction to the starburst regions, required for an accurate determination of their K luminosity, is uncertain; derived values depend on the method of derivation (near-IR colors, IR emission lines, silicate absorption feature depth, radio-IR line comparisons), and it is likely that different values of the extinction must be applied to different types of emitting sources, which implies that a high degree of spatial and spectral separation of these sources must be achieved. Fourth, the library of stellar evolutionary tracks is uncertain (see, e.g., Bernloehr 1992). Fifth, high metallicities and the presence of emission from dust and gas in starburst regions may alter the depth of the photospheric CO feature, adding ambiguity to its use as a stellar luminosity diagnostic (Gaffney & Lester 1992). If these various uncertainties can be considerably reduced, then the models can be much more tightly constrained. For example, Bernloehr (1992) has shown that, within the uncertainties, an IMF with a much shallower slope than the solar neighborhood IMF and extending from ~0.1 $M_\odot$ to 30 $M_\odot$ fits the available data for M82 (in fact, it provides the best fit).

Determination of the upper mass limit of the stellar mass function is less difficult than determining m_l because the high mass stars dominate the starburst luminosity, ionization, and excitation. Based on the ratios of the [O III] and [N III] far-IR lines to the H53α line, it appears that $m_u \approx 30 \, M_\odot$ in M82 (Bernloehr 1992; Puxley et al. 1989). Doyon et al. (1992) have used the He I λ2.06 μm/H I Brγ ratio to determine m_u for several starburst galaxies. They have assumed an ensemble of stars following the Salpeter IMF and for which the deathrate equals the birthrate (i.e., only the second term in Eqn. 1 is applicable); since the line ratio is not sensitive to stars less massive than ~10 $M_\odot$, this latter assumption is equivalent to saying that the starburst is older than ~10^7 yr. Doyon et al. (1992) infer that $m_u \approx 30 \, M_\odot$ is typical for their galaxies, but several may have $m_u > 40 \, M_\odot$.

As we have noted, Eqns. 1 and 2 permit one to estimate the star formation rate that is needed to generate L_{IR} in a starburst. For a solar neighborhood IMF extending from 0.1 $M_\odot$ or 2 $M_\odot$ up to 60 $M_\odot$, the coefficient K in the equation $M(M_\odot \, yr^{-1}) = K \times L_{IR} (L_\odot)$ has the values ~6×10^{-10} and ~3×10^{-10}, respectively. Thus, moderately luminous galaxies, with $L_{IR} = 10^{10} \, L_\odot$, are converting the gas in their central regions into stars at a rate of 3-6 $M_\odot \, yr^{-1}$, comparable to that throughout the entire Milky Way (e.g., Cox & Mezger 1987). The star formation rates for some ultraluminous galaxies, if the luminosity is in fact due to young stars, is ~100 $M_\odot \, yr^{-1}$! We see that a "typical" starburst galaxy like M82, with $L_{IR} = 3 \times 10^{10} \, L_\odot$ and $<10^8 \, M_\odot$ of gas in its central kiloparsec (Lo et al. 1987), will run out of fuel in less than a few times 10^7 yr. Since the age of the starburst in M82 is estimated to be only a few times longer than this, such a starburst essentially burns itself out in a time much shorter than the age of the Universe. Thus, the term "starburst" also conveys the brevity of this phenomenon.

4.3 Causes of Starbursts

We now briefly consider what IR observations may tell us about possible causes of the central starbursts in galaxies. We know that stars are formed from molecular clouds, so an important part of the issue about starbursts must be related to the ways in which interstellar matter is moved around in and possibly between galaxies. Studies of large samples of galaxies have shown that *barred*, especially early-type, spiral galaxies are more likely to have enhanced central star formation than are non-barred spiral galaxies (Hawarden et al. 1986; Devereux 1987; Dressel 1988). Roughly half of all disk galaxies are barred. As considered here, a bar (or oval distortion, if it is less dramatic in appearance) is an elongated stellar structure which is centered on the galactic nucleus and has a nearly constant surface brightness along its major axes and a sharp fall off in brightness along its minor axis (see, e.g., Binney & Tremaine 1987). It appears that the stars circulate through the bar, reinforcing its structure in the manner of a density wave, with the stellar and bar-pattern orbital velocities being different. N-body simulations imply that even a small bar can have a dramatic effect on the distribution of gas; Sanders & Huntley (1976) showed that an azimuthal variation in the gravitational potential of only ~3% of the mean potential at a given radius is sufficient to produce a marked response in the gas.

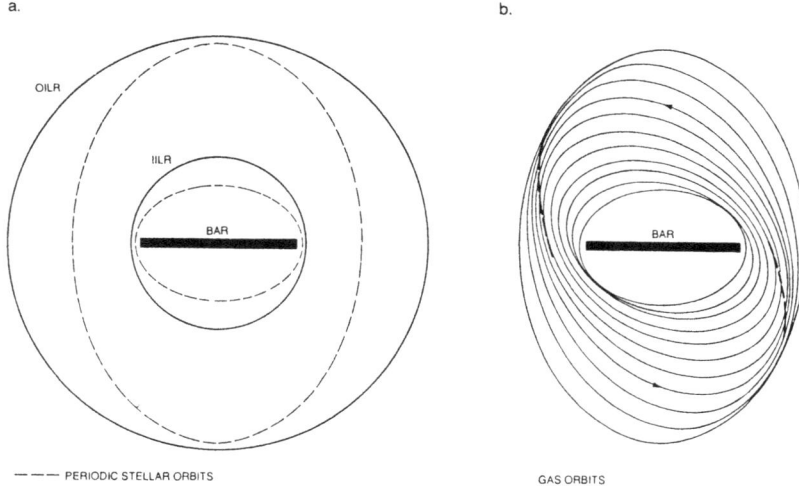

a.

OILR

IILR

BAR

--- PERIODIC STELLAR ORBITS

b.

BAR

GAS ORBITS

Figure 14: *(a) Sketch showing relationship of elongated periodic stellar orbits with ILRs in barred potential. (b) Because of viscosity, gas orbits rotate gradually, resulting in orbit crowding.*

In their study of 23 starburst and AGN galaxies, of which about two-thirds are classified as barred (many others may have undetected oval distortions), Telesco, Dressel, & Wolstencroft (1993; see also Telesco & Decher 1988) argued that the spatial distributions of the dense, neutral interstellar gas and the IR-luminous starbursts for a subset of eight galaxies for which sufficient dynamical data exist are determined by bars. To see the basis for this conclusion, we first consider the concept of the orbital resonance. Various resonances in a galaxy arise from the interplay between the bar and stellar orbits, and since we are considering central starbursts, I will focus primarily on the concept of the inner Lindblad resonance (ILR). A galaxy can have zero, one, or two ILRs depending on the rotation curve and the rotation rate of the bar. For a bar pattern rotating with angular velocity Ω_b in a galaxy with stellar rotation curve $\Omega(r)$, an ILR is located at that radius for which the bar potential perturbs the stars at their natural, or epicyclic, frequency $\kappa(r)$. Thus, $\kappa = 2(\Omega-\Omega_b)$, or $\Omega_b = \Omega-\kappa/2$, at an ILR. Those periodic stellar orbits (i.e., those closed in the reference frame of the rotating bar) located between the ILR(s) and corotation are elongated along the bar. As illustrated in Fig. 14a, if there are two ILRs, the stellar orbits between the ILRs are elongated perpendicular to the bar, and those inside the inner ILR are elongated parallel to the bar; in fact, the elongated stellar orbits change their orientation by 90° at each of the principal resonances (see, e.g., Combes 1988), namely, at each ILR, at corotation (where $\Omega = \Omega_b$), and at the outer Lindblad resonance (where $\Omega_b = \Omega + \kappa/2$). (Think of how the phase of a high-Q forced harmonic oscillator changes dramatically as the forcing frequency spans a resonance.) Unlike stars, gas clouds collide *inelastically* ($\Leftrightarrow$ viscosity), so that the orientations of the elongated gas orbits change *gradually* across each resonance (i.e., we have a low-Q oscillator). The result is that regions of orbit crowding (density waves) occur in which molecular clouds are more numerous than elsewhere in the galactic disk (Fig. 14b). Depending on the detailed bar structure and rotation rate, these density enhancements can be large-scale spirals (bar-driven spiral density waves), gas bars along the leading edge of the stellar bar, and "mini-spirals" penetrating the central regions.

Once the density enhancements have formed by this process, they are subject to rotational torques applied by the stellar bar, since the stellar bar and density enhancements are displaced azimuthally from each other. Between the ILRs and corotation the torque slows the gas, and the gas drifts toward the galactic center. Inside the ILR (if there is only one) or between the ILRs (if there are two), the torque speeds up the gas, so that it drifts out toward the ILR (or the outer ILR). Therefore, in the presence of a bar and ILRs, a ring of gas tends to form near the ILR. Another important effect is that cloud collision rates are higher at ILRs. If a ring of gas has formed at the ILR, the cloud collision rate is higher there simply because the collision rate is proportional to the square of the cloud number density. In addition, the bar perturbation can more easily transfer energy to cloud orbits at a resonance (efficient energy transfer is a universal characteristic of a resonance), and so *even a uniform distribution of clouds experiences enhanced collision rates at the ILR.* Therefore, the region near an ILR may be a preferred location for induced star formation because giant molecular clouds, where massive stars are born, tend to form there by coalescence (e.g., Casoli & Combes 1982; Roberts & Hausman 1984) or by compression-induced cloud collapse (Loren 1976).

Telesco & Decher (1988) and Telesco, Dressel, & Wolstencroft (1993) have used these ideas to interpret their observations of IR-luminous starburst galaxies. Consider the Seyfert galaxy NGC 1068 which has a luminous extended starburst and a near-IR bar (Section 4.1.2). The rotation curve $\Omega(r)$ and the calculated Lindblad precession frequency $\Omega - \kappa/2$ are shown in Fig. 15. There are zero, one, or two ILRs depending upon whether the bar pattern speed Ω_b is greater than, equal to, or less than ~70 km s^{-1} kpc^{-1}, respectively. We do not have a direct determination of Ω_b for NGC 1068, but for the reasonable value of 50 km s^{-1} kpc^{-1} the ILRs are located at ~1.3 kpc and ~1.8 kpc from the nucleus. As shown in Fig. 16, the concentrations of molecular clouds (thin lines) and intense star formation coincide with the ILRs (dotted ellipses), consistent with our notions about the longer-term evolution of gas in the presence of a bar. A similar analysis of other galaxies such as NGC 5236 and NGC 1365 (Figs. 8 and 9) shows that, even though the dense interstellar gas (dark lanes) in each galaxy is more extended than the starburst, the star formation has occurred mainly between the ILRs. Both NGC 1365 and NGC 1068 have IR-bright AGNs in addition to their extended starbursts, and possibly, because of the enhanced collision rates at the ILR, gas which has not formed stars at the ILRs will drift into the nuclei to fuel the AGNs. In contrast to these galaxies, NGC 6946, which has a marked nuclear oval distortion (Zaritsky & Lo 1986), appears not to have an ILR, since $\Omega_b > \Omega - \kappa/2$ everywhere. As a result, gas drifts inward toward the nucleus without encountering the torque reversal at an ILR, and both the molecular gas and the starburst are concentrated at the nucleus (Fig 17). From these considerations we see that bars and oval distortions and therefore probably other types of non-axisymmetric perturbations may alter the global distribution of gas in a galaxy and effect starbursts and possibly AGNs.

The weight of evidence from numerous studies implies that interactions among galaxies enhance their luminosities. IR observations lend powerful support to this conclusion. For example, Joseph et al. (1984; see also Lutz 1992) showed that 85% of the pairs in a sample of 22 interacting galaxies had IR excesses at L, as judged by their K-L colors. At far-IR wavelengths, interacting galaxies have warmer dust and higher luminosities than isolated galaxies (e.g., Xu & Sulentic 1991). Both the far-IR luminosities and the dust temperatures appear to increase with decreasing pair separation (references above), with ultraluminous IR galaxies ($L_{IR} \geq 10^{12}$ $L_{\odot}$) more likely to be actual mergers in which the component galaxies are not individually distinguishable (Joseph & Wright 1985; Sanders et al. 1988); from the large body of IRAS survey data from various studies, Rowan-Robinson (1991), in his review, shows that the percentage of interacting and merging galaxies increases steadily with system luminosity, reaching the value of 90% at the highest luminosities. The relative roles of starbursts and AGNs as a function of luminosity is still debated (e.g., Leech et al. 1989; Sanders et al. 1988). My main goal here is to consider briefly our previous discussion about the connection between bars and IR-luminous starbursts in this broader context of galaxy interactions.

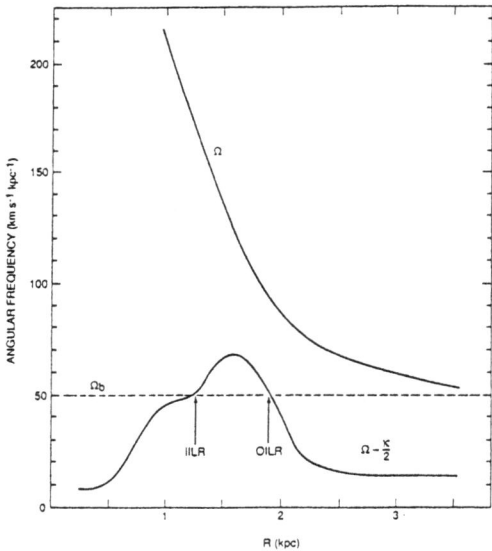

Figure 15: *Rotation curve and Lindblad precession frequency used to determine positions of ILRs for bar pattern speed Ω_b in NGC 1068.*

Generally, intense starbursts are confined to the central regions of galaxies, so that the transport of gas into the central regions, which indeed tends to be concentrated there in starburst galaxies (e.g., Wang, Scoville, & Sanders 1991), is a key issue in understanding their origin. The process responsible for this transport must somehow reduce the angular momentum of initially very extended gas by one-to-two orders of magnitude. We have seen that the non-axisymmetric gravitational potentials represented by massive bars and oval distortions can efficiently transport gas toward a galactic nucleus. An important result of N-body simulations of interacting galaxies is that tidal forces can trigger the formation of a stellar bar in a galactic disk during a close encounter (e.g., Noguchi 1988; Barnes & Hernquist 1991). Before pericenter (closest approach) the torques on the dissipative gas are due to the gravitational field of the companion (Combes 1988; Barnes & Hernquist 1991), but after pericenter, the *induced bar* removes angular momentum from the gas (Barnes & Hernquist 1991). This process evolves fairly rapidly, with the torques acting to transport the gas into the galactic center in less than a bar rotation period, or a few times 10^8 yr. Once these torques have driven the gas into the central few hundred parsecs, other ways of removing angular momentum (e.g., Shlosman, Frank & Begelman 1989) must be invoked to transport gas into the central few parsecs in order to fuel a very compact starburst or an AGN.

NGC 1068

NGC 6946

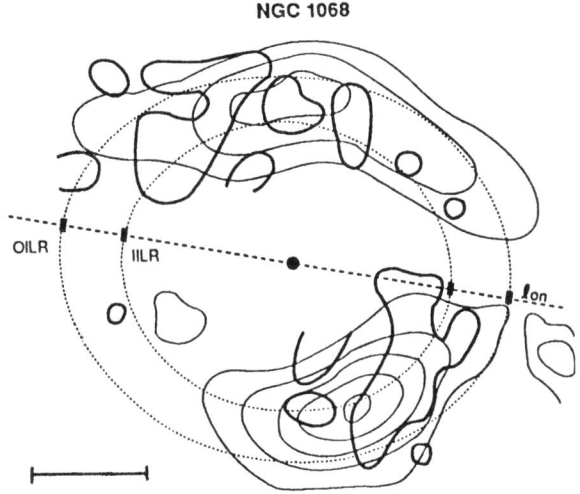

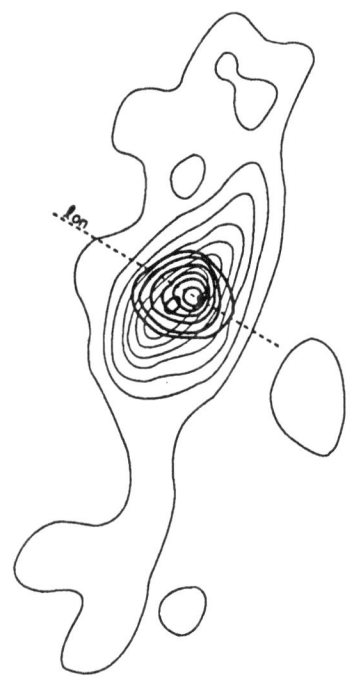

Figure 16 (left) and Figure 17 (right):
*Mid-IR maps (thick contours) of NGC 1068
and NGC 6946 overlaid on CO contours from
Myers & Scoville (1987) and Ishizuki et al.
(1990), respectively. ILRs are indicated for
NGC 1068. Bar corresponds to 10".*

5. ACTIVE GALACTIC NUCLEI AND QUASARS

Active galactic nuclei are characterized by broad emission lines which represent a wide range of excitation arising in compact nuclear regions. A supermassive black hole (the "central engine" or "monster") surrounded by an accretion disk purportedly lies at the heart of the often prodigious energy generation (but see Terlevich & Melnick 1985 for a different view). Broad permitted lines indicative of bulk motions of 2000-20,000 km s^{-1} are emitted from a high density region ($>10^9$ cm^{-3}), known as the broad-line region, that is close ($\sim 10^{17}$ cm) to the black hole (see SM92); the densities there are too high for significant forbidden, or fine-structure, line emission to occur. The broad-line region is observed only in type 1 Seyfert galaxies and quasars (it may be hidden from view in type 2 Seyferts). In all Seyferts, narrower (but still broad) permitted and forbidden lines are emitted from the so-called narrow-line region where the gas densities are 10^2-10^5 cm^{-3} and bulk motions are in the range 300-1000 km s^{-1}. Type 2 Seyferts emit only these narrower lines. AGNs are thought to generate power by accretion onto a supermassive nuclear black hole. Except for blazars, the optical-through-soft X-ray radiation (the "blue bump") is thermal emission, representative of temperatures as high as 10^5-10^6 K, from an accretion disk around the black hole; in the simplest models, the luminosity is determined by the viscosity, the mass flux, and the mass of the black hole. The hard X-ray emission (hν > 1 keV) may represent softer radiation that has been Compton scattered by hot thermal electrons close to the disk (see the review by Bregman 1990, from which much of my discussion is drawn).

Nearly all AGNs except blazars exhibit a local minimum near 1 μm in the energy distribution, and the IR emission appears as a broad hump peaking somewhere between 10 μm and 100 μm. The IR emission accounts for anywhere from 10% to 90% of the total luminosity. Type 2 Seyferts generally show a steeper rise from 1 μm to longer wavelengths than do type 1 Seyferts, which has been taken as evidence that the IR continuum in type 2 Seyferts is emisson from dust whereas that from type 1 Seyferts is nonthermal (e.g., Edelson & Malkan 1986). However, the minimum near 1 μm in the spectral energy distribution of both types of Seyferts as well as many quasars is consistent with dust emission, since dust grains evaporate at temperatures of ~1500 K and so can only emit radiation at λ > 1 μm. In at least some of these AGNs, the dust may be heated by the central nonthermal source. Evidence for this is provided by Carleton et al. (1987) who show that type 1 Seyferts with weak UV (blue bump) emission tend to have strong IR emission, consistent with there being absorption by the dust of radiation from the accretion disk. The IR emission from the Seyfert nucleus in NGC 1068, which accounts for half of the total IR luminosity of 3×10^{11} L$_\odot$, peaks in the 20-30 μm region, unlike a region of star formation (Fig. 4). A crude but useful prototype for this emission may be the IR energy distribution of the planetary nebula NGC 7027 which is powered by an extremely hot star. As a comparison of Figs 2 and 4 shows, heating of the dust by the hard UV radiation fields and the relative paucity of neutral gas compared to star-forming regions has led in both cases to a hot IR energy distribution indicative of an "ionized gas ball."

An emission component due to hot dust associated with the AGN and accompanied by far-IR emission indicative of a circumnuclear starburst, appears to be common among Seyfert galaxies, as has been shown by IRAS observations. Multi-aperture 10 μm observations of the CFA sample of Seyfert galaxies by Edelson, Malkan, & Rieke (1987) show that the emission in all of the type 2 Seyferts and about half of the type 1 Seyferts is extended on the scale of ~1 kpc; extended starbursts may account for a large fraction of the IR luminosity in these objects (see also Rodriguez Espinosa, Rudy, & Jones 1986). The large spatial extent of the modest far-IR luminosity (~3×10^9 L$_\odot$) in the type 1 Seyfert NGC 4051 implies that dust heated by young stars powers the emission (Smith et al. 1983). IR mapping and multi-aperture observations of NGC 1068 (Telesco et al. 1984, and references therein; Telesco & Decher 1988) have shown that an extended starburst disk has an IR energy distribution peaking near 100 μm (as expected for a starburst) and emitting about half of the total IR luminosity of 3×10^{11} $L_\odot$. It has also become apparent that many of the type 1 Seyferts emit in near-IR and mid-IR spectral features attributable to dust probably associated with starbursts. The type 1 Seyfert NGC 7469 emits in the 3.3 μm spectral feature, a hallmark of star formation (Section 4.1), across a region extending ~3", or 700 pc, from the nucleus (Cutri et al. 1984). Unfortunately, we cannot make general statements about how much of the IR luminosities of Seyfert galaxies is generated in starbursts and how much is powered by nuclear black holes. IR fine structure lines, as described in SM92 will alter considerably this picture.

The nature of the energy source ("monster" or starburst) in IR-luminous galaxies has been especially debated for the so-called ultraluminous galaxies, which, depending on the authors, have been defined as galaxies with IR luminosities greater than $10^{11\text{-}12}$ $L_\odot$. Sanders et al. (1988) have shown that nine out of their sample of ten galaxies with $L_{IR} > 10^{12}$ $L_\odot$ have optical emission lines characteristic of AGNs. On the other hand, Leech et al. (1989) find that only 40% of their sample of 25 galaxies with $L(60 \ \mu m) > 10^{11.5}$ $L_\odot$ have optically evident AGNs. Some ultraluminous galaxies have compact nuclear radio sources indicative of AGNs, while others have spatially diffuse radio emission, which, in view of the radio-IR spatial correlation observed for starburst regions (Section 4.1.2), is taken as evidence that starbursts are present (Condon et al. 1991; Norris et al. 1990). IR evidence bearing on this issue is scanty but includes, for example, a Paα-line detection for NGC 6240, which, after correction for extinction is ten times weaker than expected if a starburst generates the IR luminosity but is comparable to that observed for NGC 1068 (DePoy, Becklin, Wynn-Williams 1986). As we did for Seyfert galaxies, we conclude that the nature of the sources powering the IR luminosity for ultraluminous galaxies differs among the sources and will probably only be identified definitively in each source using IR spectroscopy with ISO and the new generation of groundbased facilities.

Although the thermal emission from dust may account for the IR emission in most Seyferts and radio-quiet quasars, it has been proposed that the IR spectral energy distribution for some type 1 Seyferts and unreddened quasars is nonthermal, being reasonably well fitted by a power-law and well correlated with the 2 keV X-ray emission (e.g., Edelson & Malkan 1986). The turnover in these energy distributions near 100 μm is, in this picture, due to synchrotron self absorption near the accretion disk. There are radio-loud AGNs in which the IR continuum connects smoothly with the millimeter and radio continua, strongly suggesting that the IR is indeed nonthermal, but in other of those objects with marked turnovers, most notably the radio-quiet AGNs, the slope at wavelengths longer than the turnover is inconsistent with synchrotron self-absorption, implying rather that their far-IR emission is from dust (e.g., Chini et al. 1988).

A "blazar" is an AGN distinguished from radio-quiet quasars by its smooth IR-optical-UV continuum from a star-like nucleus, high optical polarization, rapid optical variability (< days), and strong and variable radio continuum. In contrast to Seyferts and radio-quiet quasars, the blazar emission from the UV through the IR is mainly nonthermal. Most of the continuum, including the IR, is synchrotron emission generated in a plasma jet near the black-hole accretion disk. I refer the reader to the review by Bregman (1990) for a description of the observations and theoretical interpretations of these objects.

6   REFERENCES

Aannestad, P.A.: 1989, *Astrophys. J.* **338**, 162.

Aaronson, M.:1977, Ph.D. Thesis, Harvard University.

Aaronson, M., Frogel, J.A., Persson, S.E.: 1978, *Astrophys. J.* **220**, 442.

Aitken, D.K., Roche, P.F., Allen, M.C., Phillips, M.M.: 1982, *M.N.R.A.S.* **199**, 31p.

Allamandola, L.J., Tielens, A.G.G.M., Barker, J.R.: 1985, *Astrophys. J. Lett.* **290**, L25.

Andriesse, C.D.: 1978, *Astron. Astrophys.* **66**, 169.

Baldwin, J.A., Phillips, M.M., Terlevich, R.: 1981, *Pub. A.S.P.* **93**, 5.

Barnes, J.E., Hernquist, L.E.: 1991, *Astropys. J. Lett.* **370**, L65.

Beck, S.C., Beckwith, S.V., Gatley, I.: 1984, *Astrophys. J.* **279**, 563.

Becklin, E.E., Gatley, I, Matthews, K., Neugebauer, G., Sellgren, K., Werner, M.W. and Wynn-Williams, C.G.: 1980, *Astrophys. J.* **236**, 441.

Beckwith, S., Evans, N.J. III, Gatley, I., Gull, G., Russell, R.W.: 1993, *Astrophys. J.* **264**, 152.

Bernloehr, K.:1992, *Astron. Astrophys.*, in press.

Bieging J.H., Wilner, D., Thronson, H.A.: 1991, *Astrophys. J.* **379**, 271.

Binney J., Tremaine, S.: 1987, *Galactic Dynamics*, Princeton University Press. Princeton.

Block, D.L., Wainscoat, R.J.: 1991, *Nature* **353**, 48.

Boulanger, F., Perault, M.: 1988, *Astrophy. J.* **330**, 964.

Boulanger, F., Beichman, C., Desert, F.X., Helou, G., Perault, M., Ryter, C.: 1988, *Astrophys. J.* **332**, 328.

Bregman, J.N.: 1990, *Astron. Astrophys. Rev.* **2**, 125.

Bushouse, H.A., Werner, M.W.: 1990, *Astrophys. J.* **359**, 72.

Carleton, N.P., Elvis, M., Fabbiano, G., Willner, S.P., Lawrence, A., Ward, M.: 1987, *Astrophys. J.* **18**, 595.

Carico, D.P., Graham, J.R., Mathews, K., Wilson, T.D., Soifer, B.T., Neugebauer, G., Sanders, D.B.: 1990, *Astrophys. J.* **349**, L39.

Casoli, F., Combes, F.: 1982, *Astron. Astrophysics.* **110**, 287.

Chevalier, R.A., Clegg, A.W.: 1985, *Nature* **317**, 44.

Chini, R., Kreysa, E., Krugel, E., Mezger, P.G.: 1986, *Astron. Astrophys.* **166**, L8.

Chini, R., Steppe, H., Kreysa, E., Krichbaum, T., Quirrenbach, A., Schalinski, C., Witzel, A.: 1988, *Astron. Astrophys.* **192**, L1.

Churchwell, E.B., Wolfire, M.G., Wood, D.O.S.: 1990, *Astrophys. J.* **354**, 247.

Clements, D.L., Rowan-Robinson, M., Lawrence, A., Broadhurst, T., McMahon, R.: 1992, *M.N.R.A.S.* **256**, 35p.

Cohen, M., Volk, K.:1989, *Astron. J.* **98**, 1563.

Combes, F.: 1988, in *Galactic and Extragalactic Star Formation*, p. 475. Eds. Pudritz, R.E. and Fich, M. Kluwar. Dordrecht.

Condon, J.J., Anderson, M.L., Helou, G.: 1991, *Astrophys. J.* **376**, 95.

Condon, J.J., Huang, Z.-P., Yin, Q.F., Thuan, T.X.: 1991, *Astrophys. J.* **378**, 65.

Cox, P., Krugel, E., Mezger, P.G.: 1986, *Astron. Astrophys.* **155**, 380.

Cox, P., Mezger, P.G.: 1987, in *Star Formation in Galaxies*, p.23. Ed. Lonsdale Persson, C.J. U.S. Government Printing Office. Washington, D.C.

Cutri, R.M., Rudy, R.J., Rieke, G.H., Tokunaga, A.T., Willner, S.P.: 1984, *Astrophys. J.* **280**, 521.

de Grijp, M.H.K., Miley, G.K., Lub, J.: 1987, *Astron. Astrophys. Supp.* **70**, 95.

DePoy, D.L.: 1987, Ph.D. Thesis, University of Hawaii.

DePoy, D.L., Becklin, E.E., Wynn-Williams, C.G.: 1986, *Astrophys. J.* **307**, 116.

Devereux, N.: 1987, *Astrophys. J.* **323**, 91.

Devereux, N.: 1989, *Astrophys. J.* **346**, 126.

Devereux, N.A., Becklin, E.E., Scoville, N.Z.: 1987, *Astrophys. J.* **312**, 529.

Devereux, N.A., Young, J.S.: 1990, *Astrophys. J. Lett.* **350**, L25.

Doyon, R.: 1991, Ph.D. Thesis, Imperial College.

Doyon, R., Joseph, R.D., Wright, G.S.: 1990, in *Astrophysics with Infrared Arrays*, P. 69. Ed. Elston, R. Astronomical Society of the Pacific. San Francisco.

Doyon, R., Puxley, P.J., Joseph, R.D.: 1992, *Astrophys. J.* 397, 117.

Draine, B.T., Anderson, N.: 1985, *Astrophys. J.* **292**, 494.

Draine, B.T., Roberge, W.G., Dalgarno, A.: 1983, *Astrophys. J.* **264**, 485.

Dressel, L.L.: 1988, *Astrophys. J. Lett.* **329**, L69.

Duffy, P.B., Erickson, E.F., Haas, M. R., Houck, J.R.: 1987, *Astrophys. J.* **315**, 68.

Edelson, R.A., Malkan, M.A.: 1986, *Astrophys. J.* **308**, 59.

Edelson, R.A., Malkan, M.A., Rieke, G.H.: 1987, *Astrophys. J.* **321**, 233.

Engargiola, G.: 1991, *Astrophys. J. Supp.* **76**, 875.

Engargiola, G., Harper, D.A.: 1992, *Astrophys. J.* **394**, 104.

Faber, S.M., Gallagher, J.S.: 1976, *Astrophys. J.* **204**, 365.

Frogel, J.A., Whitford, A.E.:1987, *Astrophys. J.* **320**, 199.

Frogel, J.A., Persson, S.E., Aaronson, M., Mathews, K.: 1978, *Astrophys. J.* **220**, 75.

Gaffney, N.I., Lester, D.F.: 1992, *Astrophys. J.* **394**, 139.

Gaffney, N.I., Lester, D.F., Telesco, C.M.: 1993, *Astrophys. J.*, submitted.

Gallagher, J.S., Faber, S.M., Balick, B.: 1975, *Astrophys. J.* **202**, 7.

Genzel, R., Stutzki, J.: 1989, *An. Rev. Astron. Astrophysics.* **27**, 41

Gillett, F.C., Kleinmann, D.E., Wright, E.L., Capps, R.W.: 1975, *Astrophys. J. Lett.* **198**, L65.

Gonzalez-Riestra, R., Rego, M., Zamorano, J.: 1987, *Astron. Astrophys.* **186**, 64.

Graham, J.R., Carico, D.P., Mathews, K., Neugebauer, G., Soifer, B.T., Wilson, T.D.: 1990, *Astrophys. J. Lett.* **354**, L5.

Greenhouse, M.A., Woodward, C.E., Thronson, H.A., Rudy, R.J., Rossano, G.S., Erwin, P., Puetter, R.C.: 1991, *Astrophys. J.* **383**, 164.

Harris, G.L.H.: 1976, *Astrophys. J. Supp.* **30**, 451.

Hawarden, T.G., Mountain, C.M., Leggett, S.K., Puxley, P.J.: 1986, *M.N.R.A.S.* **221**, 41p.

Heckman, T.M., Armus, L., Miley, G.K.: 1990, *Astrophys. J. Supp.* **74**, 833.

Helou, G.: 1986, *Astrophys. J. Lett.* **311**, L33.

Helou, G., Khan, I.R., Malek, L., Boehmer, L.: 1989 *Astrophys. J. Supp.* **68**, 151.

Hiromoto, N., Maihara, T., Oda, N.: 1983, *Pub. A. S. J.* **35**, 413.

Ho, P.T.P., Turner, J.L., Fazio, G.G., Willner, S.P.: 1989, *Astrophys. J.* **344**, 135.

Hyland, A.R., Straw, S., Jones, T.J., Gatley, I.: 1992, *M.N.R.A.S.* **257**, 391.

Impey, C.D., Wynn-Williams, C.G., Becklin, E.E.: 1986, *Astrophys. J.* **309**, 572.

Ishizuki, S., Kaware, R., Ishiguro, M., Okumura, S.K., Morita, K.-I., Chikada, Y., Tasuga, T. and Doi, M. : 1990, *Astrophys. J.* **355**, 436.

Jennings, R.E.: 1975, in *H II Regions and Related Topics*, p. 142. Eds. Wilson, T.L. and Downes, D. Springer-Verlag. Berlin.

Joseph, R.D., Meikle, W.P.S., Robertson, N.A., Wright, G.S.: 1984, *M.N.R.A.S.* **209**, 111.

Joseph, R.D., Wright, G.S. 1985, *M.N.R.A.S.* **214**, 87.

Joy, M., Lester, D.F., Harvey, P.M.: 1987, *Astrophys. J.* **319**, 314.

Jura, M., Kim, D.-W., Knapp, G.R., Guhathakurta, P.: 1987, *Astrophys. J.* **312**, L11.

Kawara, K., Nishida, M., Gregory, B.: 1987, *Astrophys. J. Lett.* **321**, L35.

Keto, E., Ball, R., Arens, J., Jernigan, G., Meixner, M.: 1992, *Astrophys. J.* **389**, 223.

Knapp, G.R., Gunn, J.E., Wynn-Williams, C.G.: 1992, *Astrophys. J.* **399**, 76.

Leech, K.J., Penston, M.V., Terlevich, R., Lawrence, A., Rowan-Robinson, M., Crawford, J.: 1989, *M.N.R.A.S.* **240**, 349.

Leger, A., Puget, J.L.: 1984, *Astron. Astrophys.* **137**, L5.

Lester, D.F., Dinerstein, H.L., Rank, D.M.: 1979, *Astrophys. J.* **232**, 139.

Lester, D.F., Carr, J.S., Joy, M., Gaffney, N.: 1990, *Astrophys. J.* **352**, 544.

LeVan, P.D., Price, S.D.: 1987, *Astrophys. J.* **312**, 592.

Lo, K.Y., Cheung, K.W., Masson, C.R., Phillips, T.G., Scott, S.L., Woody, D.P.: 1987, *Astrophys. J.* **312**, 574.

Loren, R.B.: 1976, *Astrophys. J.* **209**, 466.

Lutz, D.: 1992, *Astron. Astrophys.* **259**, 462.

Mathis, J.S., Mezger, P.G., Panagia, N.: 1983, *Astron. Astrophys.* **128**, 212.

Meadows, V.S., Allen, D.A., Norris, R.P., Roche, P.F.: 1990, *Proc. A.S.A.* **8**, 246.

Miller, G.E., Scalo, J.M.: 1979, *Astrophys. J. Supp.* **41**, 513.

Moorwood, A.F.: 1986, *Astron. Astrophys.* **166**, 4.

Myers, S.T., Scoville, N.Z.: 1987, *Astrophys. J. Lett.* **312**, L39.

Noguchi, M.: 1988, *Astron. Astrophys.* **203**, 259.

Norris, R.P., Allem, D.A., Sramek, R.A., Kesteven, M. J., Troup, E.R.: 1990, *Astrophys. J.* **359**, 291.

Oliva, E., Moorwood, A.F.M.: 1990, *Astrophys. J. Lett.* **348**, L5.

Olnan, F.M., Baud, B., Habing, H.J., de Jong, T., Harris, S., Pottasch, S.R.: 1984, *Astrophys. J. Lett.* **278**, L41.

Panagia, N.: 1973, *Astron. J.* **78**, 929.

Piña, R.K., Jones, B., Puetter, R.C., Stein, W.A.: 1993, *Astrophys. J. Lett.*, in press.

Puget, J.L., Leger, A.: 1989, *An. Rev. Astron. Astrophys.* **27**, 161.

216

Puxley, P.J., Brand, P.W. J.L., Moore, T.J.T., Mountain, C.M., Nakai, N., Yamashita, T.: 1989, *Astrophys. J.* **345**, 163.

Rice, W., Boulanger, F., Viallefond, F., Soifer, B.T., Freedman, W.L.: 1990, *Astrophys. J.* **358**, 418.

Rice et al.: 1988, *Astrophys. J. Supp.* **68**, 91.

Rickard, L.J, Harvey, P.M.: 1984, *Astronom. J.* **89**, 1520.

Rieke, G.H.: 1988, in *Galactic and Extragalactic Star Formation*, p.561. Eds. Pudritz, R.E. and Fich, M. Kluwer. Dordrecht.

Rieke, G.H., Lebofsky, M.J.: 1979, *An. Rev. Astron. Astrophys.* 17, 477.

Rieke, G.H., Lebofsky, M.J., Thompson, R.I., Low, F.J., Tokunaga, A.T.: 1980, *Astrophys. J.* **238**, 24.

Roberts, Jr., W.W., Hausman, M.A.: 1984, *Astrophys. J.* **277**, 744.

Roche, P.F., Aitken, D.K., Smith, C.H., Ward, M.J.: 1991, *M.N.R.A.S.* **248**, 606. (R91)

Rodriguez Espinosa, J.M., Rudy, R.J., Jones, B.: 1986, *Astrophys. J.* **309**, 76.

Rowan-Robinson, M.: 1991, in *Dynamics of Galaxies and Their Molecular Cloud Distributions*, P.211. Eds. Combes, F. and Casoli, F. Kluwer. Dordrecht.

Rowan-Robinson, M., Crawford, J.: 1989, *M.N.R.A.S.* **238**, 523.

Rubin, R.H.: 1985, *Astrophys. J. Supp.* **57**, 349.

Sandage, A., Becklin, E.E., Neugebauer, G.: 1969, *Astrophys. J.* **157**, 55.

Sanders, R.H., Huntley, J.M.: 1976, *Astrophys. J.* **209**, 53.

Sanders, D.B., Soifer, B.T., Elias, J.H., Madore, B.F., Mathews, K., Neugebauer, G., Scoville, N.Z.: 1988, *Astrophys. J.* **325**, 74.

Scalo, J.M.: 1986, *Fundam. Cosmic. Phys.* **11**, 1.

Schlosman, I., Frank, J., Begelman, M.C.: 1989, *Nature* **338**, 45.

Sellgren, K.: 1984, *Astrophys. J.* **277**, 623.

Sellgren, K., Luan, L., Werner, M.W.: 1990, *Astrophys. J.* **359**, 384.

Smith, H.A., Lada, C.J., Thronson, H.A., Glaccum, W., Harper, D.A., Loewenstein, R.F., Smith, J.: 1983, *Astrophys. J.* **274**, 571.

Soifer, B.T., Rice, W.L., Mould, J.R., Gillett, F.C., Rowan-Robinson, M., Habing, H.J.: 1986, *Astrophys. J.* **304**, 651.

Spinoglio, L., Malkan, M.A.: 1992, *Astrophys. J.* **399**, 504. (SM92)

Stacey, G.J.: 1989, in *Infrared Spectroscopy in Astronomy*, p.455. Ed. Kaldeich, B.H. ESA Publications Division. Noordwijk.

Stacey, G.J., Geis, N., Genzel, R., Lugten, J.B., Poglitsch, A., Townes, C.H.: 1991, *Astrophys. J.* **373**, 423.

Telesco, C.M.: 1985, in *Extragalactic Infrared Astronomy*, p. 87. Ed. Gondhalekar, P.M. Science Eng. Res. Council. Chilton (UK).

Telesco, C.M.: 1988, *An. Rev. Astron. Astrophys.* **26**, 343.

Telesco, C.M., Becklin, E.E., Wynn-Williams, C.G., Harper, D.A.: 1984, *Astrophys. J.* **282**, 427.

Telesco, C.M., Campins, H., Joy, M., Dietz, K., Decher, R.: 1991, *Astrophys. J.* **369**, 135.

Telesco, C.M., Decher, R.: 1988, *Astrophys. J.* **334**, 573.

Telesco, C.M., Decher, R., Joy, M.: 1989, *Astrophys. J. Lett.* **343**, L13.

Telesco, C.M., Dressel, L.L., Wolstencroft, R.D.: 1993, *Astrophys. J.*, submitted.

Telesco, C.M., Gatley, I.: 1984, *Astrophys. J.* **284**, 557.

Telesco, C.M., Harper, D.A.: 1977, *Astrophys. J.* **211**, 475.

Telesco, C.M., Harper, D.A.: 1980, *Astrophys. J.* **235**, 392.

Telesco, C.M., Harper, D.A., Loewenstein, R.F.: 1976, *Astrophys. J. Lett.* **203**, L53.

Telesco, C.M., Gezari, D.Y.: 1992, *Astrophys. J.* **395**, 461.

Telesco, C.M., Decher, R., Gatley, I.: 1985, *Astrophys. J.* **299**, 896.

Telesco, C.M., Decher, R., Ramsey, B.D., Wolstencorft, R.D., Leggett, S.K.: 1987, in *Star Formation in Galaxies*, P. 497 Ed. Lonsdale Persson, C.J., US Gov. Prnting Off. Washington, D.C.

Terlevich, R., Melnick, J.: 1985, *M.N.R.A.S.* **213**, 841.

Thronson, H.A., Herald, M., Magewski, S., Greenhouse, M., Johnson, P., Spillar, E., Woodward, C.E., Harper, D.A. and Rauscher. B.J. : 1989, *Astrophys. J.* **343**, 158.

Tielens, A.G.G.M., Hollenbach, D.: 1985, *Astrophys. J.* **291**, 722.

Veilleux, S., Osterbrock, D.E.: 1987, *Astrophys. J. Supp.* **63**, 295.

Walterbos, R.A.M., Schwering, P.B.W.: 1987, *Astron. Astrophys.* **180**, 27.

Wang, Z., Helou, G.: 1992, *Astrophys. J. Lett.* **398**, L33.

Wang, Z., Scoville, N.Z., Sanders, D.B.: 1991, *Astrophys. J.* **368**, 112.

Weedman, D.W., Feldman, F.R., Balzano, V.A., Ramsey, L.W., Sramek, R.A., Wu, C.-C.: 1981, *Astrophys. J.* **248**, 105.

Werner, M.W., Beckwith, S., Gatley, I., Sellgren, K., Berriman, G., Whiting, D.L.: 1980, *Astrophys. J.* **239**, 540.

Werner, M.W., Gatley, I., Harper, D.A., Becklin, E.E., Loewenstein, R.F., Telesco, C.M., Thronson, H.A.: 1976, *Astrophys. J.* **204**, 420.

Wolfire, M.G., Tielens, A.G.G.M., Hollenbach, D.: 1990, *Astrophys. J.* **358**, 116.

Wood, D.O.S., Churchwell, E.: 1989, *Astrophys. J. Supp.* **69**, 831.

Wright, G.S., Joseph, R.D., Robertson, N.A., James, P.A., Meikle, W.P.S.: 1988, *M.N.R.A.S.* **233**, 1.

Wynn-Williams, C.G., Eales S.A., Becklin E.E., Hodapp K.W., Joseph R.D., McLean I.S., Simons D.A. and Wright G.S.: 1991, *Astrophys. J.* **377**, 426.

Xu, C., Sulentic, J.W.: 1991, *Astrophys. J.* **374**, 407.

Young, J.S., Schloerb, F.P., Kenney, J.D., Lord, S.D.: 1986, *Astrophys. J.* **304**, 443.

Zaritsky, D., Lo, K.Y.: 1986, *Astrophys. J.* **303**, 66.

Cosmology

R. D. Joseph

Institute for Astronomy
University of Hawaii
2680 Woodlawn Drive
Honolulu, HI, U.S.A.

1 INTRODUCTION

My objective in these five lectures was to introduce the fundamental physical ideas underlying cosmological models of an isotropic hot Big-Bang and the development of large-scale structure in such an expanding universe. My instructions from the Organizing Committee were to assume that the audience had not studied cosmology before, and that is what I have tried to do. My final lecture reviewed a number of results in infrared observational cosmology, but constraints of time and space have not permitted inclusion of that material here.

Much of the material on isotropic cosmological models is developed very well in any number of existing sources (although this is less so for the theory of galaxy formation) and it is with some diffidence that I offer my version. The references list a number of other developments of these topics, and I have borrowed from most of them. Perhaps the best recent text which covers these and many other topics in modern cosmology very thoroughly, and yet very readably, is that by Kolb and Turner (1990). My own research in infrared observational cosmology received an important stimulus from a series of similar lectures presented by Malcolm Longair (1977) over 15 years ago. I hope these lectures might, in a similar way, give young astronomers an introduction which will generate enthusiasm for inventing new infrared observing programs bearing on fundamental cosmological problems.

2 THE ISOTROPIC UNIVERSE

2.1 Introduction

We begin by taking a very large-scale view of the universe, and make a simplicity approximation that the universe is smooth, with no structure. In this approximation we can think of the universe as a fluid (of galaxies) of density ρ, pressure p. We will need a theory of gravitation to work out the dynamical evolution of this fluid. Here we will use Newtonian gravity, which reveals the physics quite elegantly, but we will interpret the results to conform with those reached with a proper treatment using General Relativity (GR).

It is probably helpful to begin by stating what cosmology is about, and making some distinctions about it. Cosmology is the scientific study of the structure and evolution of the large-scale universe. We note right away that cosmology is a qualitatively different scientific subject from those we normally deal with in natural science. One has, by definition only one universe to study. It is therefore not possible to distinguish those properties of the universe which are unique to this universe from those which would be characteristic of all universes. Thus it is not possible to have a theory of universes in the same way one has a theory of, say, atoms. In this sense, cosmology is at best a descriptive study. It is also useful to distinguish between cosmology and cosmogony. Cosmogony is concerned with the origin of the universe. Since scientific method is only equipped to deal with natural phenomena which are reproducible, cosmogony is therefore outside the domain of science, and should be regarded as a philosophical, religious, or aesthetic matter.

2.2   The Cosmological Principle

The starting point of standard cosmologies is assumption of what is called the Cosmological Principle (CP): the universe is *homogeneous* and *isotropic*. By isotropic, one means that the observable properties of the universe are independent of the direction in which one observes. The assumption of homogeneity means that the universe is uniform throughout. One may think of the CP as a simplicity postulate, but there is now very strong observational evidence discussed in detail later, from isotropy of the 2.7 K Cosmic Background Radiation, that the universe does seem to be isotropic to high precision.

The CP is a very strong postulate. It implies that one can define a "cosmic time" and the mass density of the universe is independent of position and can depend only on time. In other words, the universe can only expand or contract, with the mean density decreasing or increasing correspondingly, but otherwise it is uniquely determined. Thus, if $r(t)$ is distance between two test particles (*i.e.* galaxies), then the CP implies that $r(t)$ depends only on a scale factor, say $a(t)$. If t and t_o are two different cosmic times, then

$$\frac{r(t)}{r(t_o)} = \frac{a(t)}{a(t_o)}.$$

Hereafter we take the subscript "*o*" to denote a quantity evaluated at the present time, and we will set $a(t_o) = 1$, whence

$$r(t) = a(t)r(to). \tag{1}$$

With adoption of the Cosmological Principle, then, the problem is to use dynamics and a theory of gravitation to find the functional form(s) of $a(t)$. We will do this in Sections 2.5 and 2.6. But first it is instructive to investigate a bit further the consequences of the adoption of the CP.

2.3   The Hubble Law

Throughout these lectures I will usually denote a time derivative by a superior dot:

$$\frac{dr}{dt} = \dot{r}.$$

Differentiating eq. (1):

$$\dot{r} = \dot{a}r(t_o) = \frac{\dot{a}}{a}r(t),$$

or

$$v(t) = H(t)r(t),$$

where the Hubble parameter $H(t)$ is defined by

$$H(t) = \frac{1}{a}\frac{da}{dt} = \dot{a}/a. \tag{2}$$

Thus, the CP *by itself* implies a universal relation between the distance and relative velocity of test particles in the universe. Evaluating this relation at the present time becomes "Hubble's law:"

$$v(t_o) = H_o r(t_o). \tag{3}$$

H_o has dimensions of (time)$^{-1}$, but the usual units are km sec^{-1} Mpc^{-1}. Hubble originally estimated H_o to be ~ 500 km sec^{-1} Mpc^{-1}. There have been many programs to measure H_o, but the results have all tended to condense around two extreme values ~ 50 and ~ 100 km sec^{-1} Mpc^{-1}. Because of this uncertainty, it is customary to parametrize H_o by writing $H_o = 100h$ km sec^{-1} Mpc^{-1}. Extragalactic distances (derived from redshifts) and luminosities are then proportional to h^{-1} and h^2 respectively, and one can substitute one's personal best guess for h when a numerical value is required. The most recent reviews of various methodologies for determining H_o suggest a convergence toward $h \sim 0.8$.

2.4 The Redshift

Another consequence of the CP is that global properties are "local" properties. We can use a small representative piece of the universe to work out large-scale properties.

Imagine light waves travelling from a galaxy to us. Chop out a volume of the universe containing these light waves by a box with perfectly reflecting walls. As the box expands, the wavelength λ, like other lengths mustexpand proportionally. This can also be seen either from consideration of Doppler shifts off the moving walls, or as the adiabatic expansion of a gas of photons. Thus, λ is proportional to $a(t)$.

If λ_{em} is the wavelength emitted by a source at one position in the universe and λ_{obs} is the wavelength observed at a different place in the universe,

$$\frac{\lambda_{obs}}{\lambda_{em}} = \frac{a(t_{obs})}{a(t_{em})} = \frac{1}{a(t)}.$$

Then the redshift, z, defined as

$$z = \frac{\lambda_{obs} - \lambda_{em}}{\lambda_{em}} = \frac{1}{a(t)} - 1,$$

and

$$a(t) = \frac{1}{(1+z)}. \tag{4}$$

Thus, using only the implications of the CP, we have established a relation between the scale parameter $a(t)$ and an observable quantity, z. If the universe is expanding (contracting) we expect to see redshifts (blueshifts) in spectral lines of light from galaxies emitted when the universe was smaller (larger).

Early in this century galaxies weren't known to be outside Milky Way and were called "spiral nebulae." That nebulae were extragalactic was definitely established *ca.* 1925 by measuring distances to some nearby objects using the period-luminosity relation Ms. Henrietta Leavitt had discovered for Cepheid variables.

By *ca.* 1924, Slipher had discovered (mostly) redshifts in the spectra of $\sim$ 40 galaxies. In 1929, Hubble found for a small sample of galaxies the redshift of a galaxy was proportional to its distance. If z is due to the Doppler Effect,

$$\lambda_{obs} = \lambda_{em}\sqrt{\frac{1+\beta}{1-\beta}} \qquad \text{where} \quad \beta = v/c,$$

so

$$z = \sqrt{\frac{1+\beta}{1-\beta}} - 1 \simeq \sqrt{(1+\beta)^2} - 1 \simeq \beta \quad \text{for} \quad \beta \ll 1.$$

Since $cz = v$, Hubble could write his proportionality between z and r as

$$v = H_o r. \tag{5}$$

Note that this is eq. (3) evaluated at the present time, t_o. Thus, observations (*i.e.* Hubble's velocity-distance relation) show that universe is expanding. This relation is customarily plotted in the redshift-magnitude plane, where the magnitudes are for a sample of galaxies thought to be standard candles. A modern Hubble diagram is shown in Fig. 4.

2.5 Newtonian Cosmology

We now want to use conservation of energy and Newton's second law to find the equations of motion for the expansion function $a(t)$.

a) Conservation of energy. The total energy E in volume V is

$$E = \rho c^2 V.$$

As the universe expands,

$$\frac{dE}{dt} = \rho c^2 \frac{dV}{dt} + c^2 V \frac{d\rho}{dt}.$$

But dE is work done by pressure, $-pdV$, so

$$-p\frac{dV}{dt} = \rho c^2 \frac{dV}{dt} + c^2 V \frac{d\rho}{dt}.$$

227

Since

$$\frac{dV}{dt} = 4\pi r^2 \frac{dr}{dt} = 4\pi r^2 H(t) r(t) = 4\pi r^3 \frac{\dot{a}}{a},$$

we find

$$\frac{d\rho}{dt} + 3(\rho + \frac{p}{c^2})\frac{\dot{a}}{a} = 0. \tag{6}$$

Thus, with an equation of state (*i.e.* a relation between pressure and density) for the cosmic fluid, $p(\rho) = 0$, we can solve eq. (6) to find $\rho(t)$. *N.B.* this equation could also be derived directly from the field equations of GR with the assumption of the CP, so it is relativistically correct.

There are two important limiting cases for the equation of state. a) For the matter-dominated case: $p \ll \rho c^2$. This is sometimes referred to as a "dust-filled" universe. b) For a radiation-dominated universe: $p = \rho c^2/3$. As we shall soon show, this is the case early in the evolution of the universe.

Integrating eq. (6) for the matter dominated case,

$$\rho(t) = \rho(t_o) a^{-3}(t), \tag{7}$$

whereas for the radiation-dominated case,

$$\rho(t) = \rho(t_o) a^{-4}(t). \tag{8}$$

We now use Newtonian dynamics and gravitation to find the equation of motion for $a(t)$.

b) Newtonian dynamics and gravitation

$$F = dv/dt,$$

and

$$\vec{F} = -\frac{Gm}{r^2}\hat{e}_r.$$

The latter implies

$$\nabla \cdot \vec{F} = -4\pi G\rho.$$

From Hubble's Law

$$\vec{v} = \frac{\dot{a}}{a}\vec{r}.$$

Substituting this in Newton's Second Law gives

$$\vec{F} = \dot{\vec{v}} = \frac{\ddot{a}}{a}\vec{r}.$$

Then

$$\nabla \cdot \vec{F} = \frac{\ddot{a}}{a}\nabla \cdot \vec{r} = 3\frac{\ddot{a}}{a} = -4\pi G\rho,$$

or

$$\frac{\ddot{a}}{a} + \frac{4\pi G}{3}(\rho) = 0. \tag{9}$$

For a relativistic gas one gets the same equation, but with an extra term to account for the mass equivalence of energy due to pressure, p:

$$\frac{\ddot{a}}{a} + \frac{4\pi G}{3}(\rho + \frac{3p}{c^2}) = 0. \tag{10}$$

This is the equation one derives directly from the field equations using GR.

We can integrate this equation for the matter-dominated case. Substitute $\rho(t) = \rho_o a^{-3}(t)$ and multiply by $2\dot{a}$. This is obviously

$$\frac{d}{dt}(\dot{a}^2) - \frac{8\pi G}{3}\rho_o\frac{d}{dt}(\frac{1}{a}) = 0,$$

which integrates to

$$\dot{a}^2 - \frac{8\pi G}{3}\frac{\rho_o}{a} = \text{constant}.$$

We will use the result from GR to choose the constant of integration to be $-Kc^2$. Here K can take the values 1, 0, -1. We will discuss the meaning of K in a moment. Using this constant of integration:

$$(\frac{\dot{a}}{a})^2 - \frac{8\pi G}{3}\rho(t) = \frac{-Kc^2}{a^2}. \tag{11}$$

This is the Friedmann equation. With the energy equation (6) and an equation of state one can solve the Friedmann equation for $a(t)$. The resulting solutions for $a(t)$ comprise all the possible cosmological models satisfying the CP. Although we derived this equation for the matter-dominated case, the same equation follows for the radiation-dominated case; it is general.

There is a simple physical interpretation of the two terms on the left-hand-side of Eqn (11): they are kinetic energy minus potential energy: KE - PE.

If $-Kc^2 < 0$ (*i.e.* $K > 0$) $PE > KE$, which implies the expansion of the Universe will eventually halt and it will begin to collapse.

If $-Kc^2 > 0$ (*i.e.* $K < 0$) $KE > PE$, which implies unlimited expansion.

If $-Kc^2 = 0$ (*i.e.* $K = 0$) KE = PE, which implies the expansion approaches zero asymptotically. Thus, without doing any more analysis at all it is clear that there are three classes of solutions for $a(t)$; these are the homogeneous, isotropic cosmological models.

In the context of GR, K describes the geometry of space. While it may sound strange to speak of a geometry of space, all that is meant is that one can use light rays or free particles as geodesics—*i.e.* the shortest distance between two points—and then work out the corresponding geometry. This is only another way of talking about the fact that light rays are "bent" when traversing a gravitational potential gradient, but it should be clear that the geometry may or may not be Euclidean. Our adoption of the CP insures that at any given cosmic time the curvature of space will be a constant. There are just three classes of such spaces of constant curvature, and they are identified by the curvature index, K. $K = 0$ describes a flat, Euclidean space. $K = +1$ describes a space of positive curvature—in two dimensions the surface of a sphere. $K = -1$ describes a space of negative curvature—in two dimensions the surface of a saddle.

By way of historical comment: Einstein originally had another term in the field equations of GR, corresponding to a gravitational repulsion, which made eq. (11) take the form:

$$(\frac{\dot{a}}{a})^2 - \frac{8\pi G}{3}\rho(t) = \frac{-Kc^2}{a^2} + \frac{\Lambda}{3}. \tag{12}$$

The Λ in the last term on the right hand side is the "cosmological" constant, introduced in a quite *ad hoc* manner by Einstein to get a static solution, $a(t) =$ constant. (At that time Hubble's Law was not yet known.) It was abandoned with the discovery that the universe is expanding, but in recent years has been re-introduced, since such a force can arise naturally in some theories in particle physics. Since there is no compelling reason (observationally or theoretically) for including it, I will not consider the cosmological constant further.

2.6   Cosmological Models

We have now set up all the formalism for solving for $a(t)$ for the three classes of models discussed above. Before we obtain these solutions, however, it is instructive to re-write the equations for $a(t)$ in terms of some observable parameters. We define

Critical density:

$$\rho_c = \frac{3H_o^2}{8\pi G},\qquad(13)$$

Density parameter:

$$\Omega_o = \rho_o/\rho_c,\qquad(14)$$

Deceleration parameter:

$$q(t) = \frac{-\ddot{a}}{H^2(t)a(t)}.\qquad(15)$$

Evaluating the Friedman equation (11) at the present time, t_o, permits determination of Kc^2 in terms of these quantities:

$$H_o^2 - 8\pi G\rho_o/3 = -Kc^2.$$

Since

$$\rho_o = \Omega_o\rho_c = 3H_o^2\Omega_o/8\pi G$$

$$-Kc^2 = H_o^2(1 - \Omega_o).\qquad(16)$$

Thus, eq. (11) becomes, for the matter-dominated case:

231

$$\dot{a}^2 = H_o^2[(1 - \Omega_o) + \frac{\Omega_o}{a}]. \tag{17}$$

This shows explicitly that solutions for $a(t)$, *i.e.* all possible cosmological models, are uniquely determined by H_o and Ω_o. K is < 0 or > 0 as Ω_o is < 1 or > 1.

Alternatively, one can show via eq. (9) that

$$q_o = \Omega_o/2,$$

so H_o and q_o also uniquely determine world models. As would be expected from purely physical considerations, the mean mass density of the universe, through Ω_o (or q_o), determines the shape of the function $a(t)$, whereas H_o determines only the timescale for the expansion or contraction.

We are now ready to find solutions for $a(t)$ for a matter-dominated universe.

a) $\Omega_o = 1$ (K = 0). This is the Einstein-de Sitter (EdS) model. Since K = 0 the geometry is Euclidean. eq. (17) becomes:

$$\dot{a}^2 = H_o^2/a.$$

Integrating, we get

$$a_{EdS}(t) = (\frac{3H_o}{2})^{2/3}t^{2/3}. \tag{18}$$

The Einstein-de Sitter universe continues to expand without limit, but its rate of expansion goes asymptotically to zero. This solution is shown graphically in Fig. 1.

Since $a(to) = 1$, the age of an EdS universe is

$$t_o = \frac{2}{3H_o}. \tag{19}$$

b) $\Omega_o > 1$ (K = +1). This is a "closed" universe. The geometry of space is Riemannian, in two dimensions like the surface of a sphere. From eq. (17), it is evident that there is some $a(t)$ such that

$$\dot{a}^2 = 0,$$

i.e. $a(t)$ reaches a maximum. For this class of models the universe expands to some maximum and then begins collapsing. In this case one can find parametric solutions (parameter θ) of the form:

$$a(t) = \frac{\Omega_o}{2(\Omega_o - 1)}(1 - cos\theta),\tag{20}$$

$$H_o t = \frac{\Omega_o}{2(\Omega_o - 1)^{3/2}}[\theta - sin\theta].\tag{21}$$

Note that t is linear with θ, and $a(t)$ reaches a maximum when $\theta = \pi$. This solution is shown graphically in Fig. 1.

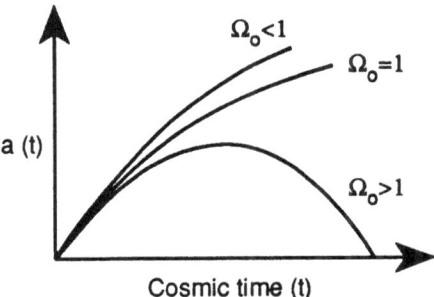

Figure 1: *Solutions for the expansion function $a(t)$ for the three classes of isotropic, homogeneous cosmological models.*

c) $\Omega_o < 1$ *(K = -1)*. This is an "open" universe. In two dimensions the geometry is like that on the surface of a saddle. In this case it is clear that $a(t)$ is always > 0. One finds parametric solutions (parameter ψ) of the form:

$$a(t) = \frac{\Omega_o}{2(1 - \Omega_o)}(cosh\psi - 1)\tag{22}$$

$$H_o t = \frac{\Omega_o}{2(1 - \Omega_o)^{3/2}}(sinh\psi - \psi).\tag{23}$$

This solution is also illustrated in Fig. 1.

Note that the slope of the function $a(t)$ at t_o is just H_o. Therefore the age of the universe, t_o, will always be less than, but of the order of H_o^{-1}.

2.7   The Hot "Big-Bang" Model

Suppose the universe had a hot early phase, with radiation in thermal equilibrium with matter. What happens to the radiation as universe expands?

We have already solved eq. (6) for both matter-dominated and radiation-dominated cases and found how the densities evolve with time. Comparing the results in Eqns. (7) and (8) we see

$$\frac{\rho_r(t)}{\rho_m(t)} = \frac{\rho_r(t_o)}{\rho_m(t_o)} \frac{1}{a(t)}.$$

Obviously, at sufficiently early time $a(t)$ will be small enough that $\rho_r \gg \rho_m$. Thus, radiation, rather than matter, dominates the dynamics of the early universe. We will show below that, as long as there are free electrons substantially present, Compton scattering will ensure that matter and radiation will be in thermal equilibrium. Thus the radiation will be blackbody in character.

In this radiation-dominated phase

$$\rho_r(t) \propto T_r^4(t) \propto a^{-4}(t)$$

so

$$T_r(t) \propto \frac{l}{a(t)} \propto 1 + z.$$

Thus,

$$T_r(t) = T_o(1 + z). \tag{24}$$

As we shall discuss below, the discovery of the cosmic background radiation has revealed that the current temperature of the radiation from the Big Bang is $T_o = 2.735$ K.

Note that temperature of the radiation, T_r, can be used as a time or expansion parameter as well as z and $a(t)$.

To find the expansion rate for the radiation-dominated epoch we need to solve eq. (11) with $\rho_r(t)$:

$$\dot{a}^2 - \frac{8\pi G}{3}\frac{\rho_o}{a^2} = -Kc^2.$$

For small $a(t)$, one can neglect the term $-Kc^2$. Substituting for ρ_o in terms of Ω_o and the critical density, we find:

$$a_r(t) = \sqrt{2H_o}\,\Omega_o^{1/4}t^{1/2}. \tag{25}$$

This describes the early evolution of a hot Big-Bang universe when radiation dominates the dynamics.

In a universe with both matter and radiation, the expansion will be controlled by the radiation energy density at early times and by the matter density at late times. The time at which these cross is t_{eq}. We want to find the corresponding redshift, z_{eq}. Substituting Eqns. (7) and (8) into the above equality gives:

$$\rho_r(t_{eq}) = \rho_m(t_{eq})$$

$$\frac{\rho_r(t_o)}{a^4(t_{eq})} = \frac{\rho_m(t_o)}{a^3(t_{eq})} \Rightarrow \frac{\rho_r(t_o)}{a(t_{eq})} = \Omega_o\rho_c.$$

Since $a(t) = 1/(1+z)$ and $\rho_r = 4\sigma T^4/c^3$,

$$1 + z_{eq} = \frac{\Omega_o\rho_c}{4\sigma T_o^4/c^3} \approx 10^4\Omega_o.$$

We will show below that Ω_o is likely to be between 0.1 and 1. Thus,

$$z_{eq} \sim 10^3 \text{ to } 10^4. \tag{26}$$

235

2.8   The Synthesis of Light Elements in the Primeval Fireball

We have adopted the hot Big-Bang model because we know there is a universal cosmic blackbody radiation background (CBR) with $T_r(t_o) \approx 2.7$K. Since $T_r(t) \propto (1 + z)$, and $\rho_m(t) \propto (1 + z)^3$, at early times (large z) the universe will have a very hot, dense phase. It turns out that the temperatures and densities are large enough to permit nuclear reactions to take place. The reactions which do occur depend on the density, temperature, and timescale for the reactions compared to the timescale for expansion. If timescales for nuclear reactions are very much shorter than the timescale for expansion, equilibrium is obtained in these reactions; otherwise reactions do not significantly affect the particle population. The expansion timescale is

$$\tau_{exp} \sim a/\dot{a} \sim [G\rho(t)]^{-1/2} \sim T_r^{-2}. \tag{27}$$

Going backward in the expansion we can identify the time and temperature and the corresponding epochs in which various physical processes take place. There are four particularly noteworthy epochs.

a) $T \sim 10^{10} - 10^{11}$ K

It is at this epoch when the ratio of neutrons to protons in the universe is determined. The relevant reaction is

$$n + \nu_e \rightarrow p + e^-,$$

and the timescale for this reaction

$$\tau_{n \leftrightarrow p} \sim T^{-5}.$$

In equilibrium, the ratio of neutrons to protons is given by:

$$\frac{N_n}{N_p} = exp[-(m_n - m_p)c^2/kT]. \tag{28}$$

This ratio decreases as T_r decreases. At $T_r \sim 10^{10}$K, the neutron/proton ratio is ~ 0.1. At this temperature the timescale for the $n \rightarrow p$ reaction is about equal to the expansion timescale, both ~ 1 sec. The neutron/proton ratio is therefore frozen out at this value. This determines the number of neutrons available to make all the elements in the periodic table.

b) $T \sim 3 \times 10^8 - 10^{10}$ K

At these temperatures one gets two-body reactions like

$$
\begin{array}{ccccccc}
p & + & n & \rightarrow & H^2 & + & \gamma \\
H^2 & + & H^2 & \rightarrow & He^3 & + & n \\
He^3 & + & n & \rightarrow & H^3 & + & \rho \\
H^3 & + & H^2 & \rightarrow & He^4 & + & n
\end{array}
$$

In this way deuterium (H^2), and isotopes of He, Li, Be, and B are built up. It turns out that more massive nuclei cannot be built up in this way because there are no stable nuclei with atomic number $A = 5$ or $A = 8$. (This can be overcome if the flux of neutrons is sufficiently intense that a mass 5 or 8 nucleus is hit by another neutron before it has time to decay. This is the situation in a supernova explosion, the site where we believe heavier elements are built up.)

Calculations show that the light element abundances (relative to hydrogen) surviving to the present can be expressed as a function of the present day mass density, ρ_o, as shown in Fig. 2. There are several interesting points to note about this figure. *i)* He^4 takes up almost all the available neutrons, and the abundance of He^4 predicted by the hot Big-Bang model, ~ 0.25 (by mass) is very close to what is observed. *ii)* Abundances of the other nuclides is expected to be $< 10^{-5}$, again consistent with what is observed. These two points give strong qualitative support to the hot Big-Bang model. *iii)* Deuterium is very sensitive to present-day mass density, and therefore it could be a useful diagnostic in placing constraints on the parameters Ω_o and H_o.

c) $T \sim 10^4 K$

At this temperature $\rho_m \simeq \rho_r$ and so this is the transition from radiation-dominated expansion to matter-dominated expansion.

d) $T \sim 4000$ K

At this temperature electrons and protons combine to form hydrogen. (This is usually called the "epoch of recombination," but of course the electrons and protons have never been bound before.) From this point on matter and radiation are no longer coupled by Compton scattering off free electrons, and radiation

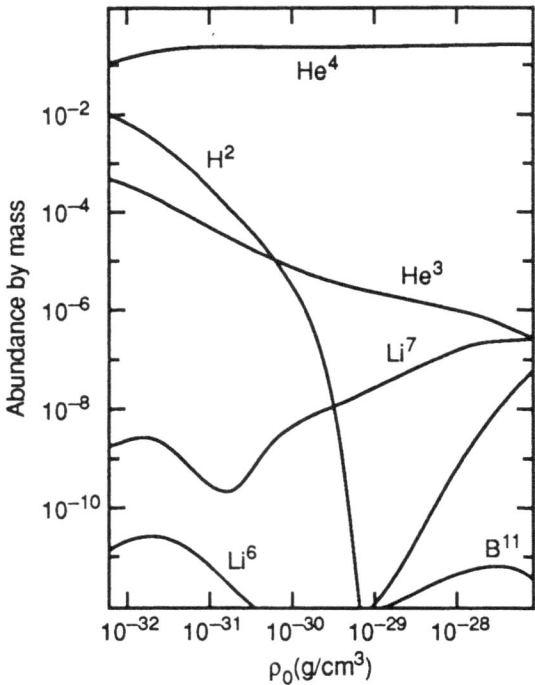

Figure 2: *Primordial light element abundances resulting from Big-Bang nucleosynthesis as a function of present-day mass density. (After Schramm and Wagoner 1974.)*

and matter evolve separately. This is called "decoupling," and the temperatures of the radiation and matter are henceforth no longer equal. Both continue to cool as the universe continues to expand. The cooling continues forever unless $\Omega_o < 1$, and in the latter case the universe would reverse the evolution we have just described once $a(t)$ has reached its maximum and contraction has begun.

This is certainly not a very interesting universe, nor is it anything like the universe we know. There is no structure (galaxies, etc.), there are no heavy elements, and therefore there are no astronomers or physicists to think about cosmology. Of course these are consequences of the simplifying assumptions we made right at the beginning. In Section 3 we will put small fluctuations into this perfectly smooth universe, and see if they can grow to the kinds of large-scale structure we see in the universe around us today. But before embarking on this next level of approximation, we will briefly review the observational constraints on this simple isotropic hot Big-Bang model, to see if we can determine, even at this very coarse level of approximation, the cosmological model parameters that best fit the universe we observe.

3 OBSERVATIONAL CONSTRAINTS ON THE ISOTROPIC HOT BIG-BANG

3.1 2.7 K Cosmic Background Radiation

As we have seen, thermal equilibrium is maintained by Compton scattering off free electrons, until decoupling . What happens to radiation after decoupling? Just before the time of decoupling the spectrum will be of a blackbody form:

$$\eta(\nu_d, t_d) = [exp(h\nu_d/kT_d) - 1]^{-1}.$$

After decoupling the photons will no longer interact with matter, and they will only experience the redshift due to expansion:

$$\frac{\lambda_d}{\lambda_o} = \frac{a(t_d)}{a(t_o)} = a(t_d) = \frac{\nu_o}{\nu_d}.$$

Then at the present time, t_o, the spectrum will be:

$$\eta(\nu_o, t_o) = [exp(h\nu_o/ka(t_d)T_d) - 1]^{-1}.$$

But $T_o = a(t_d)T_d$, so

$$\eta(\nu_o, t_o) = [exp(-h\nu_o/kT_o) - 1]^{-1}. \tag{29}$$

This is a blackbody spectrum at temperature T_o. Thus, the effect of the cosmic redshift is that the spectrum remains thermal, *i.e.* blackbody, but the temperature becomes $T_o = T_d/(1 + z_d)$. Note that this is just eq. (24) applied to the time of decoupling.

As we have noted above, $T_d \sim 4000K$, and $z_d \sim 1000$. Therefore, if one were to take the hot Big-Bang seriously, the universe should be filled with blackbody radiation at a temperature of just a few Kelvin. Such radiation would have a peak intensity at wavelengths of ~ 1 mm.

In the mid-1960's, Dicke and colleagues at Princeton thought along precisely these lines and decided to investigate the consequences of the hot Big-Bang model. They were in the process of completing construction of a receiver to search for such a radiation background, when it emerged that Penzias and Wilson at Bell Telephone Labs had detected such a radiation background by accident. They had found an irreducible noise level in an old Echo satellite horn antenna they were renovating for use in radio astronomy. The signal was distributed isotropically over the sky to a precision of $\sim 0.1\%$, and the intensity corresponded to a blackbody of temperature 3 ± 1 K. This signal was interpreted as the redshifted remnant of the primeval fireball in a hot Big Bang. The blackbody character and isotropy have since been determined more precisely and comprehensively, most recently and definitively by the Cosmic Background Explorer Satellite (COBE). The COBE spectrum for the CBR is shown in Fig. 3. The smooth curve fitting the points corresponds to a temperature of $T_o = 2.735 \pm 0.06$ K (Mather *et al.* 1990). This is obviously a most striking confirmation of the hot Big Bang model. There have been a variety of attempts to produce such a radiation background in the context of other cosmological models, but no other interpretation seems as plausible.

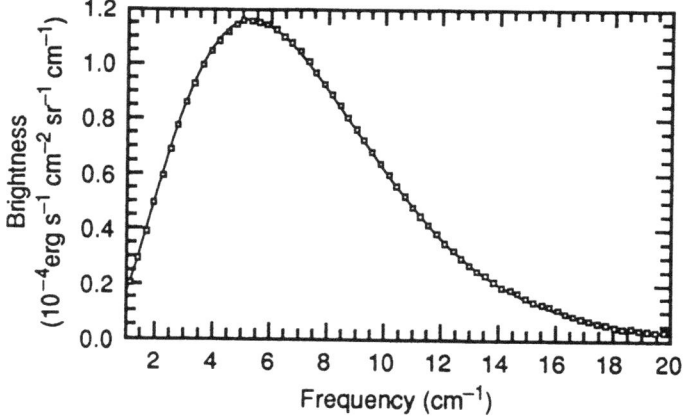

Figure 3: *The spectrum of the cosmic background radiation measured using COBE. The smooth curve is the Planck function which best fits the measured points, and it corresponds to a temperature of 2.735 K. (After Mather et al. 1990.)*

3.2 Hubble's Constant

To measure H_o one must find a way to measure distances to galaxies for which one has redshifts. A variety of distance indicators has been used—Cepheid variables, Type II supernovae, planetary nebulae, a proportionality between the absolute magnitude of a spiral galaxy and the width of the 21 cm H line (the "Tully- Fisher" relation), a relation between a characteristic size and the stellar velocity dispersion in elliptical galaxies, and other methods. The precision of the best of these should be $\sim 20\%$ in distance determination. However, as noted earlier, results have tended historically to cluster about $h \sim 0.5$ and $h \sim 1$, and so one does not know the value of H_o to within a factor of 2. The differences are due to technique and assumptions about extinction. Various biases in the samples used are also part of the reason for the discrepancies. This entire subject has been very thoroughly reviewed in Rowan-Robinson (1985). One of the major hopes for the Hubble Space Telescope is that observations with it may finally solve this problem. With its superior angular resolution it should be possible to resolve Cepheid variables in a number of galaxies at greater distance than has been possible with ground-based observations, and thereby to reveal where various potential sources of systematic error enter the distance ladder. Very recently a consensus has begun to emerge toward a value of $h \approx 0.8$ (*cf.* van den Bergh 1992). However, for the present purposes, it seems best to take $0.5 < h < 1$.

In principle, the Hubble diagram shown in Fig. 4 could provide a constraint on world models. This is a program which has been pursued most actively by Sandage and co-workers for the past 30-odd years. However, inspection of the figure shows that, while it provides strong confirmation of the Hubble law, no serious cosmological models are excluded; both closed and open models are consistent with the data.

3.3 Age of the universe

The most important consequence of the uncertainty in H_o is not how far away a galaxy with a given redshift is, but what it implies for the age of the universe. We saw in Section 2.6 that the age of an Einstein-de Sitter universe is $(2/3)H_o^{-1}$. The age of any cosmological model may be expressed as some function $f(\Omega_o)H_o^{-1}$. Thus:

$$t_o = f(\Omega_o)10^{10}/h \text{ yr.}$$

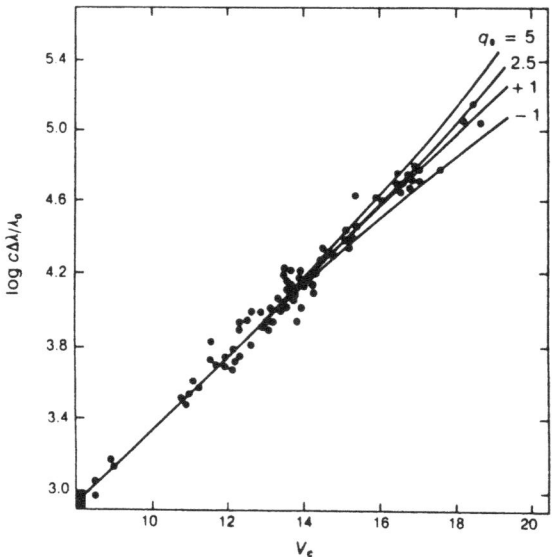

Figure 4: *The Hubble diagram, redshift plotted against corrected magnitude. Also shown are the curves expected for open and closed models. (After Bowers and Deeming 1984.)*

We know the ages of the oldest globular clusters in the Galaxy from their main sequence turnoffs; the ages of these clusters are thought to be 16 ± 3 Gyr. This is surely a lower limit to the age of the universe, and we can use it in the above expression to obtain an estimate of h.

The results are shown in Table I. Clearly the globular cluster ages favor a low value of the Hubble Constant. If h really is ~ 0.8 it is apparently just consistent with the cluster ages at the one sigma level. This may be a significant discrepancy.

Table I: Limits on Hubble Constant from ages of globular clusters.

Model	Ω_o	$f(\Omega_o)$	h
EdS	1	2/3	0.4 - 0.5
Open	$\ll 1$	1	0.5 - 0.8
Closed	10	0.35	0.2

3.4 Density parameter Ω_o

It is reasonable to try to measure the average mass density in the present-day universe by counting galaxies. The number density of galaxies times the mass of a typical galaxy should give an estimate of Ω_o. We will want to work in terms of Ω_o, and so we note that

$$\rho_o = \Omega_o \rho_c = 2 \times 10^{-26} h^2 \Omega_o \; kg \; m^{-3} = 2.9 \times 10^{11} \Omega_o h^2 \; M_\odot \; Mpc^{-3}.$$

Suppose we have space density of galaxies, n_G, each galaxy of luminosity L and mass M. Then

$$\rho_o = n_G \langle L \rangle \langle \frac{M}{L} \rangle.$$

We have written this expression in this form since we can observe the light from galaxies and use a typical M/L ratio characteristic of each type of galaxy. Note, however, that this involves the tacit assumption that light traces mass, and that is not necessarily true. However, measuring the light from galaxies we find that

$$n_G \langle L \rangle \approx 2 \times 10^8 h \; L_\odot \; Mpc^{-3}.$$

Then

$$\rho_o = 2 \times 10^8 h \langle \frac{M}{L} \rangle = 2.9 \times 10^{11} \Omega_o h^2$$

$$\Omega_o h = \frac{1}{1500} \langle \frac{M}{L} \rangle \frac{M_\odot}{L_\odot}.$$

In Table II below we list the M/L ratio per galaxy determined for different types of galaxies and in different environments, and the Ω_o implied by that value.

These imply $0.01 \leq \Omega_o \leq 1$, *if light traces mass*. The apparent trend that M/L increases with increasing scale size is taken by some to suggest that on the largest scales Ω_o may be ~ 1. On the other hand, where the data has best signal/noise it would appear that Ω_o is $\ll 1$.

Table II: Mass-to-light ratios and implied Ω_o.

Galaxy type	M/L per galaxy ($M_\odot/L_\odot$)	Ω_o
Spirals	$(9 \pm 1)h$	0.007
Ellipticals	$(10 \pm 2)h$	0.007
Galaxy pairs	$(80 \pm 20)h$	0.05
Clusters	$(500 \pm 200h$	0.3

3.5 Light Element Abundances

a) He^4 abundance. As was noted earlier in Fig. 2, $Y = m(He^4)/m(H)$ is insensitive to $\Omega_o h$, but that $Y \approx 0.25$ is consistent with hot Big-Bang nucleosynthesis. The best estimate of primordial He^4 abundance seems to be $Y = 0.230 \pm 0.004$ (Pagel 1992).

b) Deuterium abundance. $X(H^2) = m(H^2)/m(H)$ is found to vary but is always $> 2 \times 10^{-5}$. Since there are various mechanisms for destroying H^2 in stellar processing but not for making it, this abundance is taken to be at least a lower limit to the primordial value. From nucleosynthesis calculations (*cf.* Fig. 2) this implies that

$$\Omega_o h^2 < 0.035 \text{ (in baryons)}.$$

The parenthetical qualification "in baryons" is to anticipate the possibility that not all the mass in the universe is in baryonic form. In particular, there is currently much speculation that dark matter may be in the form of non-baryonic particles frozen out in the Big Bang. In any case, since the evidence cited above indicates that $h > 1/2$, this suggests that $\Omega_o(\text{baryonic}) < 0.14$.

This, along with the ages of globular clusters, is one of the most important observational constraints on cosmological models, and suggests strongly that the universe is open, unless there is a significant (dominant, actually) mass in non-baryonic form.

3.6 Conclusions

In summary, we have found:

Hubble constant: $0.5 < h < 1$.

Ages of globular clusters: $h < 0.6 \pm 0.1$.

Mass/light ratios: $0.01 < \Omega_o(\text{total}) < 0.5$ (if light traces mass).

Deuterium: $\Omega_o(\text{baryons})h^2 < 0.035$.

Taking these results at face value, it would be difficult to avoid the conclusion that we have an open universe if it were not for the evidence that Ω_o gets close to unity in measurements sensitive to dark matter on the largest spatial scales. The dynamical measurements of mass show that there is substantial dark matter which probably dominates the mass of the universe. The limits on mass in baryons from the deuterium abundance suggests that much of the dark matter must be in non-baryonic form. But there is a variety of intriguing ways to speculate about these constraints. Such speculation, and what it may imply about the actual properties of the universe is part of the fun of cosmology. One's hunches and philosophical prejudices will suggest the sorts of observation which should be pursued.

4   GALAXY FORMATION

4.1   Introduction - The Problem

Hot isotropic models are obviously a good first approximation to describing the universe we observe. They account for the Hubble law, the 2.7 K CBR and its isotropy, and even give roughly the right abundances of the light elements. However, such smooth models cannot account for the development of structure in the universe. This requires introduction of perturbations on the isotropic models, and working out how such perturbations can grow and develop in an expanding universe to form the large-scale structure (galaxies and clusters of galaxies) we observe. We will now outline how this might proceed.

But before introducing perturbations into isotropic models, it is worth asking what might be the source of the original perturbations? One possibility is Poisson statistical fluctuations in the number of particles in any given volume in the early universe. The number of baryons in a galaxy, N_b, is $\sim 10^{69}$. If Poisson statistical fluctuations are the underlying source of the fluctuations leading to galaxy formation, then in the early universe:

$$\frac{\delta N_b}{N_b} \approx \frac{N_b^{1/2}}{N_b} = N_b^{-1/2} \approx 10^{-35}$$

Comparing the density contrast of a galaxy at present, $d\rho/\rho \sim 10^6$, shows that a Poisson fluctuation must grow by a factor of $\sim 10^{41}$ in $10^{10}yr$ to make a galaxy. This is a bit daunting, as we will see below.

4.2 Growth of a fluctuation in an expanding universe

The basic idea is that the large-scale structure in the universe has grown by gravitational instability from small fluctuations present early in the evolution of the universe. To keep things simple, we will work in the Newtonian approximation and in the linear regime. Imagine a spherical fluctuation in the cosmic fluid (illustrated in Fig. 5) of uniform density

$$\rho = <\rho>(1+\delta) \tag{30}$$

and size

$$r = a(t)(1+\epsilon)r_o. \tag{31}$$

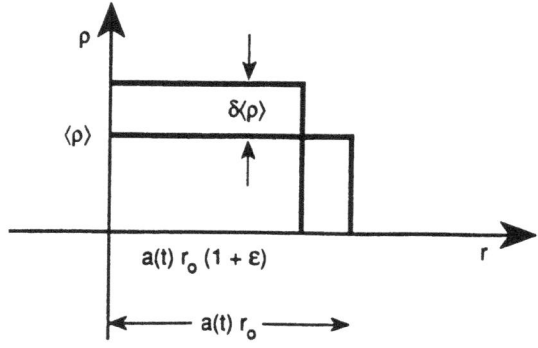

Figure 5: *A small fluctuation in the cosmic fluid.*

We wish to inquire how this density fluctuation, δ, evolves. Does it grow? Can it become a mass overdensity on the scale of a galaxy or cluster of galaxies?

To work out how δ evolves in an expanding universe we will take a linear approximation δ and $\epsilon \ll 1$, and consider the matter-dominated case, $p = 0$. Then the continuity equation (7) gives

$$\langle \rho \rangle r^3(t) = \langle \rho \rangle a^3(t) r_o^3 = \langle \rho \rangle (1 + \delta)[a(t)(1 + \epsilon)r_o]^3$$

$$\langle \rho \rangle a^3(t) r_o^3 \approx \langle \rho \rangle a^3(t) r_o^3 (1 + \delta)(1 + 3\epsilon).$$

Then

$$1 \cong 1 + \delta + 3\epsilon$$

and

$$\delta = -3\epsilon. \tag{32}.$$

As expected, if the density increases, the size decreases.

What is the dynamical evolution of such a fluctuation? For a spherical perturbation, the mass outside the perturbation does not affect the dynamics of the perturbation iself, and so the perturbation is just like in small version of the isotropic universe whose dynamics we have already worked this out. The result, of course, is the Friedmann equation (11):

$$\ddot{a} = -\frac{4\pi G}{3} \rho a.$$

Substituting $r(t) = a(t)r_o$, we get

$$\ddot{r} = -\frac{4\pi G}{3} \rho r.$$

Then substituting eq.(31):

$$\frac{d}{dt}[\dot{a}(1 + \epsilon) + a\dot{\epsilon}] = -\frac{4\pi G}{3} \langle \rho \rangle (1 + \delta)(1 + \epsilon)a$$

$$\ddot{a}(1 + \epsilon) + \dot{a}\dot{\epsilon} + \dot{a}\dot{\epsilon} + a\ddot{\epsilon} \approx -\frac{4\pi G}{3} \langle \rho \rangle (1 + \delta + \epsilon)a.$$

Robert Joseph

But

$$\ddot{a}(1 + \epsilon) = -\frac{4\pi G}{3}\langle\rho\rangle(1 + \epsilon)a$$

is just the Friedmann equation times $(1+\epsilon)$, so these two terms may be cancelled from the above expression, leaving:

$$a\ddot{\epsilon} + 2\dot{a}\dot{\epsilon} = -\frac{4\pi G}{3}\langle\rho\rangle a\delta.$$

Substituting eq. (32) this becomes

$$\ddot{\delta} + 2\frac{\dot{a}}{a}\dot{\delta} = +4\pi G\langle\rho\rangle\delta. \tag{33}$$

This is the equation we seek; it may be solved to find the time-evolution of a density fluctuation. We now solve this equation for some special cases.

a) No expansion in the universe. Then $\dot{a}/a = 0$ and we get

$$\ddot{\delta} - 4\pi G\langle\rho\rangle\delta = 0$$

which has the solution

$$\delta(t) = \delta(t_o)exp(\sqrt{4\pi G\langle\rho\rangle}t). \tag{34}$$

As we might have expected, the fluctuation grows exponentially with time. If there were no expansion, it should be easy to make galaxies from infinitesimal fluctuations early in the universe.

b) An EdS universe. Using our results for the EdS universe gives:

$$\frac{\dot{a}}{a} = \frac{H_o}{a^{3/2}} = \frac{2H_o}{3H_o}t^{-1} = \frac{2}{3}t^{-1},$$

and

248

$$\rho(t) = \frac{\rho_o}{a^3(t)} = \frac{1}{6\pi G}t^{-2}.$$

Then eq. (33) becomes

$$\ddot{\delta} + \frac{4}{3}\frac{1}{t}\dot{\delta} - \frac{2}{3}\frac{\delta}{t^2} = 0.$$

We take as a trial solution:

$$\delta = t^\alpha.$$

Substituting, we find:

$$\alpha(\alpha-1)t^{\alpha-2} + \frac{4}{3}\alpha t^{\alpha-2} - \frac{2}{3}t^{\alpha-2} = 0$$

$$3\alpha^2 + \alpha - 2 = 0$$

$$(3\alpha - 2)(\alpha + 1) = 0.$$

The solution, therefore, is

$$\delta(t) = At^{2/3} + Bt^{-1} \tag{35}$$

The first term is obviously a growing mode, whereas the second decays away. Thus an overdensity $\delta = \delta\rho/\rho$ does grow, but only at an algebraic rate, rather than at an exponential rate, due to the expansion.

Since

$$a_{EdS}(t) \propto t^{2/3} \propto \frac{1}{1+z},$$

we find

$$\delta(t) \propto \frac{1}{1+z} \tag{36}$$

for an Einstein-de Sitter universe.

c) Radiation-dominated case. This is easy to solve in the same way. The solution is

$$\delta(t) = At + Bt^{-1}. \tag{37}$$

Since

$$a_r(t) \propto t^{1/2}$$

$$\delta(t) \propto a^2(t) \propto \frac{1}{(1+z)^2}. \tag{38}$$

Growth is therefore still algebraic and slow, even in the early universe. Clearly the expansion of the universe poses significant problems for growth of fluctuations to form large-scale structure.

4.3 Jeans mass

We have so far looked at growth of fluctuations in the absence of pressure forces; these might be expected to slow growth even further. We now investigate the effects of internal pressure in an overdense region.

Consider a spherical fluctuation in the cosmic fluid of radius R. A mass element at R will feel a tendency to collapse due to self-gravity:

$$\frac{F_{grav}}{\text{unit volume}} = -\frac{GM(R)\rho}{R^2} \sim -G\rho^2 R.$$

This will be opposed by internal pressure: $\sim p/R$.

For large R, gravity will obviously win over pressure, and one gets collapse. We define the Jeans radius, R_J, as the radius when these two forces are equal:

$$\frac{p}{R_J} \approx G\rho^2 R_J.$$

Rewriting this in terms of the sound speed in the fluid,

$$c_s = \sqrt{dp/d\rho} \approx \sqrt{p/\rho}$$

gives an approximate expression for the Jeans radius:

$$R_J = c_s\sqrt{1/G\rho} \quad \text{(approximate)}. \tag{39}$$

Equating thermal energy and gravitational energy or sound-crossing time and free-fall time give same result.

A more rigorous derivation of the Jeans radius gives

$$R_J = c_s\sqrt{\frac{\pi}{G\rho}}. \tag{40}$$

The Jeans mass is then

$$M_J = \rho\frac{4}{3}\pi R_J^3$$

$$M_J \approx \frac{c_s^3}{G^{3/2}\rho^{1/2}}. \tag{41}$$

This is the minimum mass of a fluctuation which can grow by gravitational instability. Smaller masses will be dissipated by internal pressure.

4.4 Growth of fluctuations prior to decoupling

a) Types of fluctuations.

i) Isothermal fluctuations

These are fluctuations in the matter density, while the radiation density suffers no corresponding fluctuation: $\delta\rho_m \neq 0$, $\delta\rho_r = 0$. In this case a particle in an overdensity moving with velocity, v, relative to the co-moving frame feels a radiation drag proportional to $T^4 v$. Thus isothermal fluctuations are frozen due to photon drag; prior to decoupling they do not grow.

ii) Adiabatic fluctuations

In an adiabatic fluctuation matter and radiation vary in such a way that there is no net change in energy in the fluctuation:

$$\frac{\delta\rho_m}{\rho_m} = \frac{3}{4}\frac{\delta\rho_r}{\rho_r}.$$

In such a fluctuation photons will tend to have random-walked out of an overdensity by the end of the pre-decoupling era. This process, called "Silk damping," operates on mass scales less than:

$$M_D^{Silk} \sim 10^{13}\Omega_o^{-5/4}M_\odot. \tag{42}$$

Thus there is no fluctuation on mass scales less than the Silk mass which will survive the pre-decoupling era.

b) Mass in particle horizon

This is the mass in a region which is causally connected. Forces (like gravity) can't act over a larger region than $r \approx ct$.

$$M_H = \rho\frac{4}{3}\pi r^3$$

$$= \frac{\rho_c\Omega_o c^3 t^3}{a_r^3(t)}.$$

But

$$a_r(t) = \frac{1}{1+z} = \frac{T_o}{T_r}.$$

Thus,

$$M_H \approx \frac{c^3 T_o^3}{GH_o\Omega_o^{1/2}T_r^3}. \tag{43}$$

c) Evolution of Jeans mass

In a photon gas the sound speed $c_s \approx \sqrt{p/\rho} = c\sqrt{3}$. Then

$$M_J \approx \frac{c^3 T_o^3}{GH_o \Omega_o^{1/2} T_r^3}. \tag{44}$$

Putting in numerical values we find

$$M_J \approx \frac{10^{31} M_\odot}{T_r^3(K)}. \tag{45}$$

Note that the expressions for horizon mass and Jeans mass are identical. When the numerical factors are included it turns out that the baryon Jeans mass is a factor of about 30 larger than the horizon mass. The conclusion is, therefore, that baryonic fluctuations are outside the horizon and cannot grow by gravitational instability in the pre-decoupling era (at least in this simple Newtonian approximation).

4.5   Post-recombination evolution

Now the cosmic fluid behaves as a monatonic gas:

$$c_s^2 = \gamma \frac{kT_m}{m_H},$$

where

$$\gamma = c_p/c_v = 5/3.$$

In this case

$$\frac{c_s(\text{matter} - \text{dominated})}{c_s(\text{radiation} - \text{dominated})} \sim 10^{-4}.$$

Thus, at the epoch of decoupling M_J plunges by a factor of $\sim 10^{-12}$. It is easy to show, using the relations for the adiabatic expansion of an ideal gas, that the Jeans mass is now

$$M_J \approx 10 T_r^{3/2}(K) \, M_\odot. \tag{46}$$

At decoupling the Jeans mass is $\sim 10^6$ M$_\odot$. Figure 6 shows the evolution of the Jeans mass as a function of the radiation temperature pre- and post-decoupling.

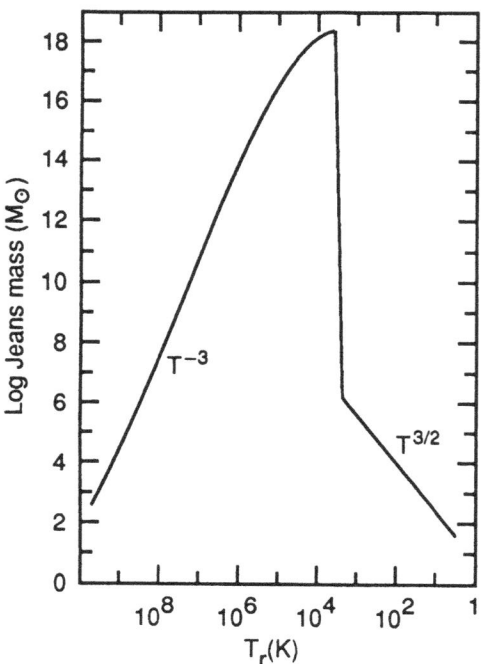

Figure 6: *Evolution of the Jeans mass as a function of the cosmic background radiation temperature. Note the steep drop at the time of decoupling (recombination). (After Narlikar 1983.)*

4.6   Scenarios for galaxy formation

Since we know how fast an overdensity grows in the post-decoupling epoch ($\delta\rho/\rho \propto 1/(1+z)$), we can work backward from any present-day galaxy and determine what its density contrast $\delta\rho/\rho$ must have been at the redshift of decoupling. For a galaxy to have formed, $\delta\rho/\rho \sim 1$ at $z = 1$ (with rapid non-linear growth thereafter), so at $z \sim 10^3$, $\delta\rho/\rho$ must have been $\sim 10^{-3}$.

This result, combined with the conclusions reached above that baryonic fluctuations cannot grow in the pre-decoupling era, suggests that the fluctuations which did lead to formation of galaxies were present at finite amplitudes at the time of decoupling and galaxies did not form from infinitesimal baryonic perturbations present early in the Big-Bang. There are a variety of proposals for such pre-galactic "seeds," all of which are highly speculative (*cf.* Kolb and Turner (1990)).

However, there is another way to solve this problem. Non-baryonic matter may dominate the mass density of the universe, as we have discussed above. Such particles will generally decouple from the radiation earlier than baryons. When such particles decouple from the radiation, they will then freely stream out of any putative overdensity, thereby damping it out. There will then be a corresponding damping mass scale which determines the smallest masses of the first fluctuations which grow in this fluid.

One general class of such particles is called "hot dark matter." Such particles decouple from the radiation very early when the temperature is high. For example, if there are neutrinos with mass of 30 eV, the corresponding damping scale is $\sim 10^{15}$ $M_{\odot}$. "Cold dark matter" particles, on the other hand, are those for which such free-streaming is not important, *i.e.* the damping scale is $< 10^8$ $M_{\odot}$. Super-symmetric particles such as axions, photinos, gravitinos, or black holes, would be examples of cold dark matter.

The importance for galaxy formation of non-baryonic matter as the dominant gravitating matter in the universe is that fluctuations in non-baryonic matter can grow during the pre-decoupling era, subject to the damping constraints mentioned. Then in the post-decoupling era baryons may rapidly fall into the now-deep potential wells produced by the growth of fluctuations in the non-baryonic material, producing large-scale structure on whatever mass scales the non-baryonic matter had. There are two general classes of such scenarios.

"Top down" scenario.

One assumes a fluctuation begins with very large mass, $M \sim 10^{15} M_{\odot}$. This is the mass scale of clusters of galaxies. The idea is that as such a fluctuation collapses and cools, it will fragment to form its constituent galaxies, which in turn collapse and fragment to form stars. Dissipative collapse of such structures makes "Zeldovich pancakes," one of the classic models in the theory of galaxy formation. This scenario for galaxy formation is associated with adiabatic fluctuations and hot dark matter. Numerical simulations of structure formation in a massive-neutrino-dominated universe have been performed (*e.g.* Centrella and Melott 1982), and these show some features similar to those observed.

"Bottom up" scenario.

In this approach, overdensities of mass $\sim 10^6 M_{\odot}$ form (*i.e.* globular cluster scale masses), and then gravitational clustering operates to make galaxies and then clusters of galaxies. Isothermal fluctuations and cold dark matter are associated with the "bottom up" scenario. The cold dark matter scenario has enjoyed considerable popularity, in part due to the rather impressive numerical simulations performed by Davis, Efstathiou, Frenk, and White ((1985). These simulations are able to reproduce a number of the observed features of large-scale structure.

4.7   Conclusions

It must be clear from this very cursory introduction that we really know very little about how galaxies formed. Of course, this makes the subject very attractive: there is so much that is fundamental still to br done. It has also become evident that there is the intimate connection between physics at the highest energies and therefore smallest spatial scales, and astrophysics at the largest spatial scales. This is one of the most beautiful intellectual results in natural science. One may actually learn about ultra-high-energy physics (for example about the sorts of particles that dominate the mass density of the universe) from astronomical observations of large-scale structure in the universe.

5   REFERENCES

Bowers, R. and Deeming, T.: 1984, *Astrophysics II: Interstellar Matter and Galaxies*, Jones and Bartlett Publishers. Boston.

Centrella, J. and Melott, A.: 1982, *Nature* **305**, 196

Davis, M., Efstathiou, G., Frenk, C. and White, S.D.M.: 1985, *Astrophys. J.* **292**, 371.

Jones, B.J.T.: 1992, in *Observational and Physical Cosmology*, p. 175. Eds. Sanchez, F., Collados, M. and Rebolo, R. Cambridge University Press. Cambridge (UK).

Kolb, E.W. and Turner, M.S.: 1990, *The Early Universe*, Addison-Wesley,. Redwood City, California.

Longair, M.S.: 1977, in *Infrared Astronomy*, p. 199. Eds. Setti, G. and Fazio, G. G. Reidel. Dordrecht, Holland.

Mather, J.C., *et. al.*: 1990, *Astrophys. J.* **354**, L37.

Narlikar, J.V.: 1983, *Introduction to Cosmology*, Jones and Bartlett Publishers. Boston.

Pagel, B.E.J.: 1992, in *Observational and Physical Cosmology*, p. 117. Eds. Sanchez, F., Collados, M. and Rebolo, R. Cambridge University Press. Cambridge (UK).

Peebles, P.J.E.: 1971, *Physical Cosmology*, Princeton University Press. Princeton.

Peebles, P.J.E.: 1980, *The Large-Scale Structure of the Universe*, Princeton University Press. Princeton.

Primack, J.R.: 1984, SLAC-PUB-3387, preprint.

Rees, M.J.: 1978 in *Observational Cosmology*, p. 261. Eds. Maeder, A., Martinet, L. and Tammann, G. Geneva Observatory. Switzerland.

Rowan-Robinson, M.: 1985, *The Cosmological Distance Ladder*, Freeman. New York.

Schramm, D.N. and Wagoner, R.V.: 1974, *Physics Today*, **27**, No. 12, p. 40.

van den Bergh, S.: 1992, *Science*, **258** 421.

White, S.D.M.: 1990 in *Physics of the Early Universe*, p. 1. Eds. Peacock, J.A., Heavens, A.F., and Davies, A.T. Scottish Universities Summer Schools in Physics. Edinburgh (UK).

G25.5+0.2, a New Ring Nebula Around a Luminous Blue Star: Case Study for the Importance of IR Observations

E.E. Becklin

Department of Astronomy,
University of California,
Los Angeles, U.S.A.

1 ABSTRACT

We present infrared images and infrared spectroscopy of the suspected very young supernova remnant G25.5+0.2 first detected at radio wavelengths. The 2.2 µm image exhibits a similar nebular structure to that seen in the radio. Spectroscopic measurements at 2.17 µm show a Brackett gamma line that has the line-to-continuum ratio expected from ionized hydrogen at a temperature near 10,000 K. At 50 km/s resolution, the line is resolved with a FWHM of 200 km/s. Ten micron photometry clearly establishes a connection between G25.5+0.2 and the IRAS source 18344-0632. Most important, the infrared image reveals a point source in the center of the nebula that has properties of a blue luminous star that could excite an ionized ring nebula with the observed radio properties. The supernova hypothesis is ruled out and G25.5+0.2 is almost certainly a ring nebula around a mass losing luminous blue star ~13 kpc distant and reddened by 20 magnitudes of visual extinction.

2 INTRODUCTION

Cowan et al. (1989) reported extensive observations of the galactic radio source G25.5+0.2. These included radio continuum images at four frequencies, 21 cm absorption measurements, and a search for radio recombination line emission. Based primarily on the absence of the H76α recombination line, Cowan et al. concluded that G25.5+0.2 is a very young galactic supernova remnant, perhaps only 25 years old. Based on 21 cm HI absorption lines, Cowan et al. give a minimum distance to G25.5+0.2 of 7.2 kpc. White and Becker (1990) and Green (1990) suggest that identification of G25.5+0.2 as a planetary nebula is more likely, based on the radio observations and IRAS fluxes. A third model is proposed by Zijlstra (1991) who suggests that it possibly belongs to a group of young outflow objects. We present infrared (1 to 2.5 µm) imaging and spectroscopy that demonstrates that G25.5+0.2 surely is not a very young supernova remnant, but is rather a ring nebula around a massive blue star that is in a mass loss stage.

3 OBSERVATIONS

3.1 Near Infrared Imaging

Infrared images of G25.5+0.2 were obtained at Mauna Kea Observatory in the photometric bands at J (1.25 µm), H (1.65 µm) and K (2.2 µm) using IRCAM on UKIRT (McLean 1987) in July 1989. The images were bias-subtracted and flat-fielded using a median sky flat. Standard stars and faint white dwarfs of known magnitude were measured at J, H and K to obtain the flux calibration. The overall uncertainty in the fluxes is estimated at 10%. Figure 1 is a contour plot of a 180-second exposure taken with a 0.62 arcsec per pixel scale. The contours show extended emission on the same scale as the radio source shown in Figures 1 and 2 of Cowan et al. (1989); the positional agreement (± 2 arcsec) and the structural agreement assure physical association. At the very center of the extended 2.2 µm emission is a very red, unresolved source that we assume is a star. Photometry at J, H, and K of the star with the nebula subtracted is given in Table 1.

Band	λ (μm)	Observed (mag)	Reddening Corrected (mag)
J	1.25	16.5	10.3
H	1.65	13.7	9.9
K	2.2	12.6	10.4

Table 1: *Photometry of Central Star of G25.5+0.2*

3.2 Infrared Spectroscopy

Spectra of G25.5+0.2 in the atmospheric windows at 1.6 and 2.2 μm wavelengths were obtained on the UKIRT in August 1989 with a circular variable filter (CVF) with a resolution of 1% and an aperture of diameter 12 arcsec on the sky. The spectral region covered contains an FeII line at 1.64 μm and the Bγ line at 2.17 μm. UKIRT spectra of G25.5+0.2 and the planetary nebula NGC 6572 are shown in Figure 2. The spectra have been calibrated using the Bright Star HR 6698. The spectrum of G25.5+0.2 shows a strong Bγ line and a red continuum. A high-resolution Bγ spectrum of G25.5+0.2 was obtained on UKIRT in September 1991 using CGS4 and a resolution of 40 km/s. The 3 arcsec wide North-South slit was placed near the center of the nebula. Bγ emission was seen in a region 15 arcsec wide in declination. A central spectrum is shown in Figure 3 along with a spectrum of the planetary nebula NGC 6572. The full width of the line from G25.5+0.2 is approximately 300 km/s. The LRS central velocity of the emission is near +20±20 km/s.

3.3 Ten Micron Photometry

In August 1989, the IRTF bolometer with a 7.2 arcsec focal plane aperture centered on the radio source, was used to measure a 10 μm flux of 1.6 Jy through a broadband N filter ($\Delta\lambda = 5$ μm). A five-point map on an 8 arcsec grid showed that the 10 μm flux peaks on the center of the radio source.

3.4 CO (3-2) Observations

Drs. N. Evans and D. Jaffe very kindly obtained CO (J=3-2) line observations of G25.5+0.2 with the 10-m CSO at Mauna Kea Observatory on 4 November 1989. A five-point map on a 30" grid with a 20" diameter beam showed a 2 to 4 K CO emission line at all five positions. The line has a width of about 15 km/s and a central LSR velocity of +106 km/s.

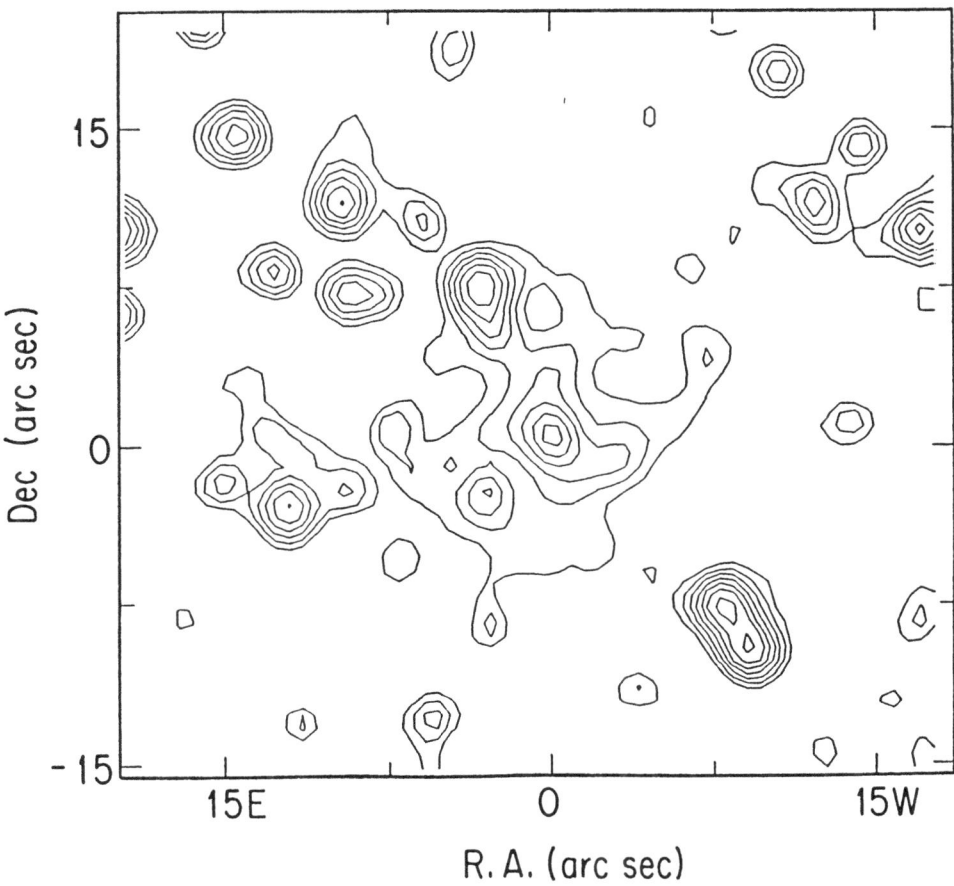

Figure 1: *A 2.2 μm (K) contour plot centered on G25.5+0.2 obtained on the UKIRT with IRCAM. The contour intervals are 0.4 magnitudes (factor of 1.45). The lowest contour is K~ 16.0 mag per square arcsec. The scale is 0.62 arcsec per pixel, the InSb detector array had 58 x 62 pixels, and the integration time was 180 seconds. Similar images were taken at 1.65(H) and 1.25(J) microns. The reddest object is a point source in the center of the extended emission at position 0,0 on the map. The position 0,0 corresponds to RA = 18h34m23.9s ± 0.2s, DEC = -06°32'13" ± 2" (1950)*

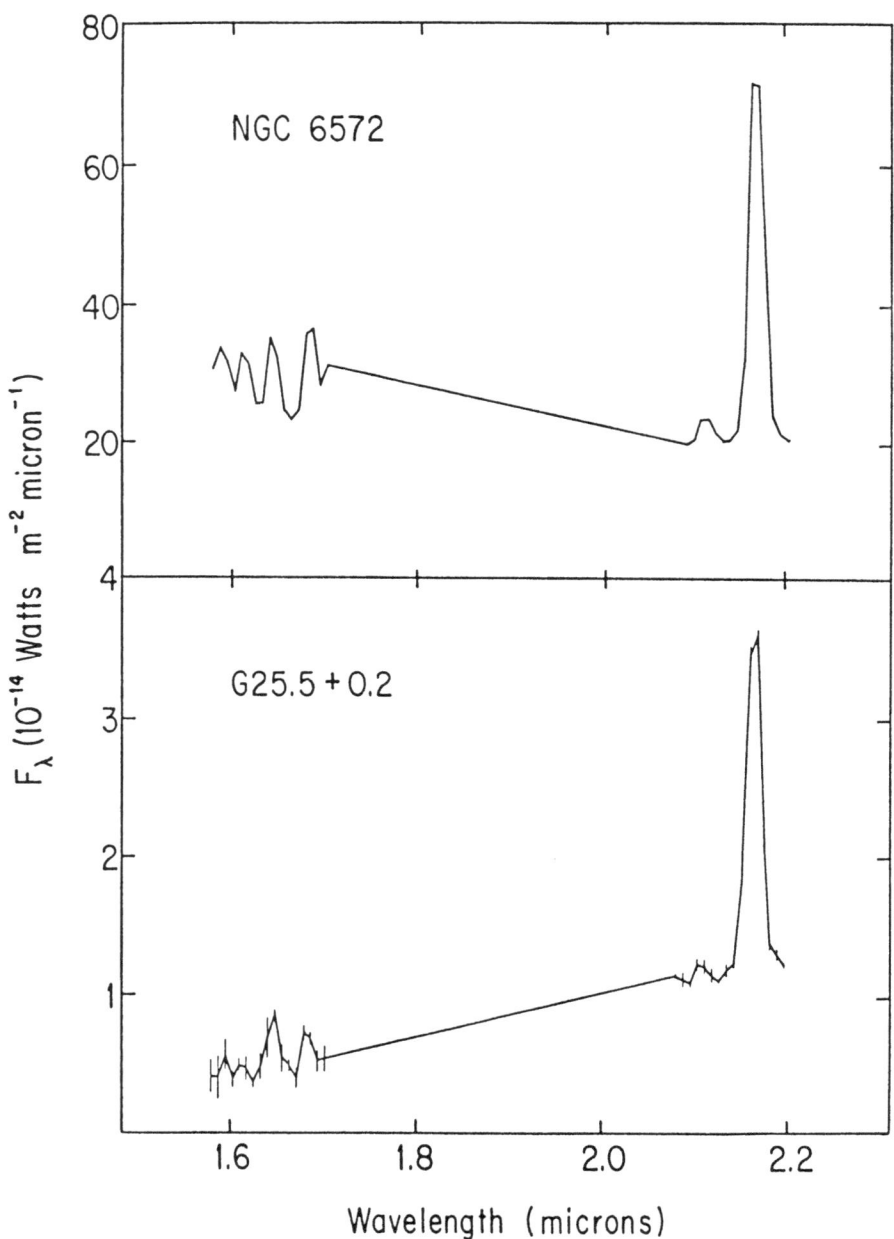

Figure 2: *A CVF spectrum of G25.5+0.2 obtained with UKT-9 on the UKIRT. The focal plane diaphragm had a 12-arcsec diameter and the spectral resolution was $\Delta\lambda/\lambda$ = 0.01. Also shown for comparison is the planetary nebula NGC 6572. Brackett lines are seen at 2.17, 1.68, 1.64, 1.61 and 1.59 μm. An FeII line seen to be very strong in supernova remnants is also at 1.64 μm.*

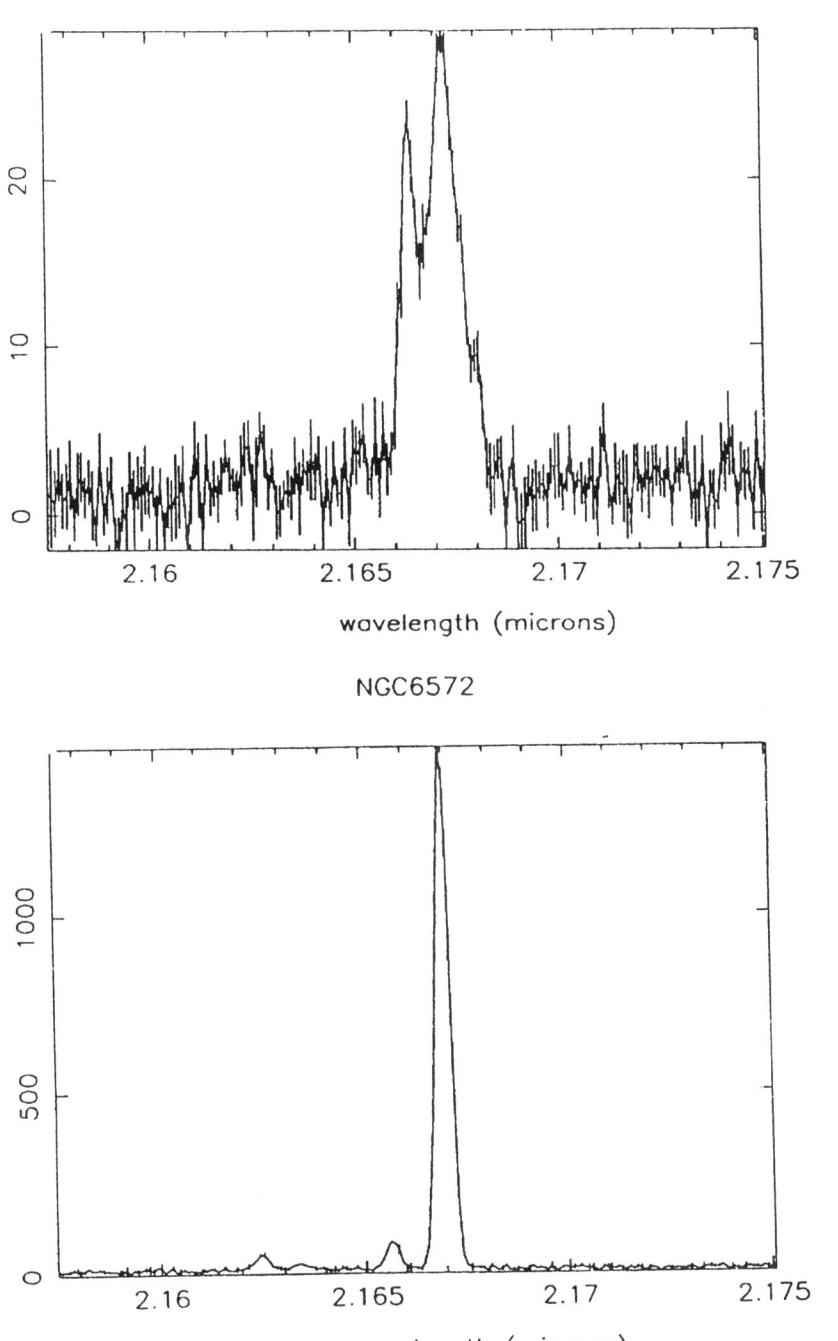

Figure 3: *A CGS4 spectrum of Brackett-γ taken at a resolution of 40 km/s. The 3 arcsec wide slit was placed near the center of G25.5+0.2 with the long axis North-South. The spectrum shown corresponds to a 3×3 arcsec region approximately 2 arcsec North of the point source. Also shown is a spectrum of the planetary nebula NGC6572.*

4 RESULTS

4.1 Distance

Cowan et al. (1989) give a minimum distance to G25.5+0.2 of 7.2 kpc based on HI absorption lines. The average LSR velocity of the Bγ lines shown in Figure 3 has a value of 20±20 km/s. Using these two facts together with the galactic rotation mode in Figure 4 of Kerr and Westerhout (1965) and a Galactic Center distance of 8.2 kpc gives a distance to G25.5+0.2 of 13.5±1 kpc. We will adopt this distance for the remainder of the paper.

4.2 Reddening Due to Interstellar Dust

Before proceeding, it is necessary to determine the interstellar extinction toward G25.5+0.2. This can be estimated in several ways. First, we assume that the point source in the center of the nebula is associated with it and has the infrared colors of a star. Below we show that the point source is most likely to be an early supergiant star. Using the reddening law of Becklin et al. (1978) given in the Table 2 and the intrinsic colors for an early type star given in Koornneeff (1983) H-K=-0.1, J-H=-0.1, we find a moderately good fit to the observed colors with A_K=2.2 mag or A_v=24 mag.

λ (μm)	A_λ/A_v
1.25	0.26
1.65	0.16
2.2	0.09
3.5	0.05

Table 2: *Interstellar Reddening*

Second, we can assume that the 1.6-2.2 μm spectral shape of G25.5+0.2 and NGC 6572 are intrinsically the same because of the similarity in the line-to-continuum ratio (Figure 2). The observed difference we ascribe to reddening; again using the Becklin et al. (1978) extinction curve, we derive A_K=1.8 mag. Finally, because we show below that the radio source is certainly a region of ionized gas, we may follow Wynn-Williams et al. (1978) and compare radio flux with the strength of the observed Bγ line in Figure 2, corrected for aperture size [Bγ = 8×10^{11} S_v(6cm)]. If T_e=10^4 K, then A_K is 1.6 mag. Below, we assume that the extinction to G25.5+0.2 is A_K=2.2 mag and that the extinction to the star is the same as to the nebula.

4.3 The Central Star

To characterize the unresolved point source in the center of the nebula, we must correct the measured magnitudes (Table 1) for reddening. With an extinction of $A_K = 2.2$ mag and the reddening curve of Becklin et al. (1978), the apparent magnitudes are given in column 3 of Table 1. From the distance of 13.5 kpc, the absolute K magnitude of the central star is $M_K = -5.3$.

4.4 The Ionized Gas

The Brackett gamma line shown in Figure 2 has the correct line-to-continuum ratio to be produced by a region of ionized gas (e.g., Wynn-Williams et al. 1978, Willner et al. 1979; and this paper, Figure 2). The measured line strength, corrected for $A_K = 1.8$ mag of reddening, can be converted to an expected free-free radio flux following Wynn-Williams et al. (1978). From Figure 2 we estimate a $B\gamma$ line strength of 5×10^{-16} Wm^{-2} in a 12 arcsec beam. Correcting for reddening of a factor of five, and flux outside the 12 arcsec diameter beam, estimated from the radio map to be about 20%, we derive a radio free-free flux of .37 Jy. This should be compared to the measured radio flux of .34 Jy (Cowan et al. 1989) and strongly suggests that the radio flux is predominantely free-free emission. Using the formulae summarized in Genzel et al. (1982) we can calculate the number of ionizing photons and the Lyman alpha line flux.

$N_L = 7.54 \times 10^{46} \, S_v D^2$ (photons/s)

L(Lyman alpha) $= 310 \, S_v D^2$ ($L_\odot$)

where S_v is the free-free 6 cm radio flux in Janskys and D is the distance in kpc.

We find that $\log N_L = 48.7$ and L(Lyman alpha) $= 2 \times 10^4 \, L_\odot$.

4.5 Dust Emission From the Nebula

The 10 μm flux of 1.6 Jy measured with a 7.2" beam centered on the nebula demonstrates that thermal emission from dust originates in and around G25.5+0.2 since free-free and line emission would be less than 0.2 Jy (Willner et al. 1972). If the emission originates from a region that has the same shape as the radio and 2.2 μm sources, then the total 10 μm emission from the nebula is about 4 Jy. This may be compared to the 12 μm flux of 6 Jy from the IRAS source 18344-0632 in the approximately one arcmin IRAS beam. Considering the shape of the spectrum and likelihood of strong interstellar silicate absorption at 9.5 μm, the agreement of the two measurements establishes that the flux from IRAS 18344-0632 is associated with G25.5+0.2. Without the benefit of ground-based 10 μm measurements, White and Becker (1990) independently deduced that IRAS 18344-0632 should be identified with G25.5+0.2.

The IRAS fluxes from G25.5+0.2 are given in Table 3. The second column gives the results from the IRAS Point Source Catalog (PSC). The third column gives the IRAS flux from "Add Scans" obtained by Dr. G. Hawkins (1990, private communication) for the best fit point source. The energy distribution is characteristic of a thermal source associated with ionized hydrogen (eg. Wynn-Williams and Becklin 1974). Integration of the total IRAS flux using an algorithm similar to that given is Appendix B of the IRAS Cataloged Galaxies (Lonsdale et al. 1985),

$$L= [6F_V(12) + 2F_V(25) + F_V(60) + 0.4F_V(100)] \times D^2,$$

where F_V is in Jy, D is in kpc and L in $L_\odot$, gives $L_{IR} \sim 10^5 \, L_\odot$.

$\lambda(\mu m)$	PSC	Point Add Scans
12	6	8
25	85	89
60	144	195
100	<466	270

Table 3: *IRAS fluxes (Jy)*

5. DISCUSSION

5.1 G25.5+0.2 is not a supernova remnant.

Four strong pieces of evidence indicate that G25.5+0.2 is not the remnant of a recent supernova. First, the infrared spectrum in Figure 2 is not what one would expect from a supernova remnant. Graham et al. (1990) have shown that the Crab Nebula and the Cygnus Loop emit very weak Bγ and strong FeII (1.64 μm) with Bγ/Fe=0.02. The spectrum in Figure 2 shows strong Bγ and FeII (1.64 μm) weak or absent with Bγ/Fe > 5. In fact, as noted in Section 4.4, the Bγ line-to-continuum ratio and the Bγ to radio flux ratio are as expected from a region of ionized gas with an electron temperature of 10^4K. Second, and even more important, the Bγ line width shown in Figure 3 is 300 km/s, whereas a supernova should have a line width of many thousands of km/s (Cowan et al. 1989).

Third, the point source seen directly in the center of the nebula would not be expected from a 25-year-old supernova remnant. The slowest decay for the central star of a supernova is ≥ 2 mag/year (Weiler and Sramek 1988). Thus, the observed M_K=-5.3 mag is inconsistent with that expected for a 25-year-old supernova. Below we show that the chance alignment of a foreground star is very small.

268

Fourth, the strong thermal dust emission associated with G25.5+0.2 is not expected from a 25-year-old supernova. Although thermal dust emission has been detected from supernovae a year after maximum (e.g., Dwek 1983), the decay rate for the dust emission should follow the decay of the optical light of at least 2 mag/year. After 25 years, this would put the expected thermal emission well below the minimum observed luminosity of 10^5 L$_\odot$. Further, as pointed out by White and Becker (1990), the ratio of infrared radiation at 60 µm to radio flux at 6 cm in G25.5+0.2 is much greater than that observed in other supernovae remnants or expected theoretically (Dwek 1987). However, the observed 60 µm to 6 cm flux ratio is the value expected for a region of ionized gas with a small amount of dust heated by trapped Lyα radiation (e.g., Wynn-Williams and Becklin 1974; Harwit et al. 1972). In conclusion, the infrared properties of G25.5+0.2 completely rule out the possibility that it is a young galactic supernova remnant.

5.2 Not a Planetary Nebula

White and Becker (1990) proposed that G25.5+0.2 is a planetary nebula based primarily on identification of the IRAS source with the radio source. As noted in Section 4.5, identification of IRAS 18344-0632 with G25.5+0.2 is now observationally secure. Our discovery of a bright near-infrared point source in the center of G25.5+0.2 makes identification of G25.5+0.2 as a planetary nebula very implausible. The central star of a planetary nebula would be more than 5 magnitudes fainter (e.g., Wood and Faulkner 1986). It is very unlikely that the central object is a field star, since the reddening to the nebula and the star are similar (Section 4.2). This is not true for the other stars in Figure 1. For example, of the 15 to 20 stars seen in Figure 1 with apparent K brighter than 14 mag, all have a J-K color much bluer than the J-K=3.9 mag observed for the central object. Thus, the chance alignment of a random field star with the correct color within a radius of 2 arcsec of the nebula center has a probability less than 10^{-2}. A conceivable way to interpret G25.5+0.2 as a planetary nebula is if the infrared point source is a physical, M giant, companion to the true central star of the planetary nebula.

There are additional arguments that make a planetary nebula unlikely, even if the central point source is an unrelated field star or a physical companion to the central star. First, the infrared luminosity of G25.5+0.2 would be larger than any of the hundreds of galactic center planetaries identified by Pottasch et al. (1988) using IRAS colors. Second, as pointed out by White and Becker (1990), G25.5+0.2 would have an ionized mass of 15 M$_\odot$ at a distance of 13.5 kpc, which is much larger than that of any known planetary nebula. We thus conclude that it is very unlikely that G25.5+0.2 is a planetary nebula. Zijlstra (1991) came to the same conclusion based on several of the above arguments.

5.3 Not an HII Region like Orion

The existence of the point source in its center makes an explanation of G25.5+0.2 as a young HII region like the Orion Nebula very natural. The absolute K magnitude of the star in the center of the nebula derived in Section 4.3 of -5.3 mag can be used to check

269

the consistency of this proposal together with the ionization rate and total thermal infrared luminosity. For an O or B star that would be necessary to produce both the ionization and luminosity the V-K color would be about -0.9 mag (Johnson 1966). Thus the absolute visual magnitude of the central object would be M_V=-6.2 with an error of at least 0.5 mag resulting primarily from the reddening correction and the distance. From the models of Panagia (1973) we find that there is no main sequence star that will fit all three parameters. An O9 to B0 supergiant could fit the visual magnitude and ionization if less than 30% of the bolometric luminosity is observed in the 12 to 100 μm region. Additional problems with the HII region explanation also exist. First, a strong CO (3-2) line would be expected to be originated from the molecular cloud associated with the HII region containing the O star (Habing and Israel 1978). However, the weak CO line observed is not at the same velocity as G25.5+0.2 but rather is a foreground cloud at a kinematic distance of 7 kpc. No molecular emission is seen at the radial velocity of G25.5+0.2 greater than 1 K. Second, the shape of the nebula is much more symmetric than would be expected from an HII region formed on the edge of a molecular cloud (Zuckerman 1973). Third, as pointed out by White and Becker (1990), the IRAS colors of G25.5+0.2 are not similar to an HII region. Finally, and most important, there is no known way to explain the broad 300 km/s Bγ line if G25.+0.2 is an HII region. HII regions typically have a hydrogen line width of a few km/s (White and Becker 1990).

5.4 A Ring Nebula Around a Massive Supergiant Star

Some Wolf-Rayet and emission O supergiant stars that have evolved from a massive main sequence Population I star (M≥40 $M_\odot$) are known to have an associated ring-like nebula of ionized gas (e.g., Chu 1982; 1991). The nebula is thought to result from the interaction of the stellar wind (V = 1400 km/s) and the interstellar medium (Chu 1982; Tutukov 1982). The mass in the ionized ring results both from interstellar gas swept up by the wind and from mass loss from the massive star itself; the former material is thought to dominate (Tutukov 1982). Ring nebulae are much less common in the Galaxy than planetary nebulae. The ionization of the nebula comes from the ultraviolet radiation from the central star. Most nebulae of this type known are excited by Wolf-Rayet stars. However, the 3.09 μm HeII line which is strong in most luminous Wolf-Rayet stars (Hillier 1982, Williams 1982) was not seen in a UKIRT spectrum. Thus we conclude that G25.5+0.2 is a ring nebula around a B0 to O9 supergiant star (Chu 1991).

Interpretation of G25.5+0.2 as a ring or bubble nebula explains naturally the broad recombination line seen in Bγ. The expansion velocity of ring nebulae is typically between 30 and 80 km/s (Georgelin and Monnet 1970; Lozinskaya and Esipov 1969; Pismis and Quintero 1982; Chu 1991).

The ring nebula interpretation also explains the general morphology. The very symmetric-appearing emission in G25.5+0.2 is typical of many ring nebula, for example, the ring nebula M1-67 (Felli and Perinotto 1979). The lack of CO emission at a kinematic velocity and the relatively large S_{25}/S_{60} flux ratio are also expected for a mass loss-generated ionized nebula (White and Becker 1990). The physical diameter of the G25.5+0.2 ring is 1.0 pc, slightly smaller than any of the nebulae discussed by Tutukov

(1982). The mass of the ionized gas is about 15 $M_\odot$ based on Equation 2 in Schneider et al. (1987). Finally, the infrared luminosity of 10^5 $L_\odot$ is about 30% of the total energy output of the central star. Other ring nebula such as M1-67 (IRAS Point Source Catalog), S308, and NGC 6888 (Van Buren and McCray 1988) are strong IRAS sources but have a much lower fraction of the stellar luminosity emitted at infrared wavelenghts.

6 CONCLUSIONS

Based on infrared imaging, spectroscopy, and photometry, G25.5+0.2 is not a young supernova remnant. Based primarily on the discovery of a central point source at 1.25, 1.65, and 2.2 μm and a broad Bγ line with a width of 300 km/s, we believe the observations are consistent with a O9 to B0 supergiant star that is lossing mass and producing a ring nebula. It is interesting to note that this is not the first ring nebula that has been identified as a supernova remnant (Johnson 1973).

Acknowledgements: This work was done in complete collaboration of Ben Zuckerman, Ian McLean and Tom Geballe. C.G. Wynn-Williams kindly pointed out to us the similarity of G25.5+0.2 and a Wolf-Rayet ring nebula. We thank Andrew J. Noymer for enthusiastic help with the imaging observations and early reduction of the data. Neal Evans and Dan Jaffe kindly made CO (3-2) observations and George Hawkins provided us with IRAS add scan data. This research was supported in part by NSF Grant AST 87-17872 to UCLA.

7 REFERENCES

Becklin, E.E., Matthews, K., Neugebauer, G. & Willner, S.P., 1978. *Astrophys. J.*, **220**, 832.

Chu, Y.-H., 1982. *Proc. IAU Symp. N°.99, Wolf-Rayet Stars: Observations, Physics. Evolution*, p. **469**, eds. de Loore, C.W. & Willis, A.J., Reidel, Dordrecht.

Chu, Y.-H., 1991. *Proc. IAU Symp. N°.143, Wolf-Rayet Stars and Interrelations With Other Massive Stars in Galaxies*, p. **549**, eds. van der Hucht, K.A. & Hidayat, B., Kluwer, Dordrecht.

Cowan, J.J., Ekers, R.D., Goss, W.M., Sramek, R.A., Roberts, D.A. & Branch, D., 1989. *Mon. Not. R. Astr. Soc.*, **241**, 613.

Dwek, E., 1983. *Astrophys. J.*, **274**, 175.

Dwek, E., 1987. *Astrophys. J.*, **322**, 812.

Felli, M. & Perinotto, M., 1979. *Astr. Astrophys.*, **76**, 69.

Genzel, R., Becklin, E.E., Wynn-Williams, C.G., Moran, J.M., Ried, M.J., Jaffe, D.I., & Downes, D., 1982. *Astrophys. J.*, **255**, 5.

Georgelin, Y.P. & Monnet, G., 1970. *Astrophys. Lett.*, **5**, 239.

Graham, J.R., Wright, G.S. & Longmore, A.J., 1990. *Astrophys. J.*, **352**, 172.

Green, D.A., 1990. *Astr. J.*, **100**, 1241.

Habing, H.J. & Israel, F.P., 1978. *Ann. Rev. Astr. Astrophys.*, **17**, 345.

Harwit, N., Soifer, B.T., Houck, J.R. & Pipher, J.H., 1972. *Nature*, **236**, 103.

Hillier, J.D., 1982. *Proc. IAU Symp. Nº.99, Wolf-Rayet Stars: Observations, Physics, Evolution*, p.**225**, eds. de Loore, C.W. & Willis, A.J., Reidel, Dordrecht.

Johnson, H.L., 1966. *Ann. Rev. Astron. Astrophys.*, **4**, 193.

Johnson, H.M., 1973. *Proc. IAU Symp. Nº.49, Wolf-Rayet and High Temperature Stars*, p. **42**, eds. Bappu, M.K.V. & Sahade, J., Reidel, Dordrecht.

Kerr, F.J. & Westerhout, G. 1965. *Stars and Stellar Systems. Vol.V. Galactic Structure.* ed. Blaauw and Schmidt, p.**167.**

Koornneef, J. 1983. *Astron. Astrophys.*, **128**, 84.

Londsdale, C., Helou, G., Good, J. & Rice, W., 1985. *Cataloged Galaxies and Quasars Observed in the IRAS Survey, prepared under the supervision of the Joint IRAS Science Working Group (JISWG), Jet Propulsion Laboratory.*

Lozinskaya, T.A. & Esipov, V.F., 1969. *Sov. Astr.*, **12**, 913.

McLean, I.S., 1987. *IR Astronomy with Arrays*, p. **180**, eds. Wynn-Williams, C.G. & Becklin, E.E., Institute for Astronomy, University of Hawaii.

Panagia, N., 1973. *Astr. J.*, **78**, 929.

Pismis, P. & Quintero, A., 1982. *Proc. IAU Symp. Nº.99. Wolf-Rayet Stars: Observations, Physics. Evolution*, p. **305**, eds. de Loore, C.W. & Willis, A.J., Reidel, Dordrecht.

Pottasch, S.R., Bignell, C., Olling, R. & Zijlstra, A.A., 1988. *Astr. Astrophys.*, **205**, 248.

Schneider, S.F., Silverglate, P.R., Altschuler, D.R. & Giovanardi, C., 1987. *Astrophys. J.*, **314**, 572.

Tutukov, A., 1982. *Proc. IAU Symp. Nº.99. Wolf-Rayet Stars: Observations, Physic. Evolution*, p. **485**, eds. de Loore, C.W. & Willis, A.J., Reidel, Dordrecht.

Van Buren, D. & McCray, R., 1988. *Astrophys. J. Letters*, **329**, L93.

Weiler, K.W. & Sramek, R.A., 1988. *Ann. Rev. Astr. Astrophys.*, **26**, 295.

White, R.L. & Becker, R.H., 1990. *Mon. Not. R. Astr. Soc.*, **244,** 12p.

Williams, P.M., 1982. *Proc. IAU Symp. No. 99, Wolf-Rayet Stars: Observations, Physics, Evolution,* p. **73**, eds. de Loore, C.W. & Willis, A.J., Reidel, Dordrecht.

Willner, S.P., Becklin, E.E. & Visvanathan, N., 1972. *Astrophys. J.*, **175**, 699.

Willner, S.P., Jones, B., Puetter, R.C., Russell, R.W., & Soifer, B.T., 1979. *Astrophys. J.*, **234**, 596.

Wood, P.R. & Faulkner, D.J., 1986. *Astrophys. J.,* **307**, 659.

Wynn-Williams, C.G., Becklin, E.E., Matthews, K. & Neugebauer, G., 1978. *Mon. Not. R. Astr. Soc.*, **183**, 237.

Wynn-Williams, C.G. & Becklin, E.E., 1974. *Publ. Astr. Soc. Pacif.*, **86**, 5.

Zijlstra, A.A., 1991. *Mon. Not. R. Astr. Soc.*, **248**, 11p.

Zuckerman, B., 1973. *Astrophys. J.*, **183**, 863.

Cosmic Grains: 1. Basic Constraints and Standard Models

N. C. Wickramasinghe

School of Mathematics
University of Wales College of Cardiff
Cardiff, U.K.

1   INTRODUCTION

Amongst the most startling pictures of the Milky Way are those that involve interstellar dust clouds.   They show up characteristically in long exposure photographs as dark patches and striations against more or less uniform starfields. Fig. 1 shows a typical example of such a picture depicting the Milky Way in the direction of the constellation of Sagittarius.   The patches of Fig. 1 are not recesses in the distribution of stars, as they were once thought, but rather are caused by the very effective obscuration of background optical radiation by cooler clouds containing cosmic dust grains.   Nowadays a variety of observational techniques are deployed to study these grains, involving the measurement of the radiation they emit, scatter and absorb at wavelengths ranging from the far infrared to the far ultraviolet.

Figure 1:   *Milky Way in the direction of Sagittarius showing conspicuous obscuring clouds (Neb. Barnard 92 Sag.) (Courtesy of the Mt Wilson and Palomar Observatories)*

Early measurements of the interstellar extinction curve at visual wavelengths (Stebbins *et al.*, 1939) established a broad result which has survived unchanged, namely that the amount of the extinction on a magnitude scale has an approximately linear dependence on wavenumber, i.e. $A_\lambda \propto 1/\lambda$. This result was later refined (Nandy 1964a,b, 1965) to show that in a second order of approximation A_λ could be represented in a plot against $1/\lambda$ by two straight line segments intersecting at $1/\lambda \cong 2.4$ (μm^{-1}), with the segment corresponding to blue wavelengths being somewhat shallower in slope than the segment corresponding to red wavelengths. The shape of the interstellar extinction curve over the waveband $1 < \lambda^{-1} < 3 \mu m^{-1}$ remained more or less invariant from star to star. Grain models were thus constrained to possess a wavelength dependence of extinction that accorded with this general result.

Another observational criterion that was available from the 1930's related to the amount of the interstellar extinction per unit path length. For instance at the visual wavelength, corresponding to $1/\lambda = 1.8$ $(\mu m)^{-1}$, the mean extinction of starlight in directions close to the galactic plane is about 2 mag/kpc.

A further result of relevance is the so-called Oort limit for the total mass density of interstellar material - gas and dust - which amounts to $\sim 3 \times 10^{-24}$ g cm^{-3} (Oort, 1932, 1952). The dust grain density had certainly to be less than this value, probably considerably less, in view of the fact that the bulk of cosmic material is made up of H which cannot by itself condense into solid grains at a temperature above that of the cosmic microwave background, $\sim 2.7°K$.

The composition of interstellar grains has been the subject of extensive discussion for over 4 decades, but no final solution has yet emerged. In chronological sequence the following types of model have been proposed by various authors at various times:

(1) Iron grains (Schalen, 1939), 1939-1946
(2) Ice grains (van de Hulst, 1946, 1949), 1946-1962
(3) Graphite grains (Hoyle and Wickramasinghe, 1962); Composite core-mantle grains, 1962-70
(4) Refractory grain mixtures - silicates, iron, graphite (Hoyle and Wickramasinghe, 1969), 1969-
(5) Organic polymers (Wickramasinghe, 1974), 1974-
(6) Biological grains (Hoyle and Wickramasinghe, 1979), 1979-

We shall discuss models (1)-(4) in relation to observational constraints in the present contribution; models (5) and (6) in the next.

We can obtain upper limits to the mass densities of various types of grain materials proposed thus far using a remarkably simple argument. There is good evidence to assert that the nuclear composition of the outer layers of a star is closely similar to that of the interstellar cloud from which it condensed. This is particularly true for the case of the Sun. Table 1 shows a compilation of the relative abundances in the Sun (Cameron, 1970). Assuming these to be the relative abundances in interstellar matter as a whole, and fixing the value of the hydrogen density, we can then calculate the maximum density permitted for a given grain composition. Table 1 together with the Oort limit $\rho_H = 3 \times 10^{-24}$ g cm^{-3} (which is corroborated from neutral hydrogen measurements) gives the values set out in Table 2.

Table 1: *Solar Abundance Ratios*

Element	Atomic Weight	Relative Abundance
H	1	3.18×10^{10}
O	16	2.14×10^{7}
C	12	1.18×10^{7}
N	14	3.64×10^{6}
Mg	24	1.06×10^{6}
Si	28	$1.0 \ \times 10^{6}$
Fe	56	$8.3 \ \times 10^{5}$
Al	27	8.51×10^{4}

Table 2: *Upper limits to grain densities*

Species	Max. density (g cm^{-3})
H_2O (ice)	$3.6 \ \times 10^{-26}$
C	1.33×10^{-26}
Fe	0.44×10^{-26}
$MgSiO_3$	0.94×10^{-26}
Al_2O_3	0.04×10^{-26}
SiO_2	0.57×10^{-26}
Organic polymers	3.34×10^{-26}

2 FORMATION THEORIES

Theories for the formation of grains of various selected types have been developed mostly on a *post-hoc* basis. When iron grains were thought to be a plausible model in the 1930's the connection was made between interstellar grains and iron meteorites. This connection was subsequently shown to be spurious.

The ice grain model was developed partly to take account of the observed correlation between the extinction of starlight and the gas density in the line of sight. This observation is, however, open to other interpretations, notably the idea that grains expelled from stellar sources are preferentially stopped in gas clouds of high density. Particles of an icy composition were supposed to condense from gas clouds, with their eventual sizes being limited by destruction processes. It was assumed that the formation of icy grains from pure gaseous material involved three processes: (a) the formation of molecules through gas phase reactions, (b) the growth of condensation nuclei comprised of 30-50 atoms and (c) the accretion of molecules onto the nuclei. Recent arguments show that processes (a) and (b) are virtually impossible for low density interstellar gas clouds.

The formation of refractory grains is thought to accompany mass flows from cool stars as well as from novae and supernovae. Carbon stars produce carbon particles (possibly graphite) and oxygen-rich Mira variables produce particles comprised of silica, silicates and iron. Carbon, silicate and iron grains can form in ejecta from novae as well as from supernovae. These ideas were developed by Hoyle and Wickramasinghe through the 1960's and 70's. Direct observational evidence for grain formation exists as infrared emission from all these types of objects. There can be no doubt that these are sources of interstellar grains. The only doubts relate to the relative proportions of the various chemical species and the total mass of grains thus supplied. There is also the possibility that grains ejected from stellar sources would subsequently accrete mantles of organic or inorganic material in dense interstellar clouds.

The formation theory for organic and biological grains will be discussed in the following contribution.

3 EXTINCTION OF STARLIGHT

An important aspect of the modelling of interstellar grains involves the computation of extinction, absorption and scattering cross-sections of particles comprised of candidate materials, with a view to comparisons with astronomical data. Since interstellar particles would in general be expected to have non-spherical, irregular shapes a necessary simplification is the assumption of spherical grains. It could be argued, with some empirical justification, that randomly (or nearly randomly) oriented irregular grains can be represented by an equivalent set of particles of spherical shape for the purposes of scattering calculations. In this case the computational procedures are straightforward and are readily available in standard texts (van de Hulst, 1957, Wickramasinghe, 1973). For a given value of the refractive index m (= n-ik), particle radius a and wavelength λ the efficiency factors (cross-sections/πa^2) for extinction (Q_{ext}), absorption (Q_{abs}) and scattering ($Q_{sca} = Q_{ext}-Q_{abs}$) could be calculated. For non-absorbing particles $Q_{abs} = 0$ and $Q_{ext} = Q_{sca}$. Related to these various Q-values are mass absorption/extinction/scattering coefficients given by $\kappa = 3Q/4as$, where s is the density of grain material.

Fig. 2 gives Q_{ext} curves for several values of m plotted as functions of x = $2\pi a/\lambda$. Fig. 3 shows the $Q_{ext}(1/\lambda)$ curves for graphite spheres of various radii. For an ensemble of identical grains, each with extinction efficiency Q_{ext}, the solution of the equation of transfer for radiation along the line of sight can be shown to yield

$$\Delta m = 1.086 \, N \, \pi a^2 \, Q_{ext} \tag{1}$$

where N is the column density of grains (cm^{-2}), Δm is the extinction in magnitudes, and a is the grain radius (Wickramasinghe, 1967). We first consider extinction at one particular wavelength. Using our earlier result that the interstellar extinction at $\lambda^{-1} =$ 1.8 (μm)$^{-1}$ amounts to ~ 2 mag/kpc equation (1) now readily leads to the condition

$$\rho = 3.0 \times 10^{-22} \, as/Q_{ext} \qquad g \, cm^{-3} \tag{2}$$

for the smeared-out density of spherical grains, where s is the density of the material comprising the grain. With Q_{ext} values calculated for various models (as from Figs. 2 and 3) for $\lambda^{-1} = 1.8$ (μm^{-1}) we have the following table for the average densities required in each of the models listed.

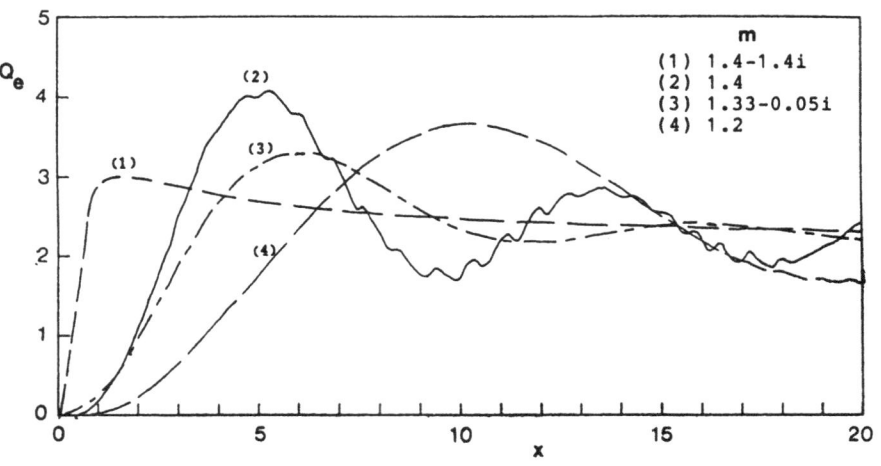

Figure 2: *Extinction efficiencies as functions of x for various values of m*

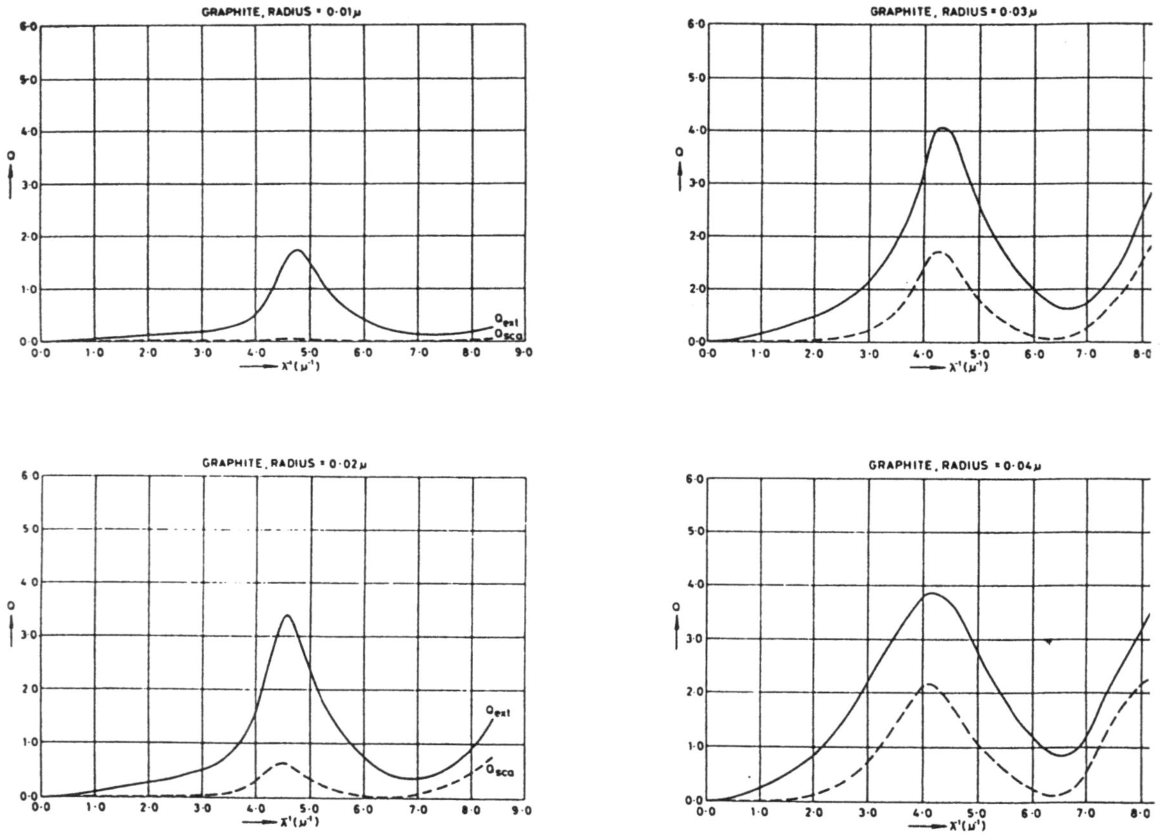

Figure 3: *Extinction efficiencies of graphite spheres of radii 0.01, 0.02, 0.03 and 0.04 μm*

Table 3: *Density requirements*

Grain Model	ρ (g cm^{-3})
ice (a = 3 × 10^{-5} cm)	10^{-26}
graphite (a = 2 × 10^{-6} cm)	6 × 10^{-27}
iron (a = 2 × 10^{-6} cm)	~ 4.5 × 10^{-26}
silicate (a = 2 × 10^{-5} cm)	1.6 × 10^{-26}
organics (a = 3 × 10^{-5} cm)	~ 10^{-26}

A comparison with Table 1 shows that a grain model based solely or primarily on iron is excluded for the reason that not enough of this element is available in interstellar space. Likewise, the case for silicates appears to be balanced on a razor's edge, if silicates are to make up the large bulk of the grains. The possibility of silicates, silica or even iron making up a small fraction of mass of the interstellar dust cannot, however, be ruled out on the basis of the present argument. What is amply clear is that solid particles with optimal efficiency for extinction at $\lambda^{-1} = 1.82$ (μm^{-1}) are required for producing the observed total amount of visual extinction, and for this purpose the main contributor to the grain mass can only come from the elements C, N, O. Two possibilities remain open:

(1) The C, N, O may be combined with H as volatile inorganic ices, much in the same way as discussed by van de Hulst, but including a possible major contribution from C in the form of graphite.

or

(2) The C, N, O may be combined with H as organic polymers of low to moderate volatility.

At the present time the wavelength base for interstellar extinction observations extends from the far infrared to ~ 1000 Å in the ultraviolet. Extinction data are available for hundreds of stars distributed throughout the galaxy. Measurements of colour excess ratios give extinction curves in the ultraviolet as shown in Fig. 4.

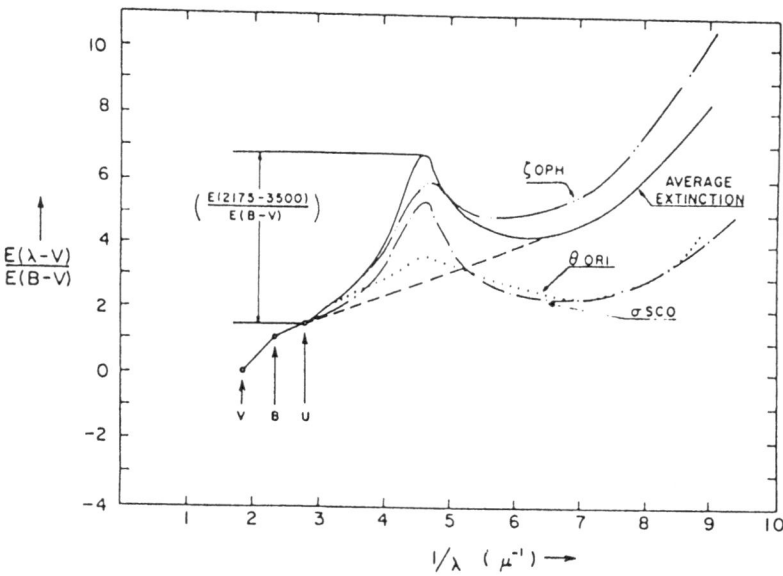

Figure 4: *Normalised ultraviolet extinction data for several stars combined with appropriate optical data (Bless and Savage, 1970)*

We note that the ultraviolet extinction curve shows considerable variability whilst maintaining a more or less constant wavelength of peak extinction at $\lambda \cong 2175$ Å. The extinction in the visual spectral region also maintains an approximately constant shape for the majority of stars. Colour excess ratios are readily converted to estimates of absolute extinction values $\Delta m(\lambda)$ if we know the ratio of total to selective extinction A_V/E_{B-V}. In most cases this ratio is close to 3.0, but exceptionally, as in the case of θ Ori, the ratio could be as high as 6. An average extinction curve from data compiled by Sapar and Kuusik (1978) is shown in Fig. 5. More recent compilation including a larger number of stars still retains almost the precise shape of the curve plotted in this figure. This extinction curve will be used as a discriminant of grain models that we shall discuss later in this contribution.

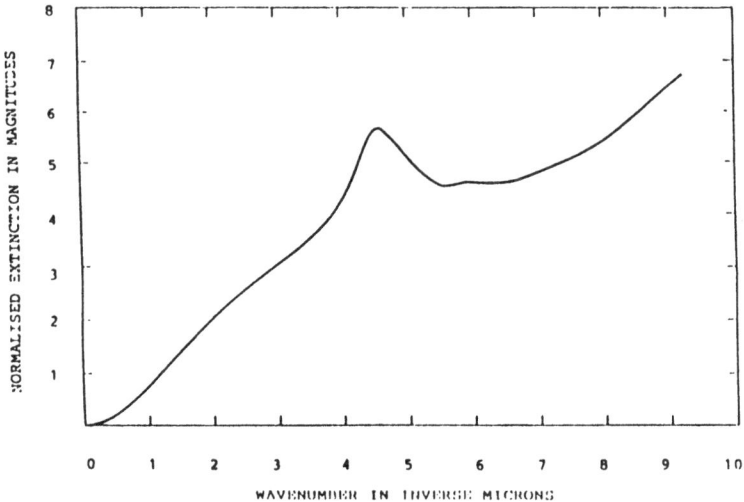

Figure 5: *Normalised mean extinction curve in the galaxy from the compilation by Sapar and Kuusik (1978). Normalisation is to $\Delta m = 1.8$ at $\lambda^{-1} = 1.8 \; \mu m^{-1}$*

4   INTERSTELLAR POLARIZATION

Starlight passing through interstellar dust clouds suffers extinction as well as partial linear polarizatiion. The polarization measured as the difference in extinction for two orthogonal directions of the electric vector normal to the line of sight, and expressed in the form

$$\Delta m_p = \Delta m_1 - \Delta m_2 \; , \tag{1}$$

is, in general, less than a few percent of the total extinction

$$\Delta m = \Delta m_1 + \Delta m_2 \quad . \tag{2}$$

285

Fig. 6 shows the average behaviour of $\Delta m_p(\lambda)$ normalised to unity at its peak. It is customary to discuss polarisation data using models of particles in the shape of cylinders of infinite length, just as particles of spherical shape were considered in relation to the total extinction. The justification for this duality of approach is based on the near coincidence between the extinction behaviour of randomly oriented cylinders and spheres, as seen for instance in Fig. 7.

Polarization of starlight results from the alignment effectively of only a few percent of elongated grains. Because many types of molecules and polymers that condense from a gas show whisker-type growth the presence of highly elongated interstellar grains might be expected.

Although Fig. 6 shows the mean wavelength dependence of polarization, it is known that a considerable degree of variability occurs from star to star. Attempts to model interstellar polarization have led to two general results:

(1) Metallic grains (grains with high values of k) do not give the correct mean wavelength dependence of polarization.

(2) Dielectric grains with refractive index in the range n = 1.4 to 1.6 are best suited to reproduce the observed data.

In view of the fact that only a small subset of interstellar grains are selectively involved in the phenomenon of polarization, the above constraints do not turn out to be limiting in any significant way.

5   DIFFUSE GALACTIC LIGHT

Two important parameters characterising the interstellar grains are the albedo γ and the phase parameter g = <cosθ>. The albedo for a single particle is defined by

$$\gamma = Q_{sca}/Q_{ext} \tag{3}$$

and the phase parameter is defined by

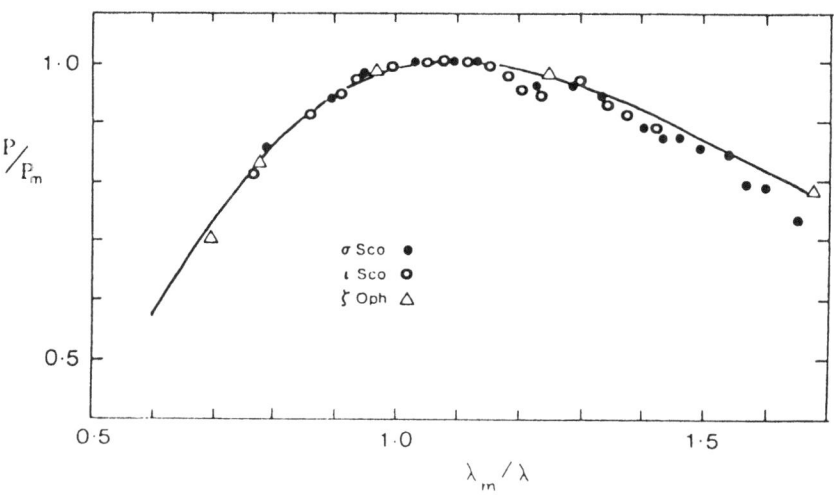

Figure 6: *Wavelength dependence of polarisation for several stars*

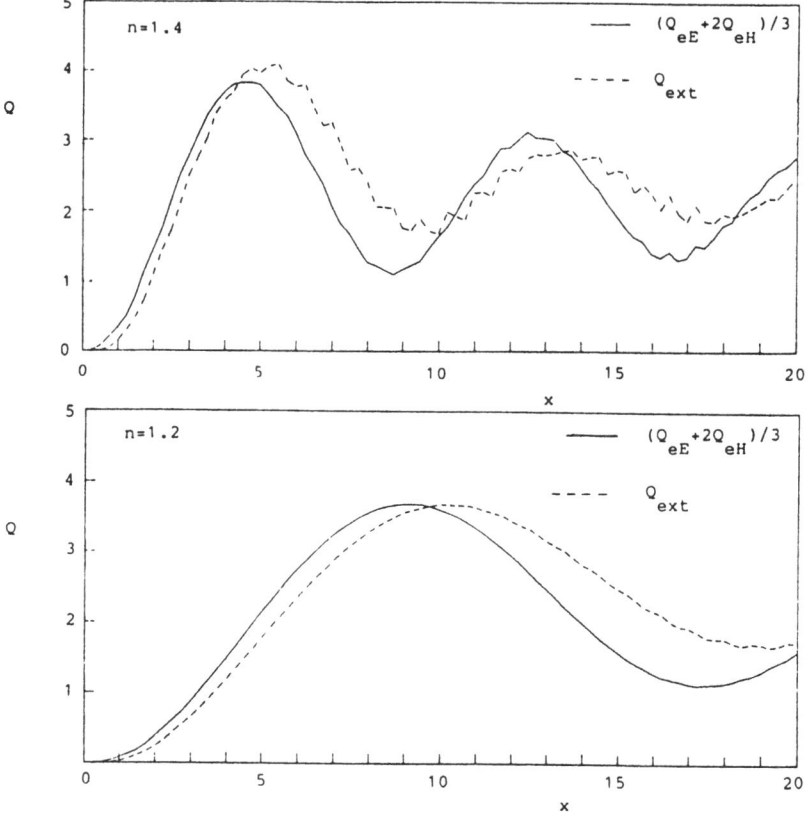

Figure 7: *Comparison of $Q_{ext}(x)$ for a sphere with the average extinction efficiency of cylinders in random orientation $(Q_{eE}+2Q_{eH})/3$.*

287

$$g = <\cos\theta> = \frac{\int_{o}^{\pi} S(\theta)\,\sin\theta\cos\theta\,d\theta}{\int_{o}^{\pi} S(\theta)\,\sin\theta\,d\theta} \tag{4}$$

where $S(\theta)$ is the fraction of incident energy scattered into a unit solid angle about a direction that makes an angle θ with the direction of the incident beam. A study of light scattered by interstellar grains in the plane of the galaxy may be expected to yield important information concerning the average values of both γ and g.

The diffuse galactic light (DGL) is the starlight that has been scattered diffusely by the interstellar dust grains. It is evident as a background sky brightness in the galactic plane. Measurements of this radiation, even with modern techniques, are difficult due to spurious effects - airglow, zodiacal light and light from unresolved faint stars which inevitably add to the sky brightness. Uncertain corrections have to be introduced to take account of these factors. The best hope for accurate results is to make measurements above the atmosphere and only a few such observations are available at the present time. The interpretation of diffuse galactic light data once obtained is also difficult. A radiative transfer model is required in order to deduce properties of grains, in particular to deduce values of the albedo γ and the phase parameter g (= $<\cos\theta>$).

Some idea of the asymmetry of the scattering diagram of individual grains may be obtained if observations are available of the diffuse light as a function of galactic latitude. Lillie and Witt (1973, 1977) have obtained such data in the waveband 1500-4200 Å using equipment carried on the Orbiting Astronomical Observatory OAO 2. Using this data in conjunction with the computations of van de Hulst and de Jong (1969), Witt and Lillie have obtained a wavelength dependence for the grain albedo and also for the phase parameter g = $<\cos\theta>$. Fig. 8 shows their results for the albedo which indicate a sharp minimum at 2200 Å, coinciding with the peak in the extinction curve. The scattering phase function is found to be strongly forward throwing at all wavelengths. The excessively high values of g ~ 0.9 found near the wavelength λ ~ 2200 Å would militate against graphite spheres of radius 0.02 μm, for which it can be shown that g << 0.5.

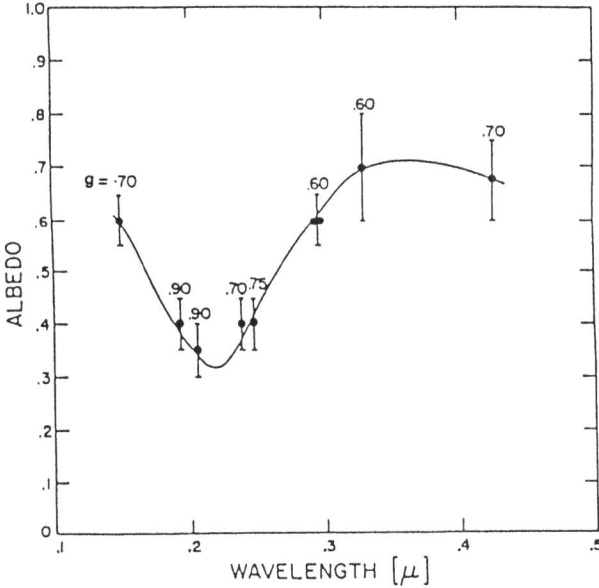

Figure 8: *The albedo and phase parameter g of interstellar dust as a function of wavelength based on OAO-2 measurements of the diffuse galactic light (Lillie and Witt, 1973, 1977)*

6   ICE GRAIN MODEL

From the early 1940's to the early 1960's astronomers accepted the idea of ice grains almost without dissent. In this section we shall review the progress of this model from its first publication to the present day.

Thus far we have taken the average radius of grain as given for each of the grain models considered (e.g. Table 3). The average radius, however, is not always a free parameter and must be fixed in such a way that the wavelength dependence of Q_{ext} approximates to the linear variation with $1/\lambda$ over the visual waveband, as we have earlier discussed. The points in Fig. 9 show the average wavelength dependence of extinction for stars in the Cygnus region (Nandy, 1964a). The curve shows the best fitting theoretical extinction for pure H_2O-ice grains with measured values of the refractive index. This calculation is not for grains of a single size but takes account of a distribution of sizes resulting from a balance between growth and destruction processes. The effective mean particle radius that contributes most to the extinction is 0.33 μm. The details of this 'fit' could be improved by assuming an arbitrarily high value of the imaginary component, k = 0.1, of the refractive index, possibly corresponding to iron impurities, but the amount of iron implied by such a value of k is in excess of the available proportion of iron in the interstellar medium and cannot therefore be justified.

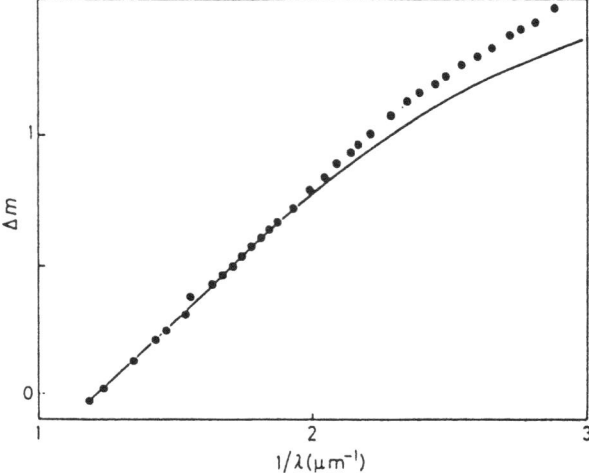

Figure 9: *Theoretical extinction curve for a size distribution of pure ice grains compared with Nandy's observations (•) for stars in the Cygnus direction. The curve is for ice grains with refractive index n = 1.33 with a mean radius r_1 = 0.34 μm. Normalization is to Δm = 0.409 at $λ^{-1}$ = 1.62 $μm^{-1}$, Δm = 0.726 at $λ^{-1}$ = 1.94 $μm^{-1}$*

The evident defects in a pure ice grain model became further exacerbated when the interstellar extinction curve was extended into the ultraviolet spectral region. Moreover, the appearance of a broad mid-UV extinction peak centred on λ = 2175 Å as seen in Fig. 5 pointed strongly to the predominance of grain materials unrelated to ice. No distribution of ice grain sizes could reproduce even remotely a feature such as this. Fig. 10 shows the distribution in wavelength of the peak of the ultraviolet extinction for a representative number of stars observed by the IUE satellite. The maximum frequency is at λ = 2175 Å with only a shallow spread towards shorter and longer wavelengths. As we have noted earlier, independent observations of the diffuse galactic light suggest that interstellar grains are mostly scatterers (rather than absorbers) at both visual and far ultraviolet wavelengths, but near the extinction peak at 2175 Å they have strongly absorbing properties (see Fig. 8). The 2175 Å extinction cannot therefore be caused by dielectric grains.

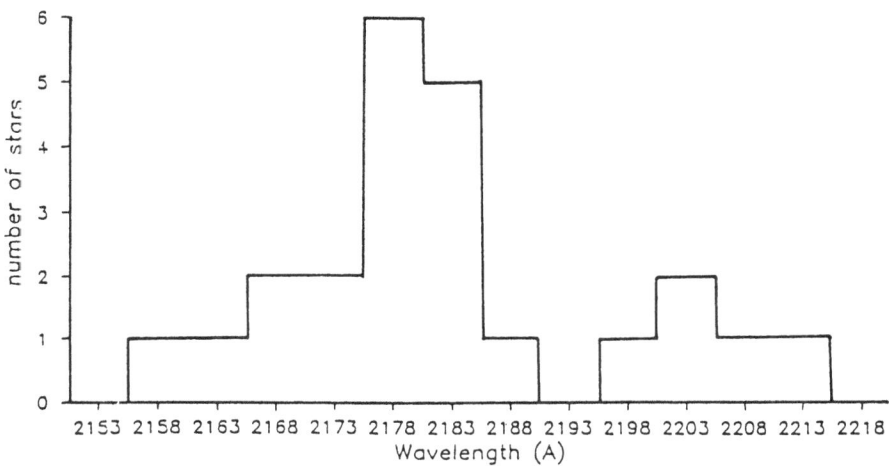

Figure 10: *Histogram of the distribution of ultraviolet extinction peaks for 26 reddened stars from IUE spectra*

From about 1965 when infrared stellar spectroscopy came to the fore, it emerged that ice grains could not form a substantial component of interstellar dust. Ice grains have a strong infrared absorption feature near $\lambda = 3.1$ μm with a peak absorption coefficient of $\kappa_{abs} \sim 30,000$ cm^2 g^{-1}. Because the mass coefficient for scattering by ice grains of radii $\sim 10^{-5}$ cm is of the same order, we have a ratio Δm (3.1 μm)/Δm (5000 Å) ~ 1. A star with one magnitude of visual extinction would thus have about one magnitude of absorbtion at 3.1 μm , on the basis of the ice grain model, and this was clearly not evident even from the early work of Danielson *et al.* (1965). Ice was thus relegated to be a minor component of the interstellar grains.

7   GRAPHITE MODEL AND ITS INADEQUACIES

The 2175 Å absorption peak was provisionally identified with spherical graphite particles from the early Mie calculations of Wickramasinghe and Guillaume (1965). Graphite spheres with an **assumed** optical isotropy were shown to possess extinction curves as shown in Fig. 3. Particles of radii 0.02 μm are seen to match the observations (see Figs. 5 and 10) to a remarkably precise degree. Because a graphite particle model (which involved solid particle condensation in the photospheres of cool carbon stars) had been discussed a few years earlier by Hoyle and Wickramasinghe (1962), the subsequent Mie calculations for spheres and the mid-UV data were taken as strong confirmation of this model. Even to this day graphite spheres are the favoured model for accounting for the mid-ultraviolet extinction of starlight.

It is, however, well-known that graphite is strongly anisotropic in its optical and electrical properties and tends to form flakes rather than spheres. Graphite flakes are comprised of planar linkages of hexagonal carbon rings, stacked in parallel planes. The electrical conductivity along the planes is high, similar to a metal, but in directions transverse to the planes the conductivity is about 100 times lower and the material behaves essentially like a dielectric.

The best determinations of the complex refractive index of graphite are available for the case of light with electric vector parallel to the basal planes (Taft and Phillipp, 1965; see also Hoyle and Wickramasinghe, 1991). For hypothetical graphite spheres, in which a complex refractive index equal to that determined by Taft and Phillipp is assumed to hold for all directions of the electric vector of incident light, the calculated normalised extinction efficiency in a Rayleigh approximation is shown by the long-dashed curve of Fig. 11. We note that the peak of extinction occurs near 2050 Å, about 125 Å short of the astronomical observations. Mie computations for a radius $a = 0.02$ μm (as we saw earlier) show that the peak moves near to the correct value $\lambda = 2175$ Å, as indicated in the dashed curve of Fig. 11. A *post hoc* justification of a spherical isotropic graphite model is that it might reasonably be thought to represent the behaviour of a randomly oriented set of graphite flakes. The equivalence of these two quite distinct optical systems could, however, be tested in a rigorous way (Wickramasinghe *et al.*, 1992). If the equivalence can be disproved, the validity of the graphite particle model will be in serious doubt.

A graphite flake comprised of a stack of planar hexagonal layers may be well represented by a thin circular disk of radius a, thickness t $\ll$ a. The complex dielectric constant for light with electric vector in the plane of the disk is obtained from measurements of $n_{//}$ and $k_{//}$ by Taft and Phillipp (1965):

$$\epsilon_{//} = m_{//}^2 = (n_{//} - ik_{//})^2. \tag{5}$$

The complex dielectric constant for electric vector transverse to the planes may be approximated by

$$\epsilon_{\perp} = m_{\perp}^2 = K - 2i\sigma_{\perp}\,\lambda/c \tag{6}$$

where $K \cong 4$ and $\sigma_\perp \cong \sigma_{//}/100$, $\sigma_{//}$ being the conductivity for light with electric vector along the planes. Since $\sigma_{//} \lambda/c = n_{//}k_{//}$ equation (5) yields

$$\epsilon_\perp \cong 4 - 2i \, n_{//}(\lambda) \, k_{//}(\lambda)/100. \tag{7}$$

In the Rayleigh limit where $\lambda \gg 2\pi a$, the extinction cross-section for light with electric vector parallel to the plane of the disk is

$$C_{//} = -\frac{2\pi V}{\lambda} \, \text{Im} \, (\epsilon_{//} - 1), \tag{8}$$

and that for light with electric vector perpendicular to the plane is

$$C_- = -\frac{2\pi V}{\lambda} \, \text{Im} \left(\frac{\epsilon_\perp - 1}{\epsilon_\perp} \right), \tag{9}$$

V being the particle volume (Wickramasinghe, 1973). For a randomly oriented ensemble of graphite flakes in the Rayleigh limit we thus have an average cross-section for extinction given by

$$C(\lambda) = (C_{//} + 2C_\perp)/3. \tag{10}$$

From equations (5) to (10) we can calculate $C(\lambda)$ and hence a normalised extinction efficiency defined by $Q(\lambda) = C(\lambda)/C(\lambda_o)$ where λ_o is the wavelength of maximum extinction. The solid curve in Fig. 11 shows the function $Q(\lambda)$. We find that λ_o is close to 2700 Å, considerably longward of 2175 Å, the wavelength of the interstellar absorption feature. This is for graphite disks that are small enough for the Rayleigh formulae (9) and (10) to apply. With increasing flake (disk) radius a, the resonance would be expected to move to wavelengths longward of 2700 Å, analogous to the trend in the spherical case (see Fig. 3). The discrepancy with observation therefore worsens with increasing particle size.

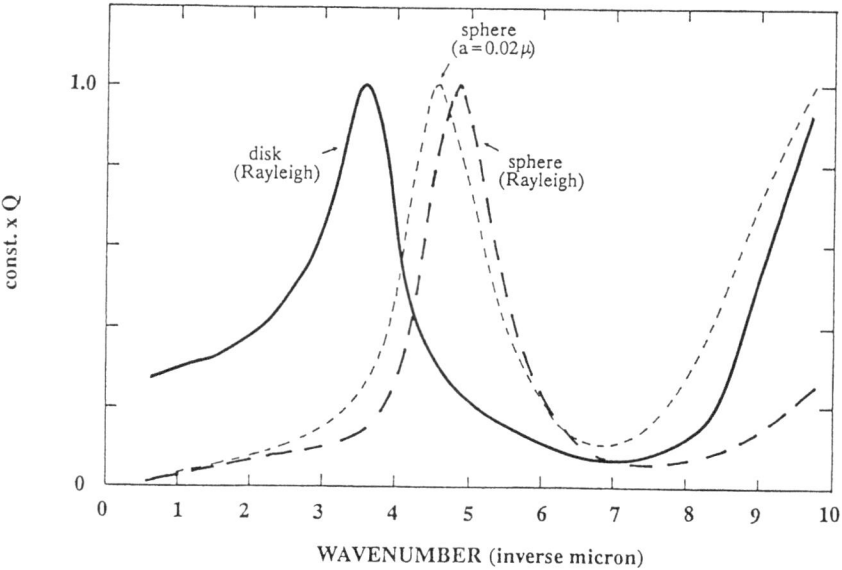

Figure 11: *Normalised extinction efficiencies for graphite flakes in the Rayleigh limit (solid curve); normalised extinction efficiency for graphite spheres in the Rayleigh limit (long-dash curve); normalised extinction efficiency of a graphite sphere of radius 0.02 μm (dashed curve). The normalisation is to unity at the peak of the mid-ultraviolet extinction profile*

We conclude this section by noting that graphite flakes in random orientation cannot be represented by Mie calculations for spherical particles. Accordingly, the coincidence of the interstellar absorption feature at $\lambda = 2175$ Å with calculations for spherical graphite grains of radius 0.02 μm must be dismissed as being fortuitous.

8   EVIDENCE OF SILICATES?

The first positive detection of any spectral feature of grains in the infrared waveband was made in 1969. Woolf and Ney (1969), Knacke *et al.* (1969), Ney and Allen (1969) and Stein and Gillett (1969) detected a strong infrared excess (above a thermal continuum) in the 8-12 μm waveband in the spectra of several oxygen-rich Mira-type stars. In view of broad similarities between the observations and the spectra of terrestrial silicates in this waveband the 8-12 μm feature of Mira stars has widely been attributed to silicate grains that condense in stellar mass flows. It is certainly true that a spectral feature similar to this could arise from Si-O bending vibrations as is present in silicates, but the identification cannot be unique.

At the present time the 8-12 μm feature has been observed both in emission and in absorption in a variety of types of astronomical object - these include the Trapezium nebula, planetary nebulae, compact HII regions, OH/IR sources, the galactic centre and comets. It is significant that in a small fraction of cases the same feature has also been seen in the spectra of carbon stars (Little-Marenin, 1986). The shapes of the bands in all these instances appear fairly constant. It would seem unlikely that the carbon star 10 μm band is due to silicates - thus casting doubt on the usual 'silicate only' hypothesis for this spectral feature.

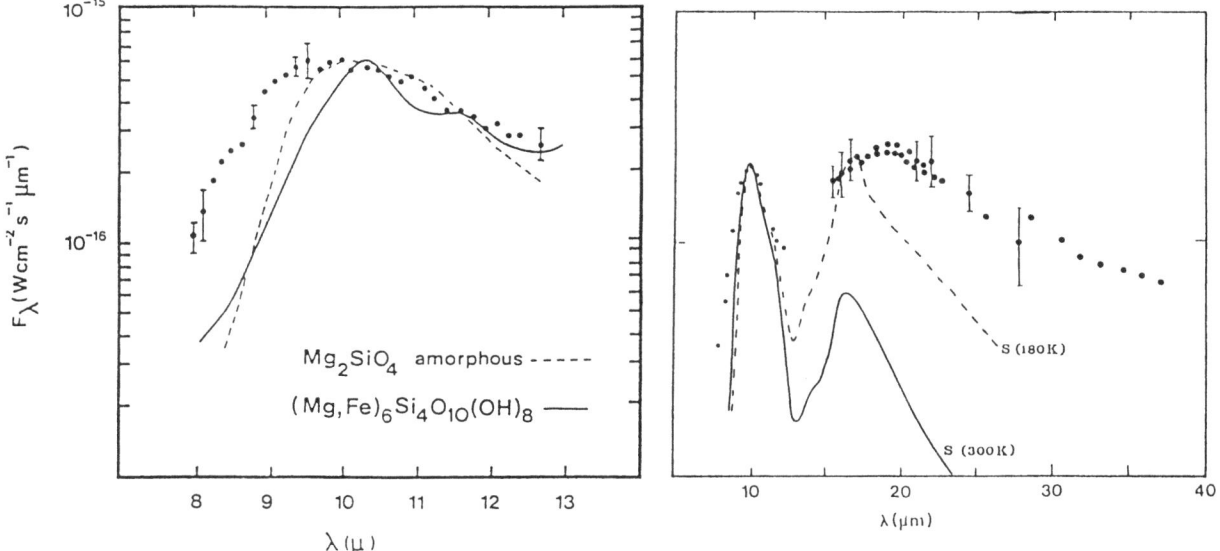

Figure 12: *Left: Flux data for the Trapezium nebula (dots) compared with the theoretical flux for models of magnesium silicate grains at a temperature of 175 °K. Normalisation is to F = 6 × 10⁻¹⁶ W cm⁻² s⁻¹ μm⁻¹ at the wavelength of maximum flux. The data is taken from Forrest et al., (1975a,b) Right: The infrared spectrum of the Trapezium nebula compared with amorphous silicate models at two temperatures. The data is from Forrest et al. (1976)*

Modelling the 8-12 μm feature as well as the subsequently observed 14-40 μm spectrum is relatively straightforward for the case of optically thin dust emission as in the Trapezium nebula. The points in Fig. 12 show the data for the Trapezium emission obtained by Forrest *et al.* (1975a,b, 1976) compared with calculated flux curves for small particles comprised of amorphous and hydrated silicates. The agreement between the data and the silicate models is seen to be poor, and this evident mismatch is not rectified even by considering the various ways in which silicates might be modified under astronomical conditions - e.g. irradiation by energetic photons and particles. The best 'mineral' fit to the 10 μm feature is for grains comprised of material derived from carbonaceous meteorites. Fig. 13 shows a comparison between the data and a calculation based on optical constants of finely crushed Murchison meteorite material. It should be remembered, however, that this material includes quantities of carbonaceous organic matter, which can also contribute to absorptions in the 8-12 μm band through C-O, C-O-C bending vibrations.

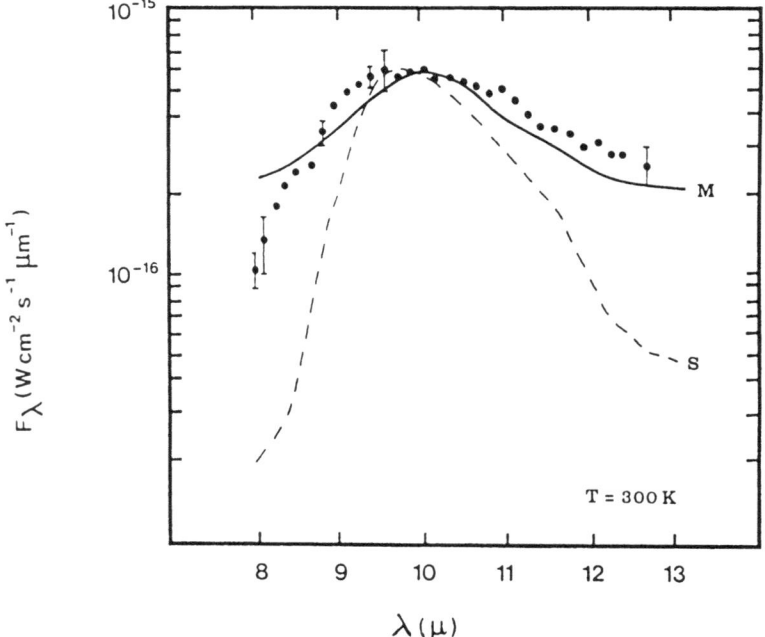

Figure 13: *Normalised flux over the waveband 8-13 μm calculated for amorphous silicate particles and for Murchison material at 300 °K compared with data points for the Trapezium nebula from Forrest et al. (1975a,b)*

9 THE THREE COMPONENT GRAIN MODEL

Despite the problems of spectroscopic identifications which remain unsolved, an analysis of available extinction and polarization data in terms of Mie-type calculations yields positive results. A study of the variations of the ultraviolet extinction law from star to star (as shown in Fig. 4) combined with the relative constancy of the visual extinction curve (Fig. 9) leads almost inevitably to a model with 3-components:

Component 1: Dielectirc spheres or needles with typical refractive index $m = n-ik$ ($n \sim 1.2$-1.4; $k \cong 0.01$-0.03), radius 0.3 μm, contributing to the visual and near IR extinction. A fraction of aligned cylinders would explain the data relating to interstellar polarisation.

Component 2: An absorber responsible for the 2175 Å absorption profile. The popular view is that this is identifiable with graphite, but the identification is dubious as we saw in Section 7.

Component 3: Small dielectric grains of typical refractive index $n = 1.4$-1.5 and radius 0.03 μm which contribute to the far ultraviolet extinction.

The problem of identifying components 1,2,3 will be addressed in the following contribution. Fig. 14 shows how the three extinction components could add up to yield the total average extinction curve.

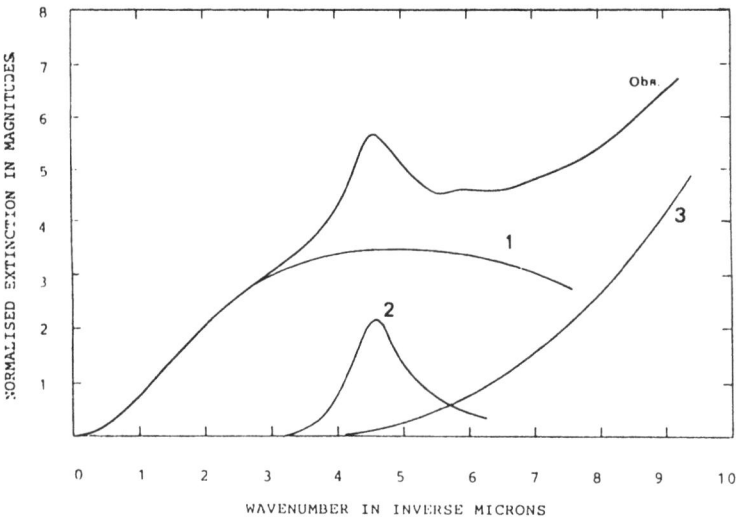

Figure 14: *The observed interstellar extinction curve schematically decomposed into 3 components*

REFERENCES

Bless, R.C. and Savage, B.D.: 1970, in L. Houziaux and H.E. Butler (eds.) *Ultraviolet Stellar Spectra and Groundbased Observations*. D. Reidel.

Cameron, A.G.W.: 1970, *Space Science Reviews*, **15**, 121.

Danielson, R.E., Woolf, N.J. and Gaustad, J.E.: 1965, *Astrophys. J.*, **141**, 116.

Forrest, W.J., Gillett, F.C. and Stein, W.A.: 1975a, *Astrophys. J.*, **192**, 351.

Forrest, W.J., Gillett, F.C. and Stein, W.A.: 1975b, *Astrophys. J.*, **195**, 423.

Forrest, W.J., Houck, J.R. and Reed, R.A.: 1976, *Astrophys. J.*, **208**, L133.

Hoyle, F. and Wickramasinghe, N.C.: 1962, *Mon. Not. Roy. Astr. Soc.*, **124**, 417.

Hoyle, F. and Wickramasinghe, N.C.: 1969, *Nature*, **223**, 459.

Hoyle, F. and Wickramasinghe, C. : 1979, *Astrophys. Sp. Sci.*, **66,** 77.

Hoyle, F. and Wickramasinghe, N.C.: 1991, *The Theory of Cosmic Grains*, Kluwer Academic Press).

Knacke, R.F., Cudaback, D.D. and Gaustad, J.E.: 1969, *Astrophys. J.*, **158**, 151.

Knacke, R.F., Gaustad, J.E., Gillett, F.C. and Stein, W.A.: 1969, *Astrophys. J.*, **155**, L189.

Lillie, C.F. and Witt, A.N.: 1973, in J.M. Greenberg and H.C. van de Hulst (eds.) *Interstellar Dust and Related Topics*, D. Reidel.

Lillie, C.F. and Witt, A.N.: 1977, *Astrophys. J.*, **208**, 64.

Little-Marenin, I.R.: 1986, *Astrophys. J.*, **307**, L15.

Nandy, K.: 1964a, *Publ. Roy. Obs. Edin.*, **4**, 57.

Nandy, K.: 1964b, *Publ. Roy. Obs. Edin.*, **3**, No. 6, 142.

Nandy, K.: 1965, *Publ. Roy. Obs. Edin.*, **5**, 13.

Ney, E.P. and Allen, D.A.: 1969, *Astrophys. J.*, **155**, L193.

Oort, J.H.: 1932, *B.A.N.*, No. 238.

Oort, J.H.: 1952, *Astrophys. J.*, **116**, 233.

Sapar, A. and Kuusik, I.: 1978, *Publ. Tartu Astrophys. Obs.*, **46**, 71.

Schalén, C.: 1939, *Uppsala Obs. Ann.*, **1**, No. 2.

Schalén, C.: 1945, *Uppsala Obs. Ann.*, No. 9.

Stebbins, J., Huffer, C.H. and Whitford, A.E.: 1939, *Astrophys. J.*, **90**, 209.

Stein, W.A. and Gillett, F.C.: 1969, *Astrophys. J.*, **155**, L197.

Taft, E.A. and Phillipp, H.R.: 1965, *Phys. Rev.*, **138**A, 197.

van de Hulst, H.C.: 1946 and 1949, *Rech. Astron. Obs. Utrecht,* XI,. Parts 1 and 2.

van de Hulst, H.C.: 1957, *Light Scattering by Small Particles*, John Wiley.

van de Hulst, H.C. and de Jong, T.: 1969, *Physica*, **41**, 151.

Wickramasinghe, N.C.: 1974, *Nature*, **252**, 462.

Wickramasinghe, N.C.: 1973, *Light Scattering Functions for Small Particles with Applications in Astronomy*, J. Wiley, N.Y.

Wickramasinghe, N.C.: 1967, *Interstellar Grains*, Chapman & Hall, Lond.

Wickramasinghe, N.C. and Guillaume, C.: 1965, *Nature*, **207**, 366.

Wickramasinghe, N.C., Wickramasinghe, A.N. and Hoyle, F.: 1992. *Astrophys. Sp. Sci.*, **196**, 167.

Woolf, N.J. and Ney, E.P.: 1969, *Astrophys. J.*, **155**, L181.

Cosmic Grains: 2. Spectroscopic Identifications and the Organic Model

N. C. Wickramasinghe

School of Mathematics
University of Wales College of Cardiff
Cardiff, U.K.

1   INTRODUCTION

We pointed out in the preceding contribution that the measured total amount of the visual extinction in directions close to the galactic plane, combined with the known distribution of chemical elements in the interstellar medium, implied that C, N, O must be the main contributors to the mass of the grains.   Since inorganic combinations of these elements are evidently ruled out for the main mass of the grains from considerations in the previous contribution, we are left now with the possibility that C, N, O combined with H exist mainly as organic molecules and polymers in condensed form.

From the mid-1960's onwards it became clear that interstellar molecules included a component that was organic. This is clearly seen in Table 1. These molecules were mostly detected by their spectral features in the radio and millimetre waveband. The list continues to grow as new transitions are predicted and new instruments and techniques are deployed to search for these predictions.

Table 1: *Interstellar molecules listed according to number of atoms*

2	3	4	5	6	7	8	9	10	11	13
H_2	H_2O	H_2CO	H_2C_2O	$HCONH_2$	HC_5N	$HCOOCH_3$	HC_7N	CH_3C_5N	HC_9N	$HC_{11}N$
CH	H_2	H_2CS	H_2CNH	CH_3CN	$HCOCH_3$	CH_3C_3N	$(CH_3)_2O$	CH_3COCH_3		
CH^+	HCN	$HCNH^+$	H_2NCN	CH_3NC	CH_3C_2H		CH_3CH_2CN			
C_2	HNC	$HNCO$	HC_3N	CH_3OH	CH_2CHCN		CH_3CH_2OH			
CN	HCO	$HNCS$	$HCOOH$	CH_3SH	NH_2CH_3		CH_3C_4H			
CO	HCO^+	$HOCO^+$	CH_2CN	C_5H	C_6H					
CS	HOC^+	C_3H	C_3H_2							
NO	HCS^+	C_3N	C_4H							
NS	HNO	C_3O	SiH_4							
OH	N_2H^+	C_3S	CH_4							
SiO	C_2H	NH_3								
SiS	OCS	H_3O^+								
SO	SO_2	C_2H_2								
SO^+	SiC_2	C_2H_4								
HCl	C_2S									

Whilst the simpler molecules in this table could form either through grain surface reactions or ion-molecule reactions in the gas phase, the larger and more complex molecules pose a serious problem as to how they are formed. A possible connection between these complex molecules and the nature of interstellar grains was first suggested in the 1970's (Wickramasinghe, 1974, 1975). Polymerisation reactions occurring on grain surfaces could be argued as providing a route to the formation of polyformaldehyde or similar polymeric mantles on grains. An immediate application of this class of model was seen to be its possible contribution to the 8-12 μm and 16-25 μm emission features in sources such as the Trapezium nebula. Because C-O, C-C and C-N linkages have absorptions in the 7.7-12.5 μm wavelength interval, the deficiencies of the silica/silicate models discussed in our earlier contribution could be attributed to the neglect of an organic component of the dust.

Fig. 1 shows the improvement to the earlier silicate models that is obtained for the polyformaldehyde grain model (Wickramasinghe, 1974), and Fig. 2 shows a further refinement using data based on laboratory measurements of the transmittance of cellulose (Hoyle and Wickramasinghe, 1977). In each case the radiation flux is calculated according to the relation

$$F_\lambda \propto \varepsilon_\lambda \, B_\lambda(T) \tag{1}$$

where ε_λ is the emissivity of the candidate material, and $B_\lambda(T)$ is the Planck function at a temperature T. The value of T is chosen to provide the best fit to the data.

Fig. 2 does not prove that polysaccharides actually exist in interstellar clouds. However, it is clear that the quality of fit to at least one astronomical data set improves dramatically as the postulated organic component becomes more complex. Polysaccharides (of which cellulose is an example) is a most important class of biochemical, so any hint of its widespread occurrence in the galaxy has an immediate bearing on the problem of the origin of life. Many scientists now believe that the general character of even the molecules listed in Table 1 implies that the first steps towards an origin of life are occurring on an extraterrestrial scale. A more radical hypothesis that we shall develop in this contribution is that the molecules of Table 1 are only the tip of an iceberg, representing fragments of an ensemble of much more complex organic/biotic material condensed in the form of interstellar grains.

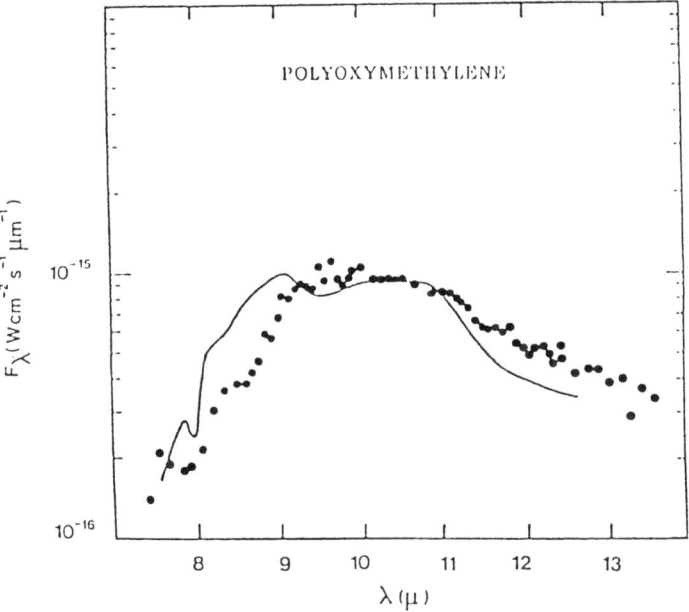

Figure 1: *The infrared spectrum of the Trapezium nebula compared with the prediction for a 445 °K polyformaldehyde grain model (Data points: Forrest et al., 1975a,b)*

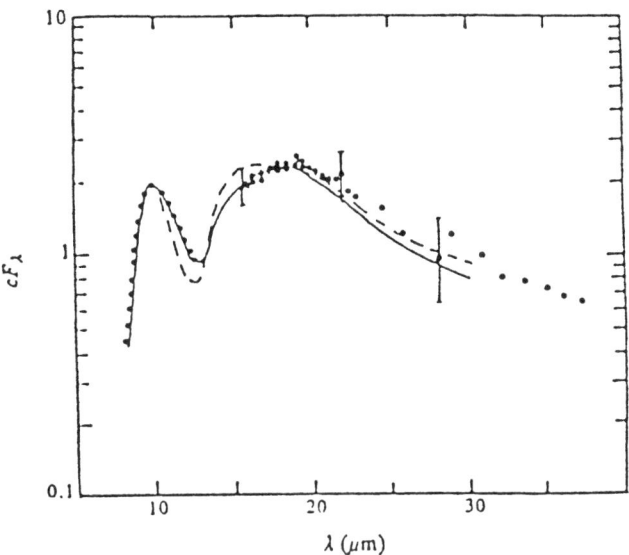

Figure 2: *Normalised flux from the Trapezium nebula (Forrest et al., 1976) (points) compared with emission from polysaccharide grains with temperature 175 °K (Hoyle and Wickramasinghe, 1977)*

2 PRODUCTION OF INTERSTELLAR ORGANICS

All the inorganic processes that seem to be available for converting inorganic material into organics in interstellar clouds - e.g. the Fischer-Tropsche reaction - are self-limiting and could yield only miniscule fractions of organics in the final outcome.

One could offer general arguments to the effect that it is impossible to produce cosmically by non-biological means adequate quantities of organic material from inorganics like water, carbon monoxide and dioxide, and nitrogen gas, and therefore the immense amount of interstellar organics must be biological in origin. The total quantity of interstellar organics implies that perhaps as much as 50% of all the carbon in the space is tied up in the form of complex organic matter. Under carefully controlled laboratory conditions, using the best prepared catalysts, no more than trace quantities of organics can be produced from inorganic materials. Vast quantities of interstellar organics will not be produced in uncontrolled conditions where every prospective catalyst becomes poisoned on very short timescales.

Ultraviolet processing of frozen inorganic ices such as H_2O, CH_4, NH_3 into complex molecules also appears to be limited in scope. The process requires bond-breaking events that are cumulative, leading to re-arrangements of neighbouring molecular structures. Under interstellar conditions in which the time interval between successive photon encounters is ~ 10 min, considerably longer than a relaxation time, the required re-arrangements will not occur. Another general objection to the production of organics by non-biological means is that production processes as we encounter them in chemical laboratories are all of a linear kind. Biological systems, on the other hand, have a crucially different mode of expansion. Instead of being linear they are exponential in character: one makes two, two makes four, four makes eight, and so on. Biological expansion is in effect explosive.

If we start from a single bacterium and suppose that this bacterium and its progeny is supplied with suitable nutrients, a typical time for replication under favourable conditions would be two or three hours. In a single day the initial bacterium would have expanded through sequential doublings into a colony of about 1000 bacteria. In two days there would be a million bacteria which would be just about visible to the naked eye. In four days we would have a thousand billion bacteria, weighing together about a gram. In five days the colony would be approaching a kilogram in weight, and so on, adding three zeros to the numbers for every day that passed - a tonne after six days, 1000 tonnes after a week, the mass of the Earth after thirteen days, our galaxy after nineteen days, and the whole of the visible universe in a mere twenty-two days. This timespan is of course unrealistic for the reason that it assumes an instantaneous propagation of the colony in space, as well as continued and uninterrupted access to nutrients. These numbers illustrate the inherent potential for biological expansion on a cosmic scale, a potential that would be limited only by the finite travel speeds of bacteria between nutrient sites.

3 BACTERIAL GRAIN MODEL

It would seem obvious that the most inevitable logic for replicating a whole range of complex molecules, and for reproducing structures of almost invariable size, is biology itself. Bacteria have the right sizes and shapes to be *prima facie* candidates for grains. Under astronomical conditions, where bacterial particles can be propelled around the Galaxy by radiation pressure, limitations to growth arise only from limited access to molecular feedstock. We estimate a regeneration time for all the 10^{40} g of interstellar bacteria to be 10^8-10^9 years, most of the regeneration occurring in cometary interiors as we shall see in Section 8.

Figure 3 shows the observed size distribution of a representative sample of spore-forming bacteria (cross-sectional diameters of rods and spheres). Terrestrial bacteria are bound by *rigid* outer cell walls comprised mainly of polysaccharides and lipoproteins, with an interior that is comprised of a rich variety of biochemicals and water. Water makes up about two-thirds by volume of a bacterium. We estimate that interstellar cloud conditions lead to evaporation of water so that the mean refractive index of a freeze-dried bacterium is made up of one-third organic material with m = 1.48, two-thirds vacuum with m = 1, giving a mean value

$$\overline{m} = \frac{1}{3}(1.48+2) = 1.16.$$

Figure 4 shows the normalized extinction properties of bacteria thus freeze-dried under interstellar conditions. If one accepts the laboratory data on size distribution as being representative, there are *no* free parameters left. The calculated curve, and consequently this agreement with the astronomical data, are then unique. The fit is a consequence of the size distribution $\propto da/a^3$ and the property that $n \cong 1.16$ for freeze-dried bacteria. The agreement is better than for any model that has been previously considered.

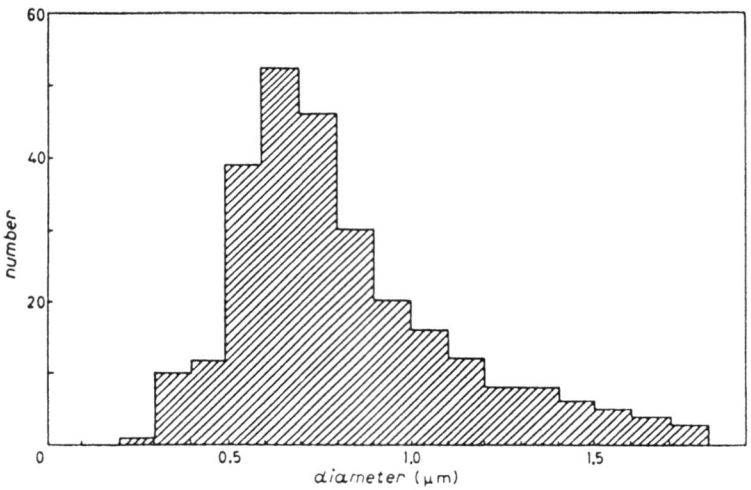

Figure 3: *Size distribution of spore-forming terrestrial bacteria*

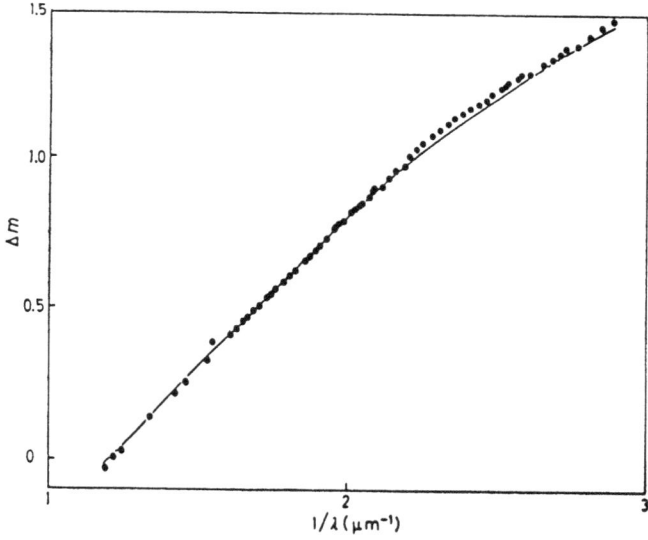

Figure 4: *Calculated extinction curve for the size distribution of bacteria prescribed by the histogram in Fig. 3 assumed to be freeze-dried under interstellar conditions. Normalization is to $\Delta m = 0.409$ at $\lambda^{-1} = 1.62~\mu m^{-1}$, $\Delta m = 0.726$ at $\lambda^{-1} = 1.94~\mu m^{-1}$. Points are the mean extinction data of Nandy for the Cygnus region*

4   THE 3.4 μm BAND: PROOF THAT GRAINS ARE MAINLY ORGANIC

We have already seen that the 10 and 20 μm emission features in the Trapezium Nebula provide suggestive evidence of a major organic contribution to interstellar dust. Clearly the deficiencies of emissivity in the 8-9 μm spectra for most silicates has to be augmented with an appropriate non-siliceous material. It is natural to seek a contribution in the 8-9 μm region from organic solids, which also possess spectral features over a wider infrared waveband. In particular, the observation of a broad CH stretching feature near $\lambda \sim 3.4$ μm (in various aromatic/aliphatic configurations) would be conclusive evidence of organic grains.

When the organic/biological grain model was first proposed there were no suitably calibrated transmittance spectra of desiccated bacteria at infrared wavelengths. We set out to obtain such spectra with a view to testing the bacterial grain hypothesis. An experimental procedure was designed so that the measured transmittance curves were calibrated to give an optical depth $\tau(\lambda)$ on an absolute scale. Our results were not contaminated by scattering because of the choice of KBr as the dispersing agent, a material that possesses the same $n(\lambda)$ values in the 2.5-15 μm band as bacteria to within a small margin (Hoyle *et al.*, 1982a).

Fig. 5 shows the measured transmittance curve over the wavelength range 2.6-3.9 μm for three systems: (a) *E. coli* at room temperature, (b) *E. coli* at 350°C and (c) dehydrated vegetative yeast cells.   Whilst the transmittance curve is seen to be generally the same for the three cases over the whole wavelength range shown in Fig. 5, we note that there is a detailed invariance of shape between 3.3 μm and 3.5 μm. Still more strikingly, the same invariance is seen to hold for eukaryotes (cells with nuclei) as for prokaryotes (bacteria and blue green algae).

When these spectra were first obtained on 21 May 1981 it was noticed that the 3.3-3.9 μm opacity behaviour bore a general resemblance to the spectrum of GC-IRS 7 published by D.T. Wickramasinghe and D.A. Allen (1980).   The significance of this particular comparison arises because GC-IRS 7 happens to be ideally placed for studying the properties of interstellar dust over a 10 kpc path length from the solar system to the galactic centre.   In May 1981 Allen and D.T. Wickramasinghe observed the same object at a higher spectral resolution than before and over a slightly more extended wavelength region. Their spectrum is shown in Fig. 6. The source of the primary radiation in which absorption takes place is thought to be a late-type supergiant. After experiencing general interstellar reddening, the radiation over a limited wavelength region centred at 3.4 μm would, if specific absorptions were absent, approximate to a black-body distribution for a lower temperature than the supergiant itself. The observations actually have an envelope that is close to a black-body distribution at 1100°K.

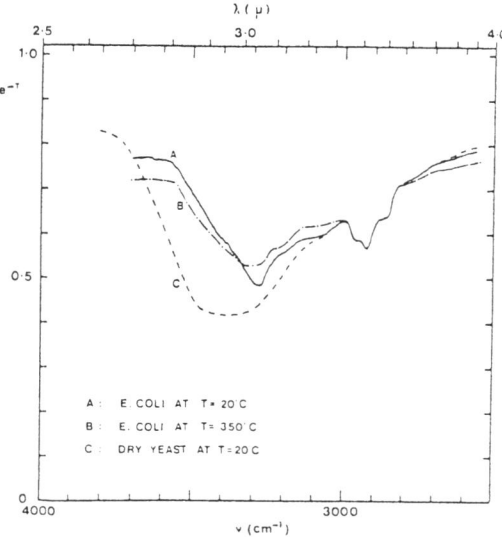

Figure 5: *The measured transmittance curves of microorganisms. For E. coli a dry mass of 1.5 mg was dispersed in a KBr disc of radius 0.65 cm. The transmittance data for yeast was normalised to agree with the E. coli curves at λ = 3.406 μm*

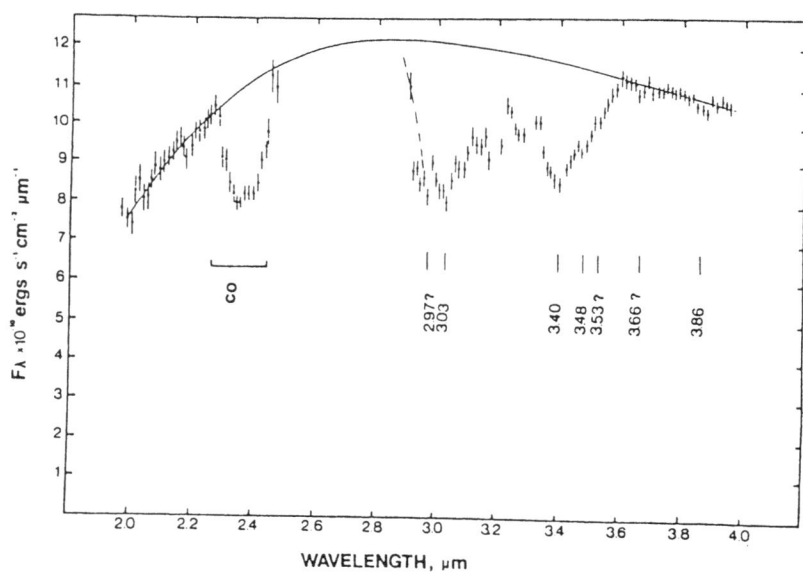

Figure 6: *Observed flux from GC-IRS 7 (from Allen and D.T. Wickramasinghe, 1981)*

It is against this background continuum that absorptions by both gas and dust can be seen. The 2.4 μm absorption band seen here is due to CO, and the 3.4 and 2.95 μm absorptions are most probably due to organic grains with OH, CH and NH linkages. We note that there is no evidence whatsoever for a 3.1 μm water-ice absorption in this source.

From Fig. 6 it is seen that the depth of the 3.4 μm absorption peak below the continuum amounts to ~ 0.3 mag. For grains with a measured absorption coefficient 500 cm^2 g^{-1} distributed along the 10 kpc path length to the galactic centre, we require a smoothed-out organic grain density of ~ 2×10^{-26} g cm^{-3}, essentially the bulk of the interstellar carbon. This conclusion cannot be relaxed unless one adopts higher values for the mass absorption coefficient. Even $\kappa \cong 1000$ cm^2 g^{-1}, which is high for a band of this width, leads to $\rho_g = 10^{-26}$ g cm^{-3}.

This conclusion can only be avoided if a large part of the 3.4 μm absorption in the GC-IRS 7 spectrum arises within a circumstellar environment close to the source. Recent observations by Okuda *et al.* (1990) of 6 other obscured supergiants all lying within a few parsecs radius of GC-IRS 7 can be used to discount this possibility, however. Fig. 7 shows the relative intensity curves of these sources along with the spectrum of GC-IRS 7. The solid lines are the best-fitting background continua represented by Planck curves with temperatures ranging from 450-1100°K. Against such black-body continua we note that absorptions near 3.4 μm for each source correspond to optical depths that agree to within about 10% with the value for GC-IRS 7. These results confirm beyond reasonable doubt that the bulk of the 3.4 μm absorption does not arise from grains local to the sources, but arises instead from grains in the general interstellar medium.

We shall now model this data more carefully with a view to identifying the organic materials of the grains, starting with the calibrated laboratory spectrum of desiccated *E. coli* in the 2.6-3.6 μm waveband measured in the laboratory by S. Al-Mufti and plotted in Fig. 8 (Hoyle *et al.*, 1982a). Imagine the radiation from GC-IRS 7 to be collimated to give a beam with intensity distribution $I(\lambda)\, d\lambda$ directed towards the Earth. Because of the scattering and absorption of the radiation which occurs *en route* to the Earth, a terrestrial observer determines the spectrum

$$\exp[-Q_{sca}(\lambda)]\, \exp[-Q_{abs}(\lambda)]\, I(\lambda)\, d\lambda, \tag{2}$$

$Q_{sca}(\lambda)$ and $Q_{abs}(\lambda)$ being the wavelength dependent scattering and absorption integrated along the line of sight. The source of $I(\lambda)$ is inferred from studies of CO absorption at $\lambda \cong 2.4$ μm and from near-infrared filter photometry to be an M supergiant with an effective temperature near 3200°K, so that $I(\lambda)\, d\lambda$ is much like

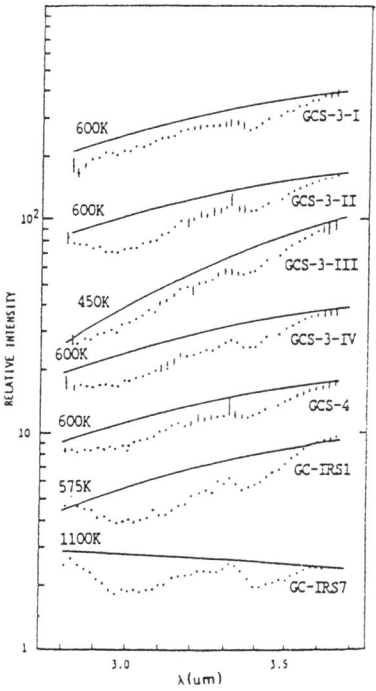

Figure 7: *Spectra of several discrete infrared sources in the galactic centre region distributed within a volume of 3 cu.pc. The solid lines are proposed underlying black-body continua. The data are from Okuda et al., 1990*

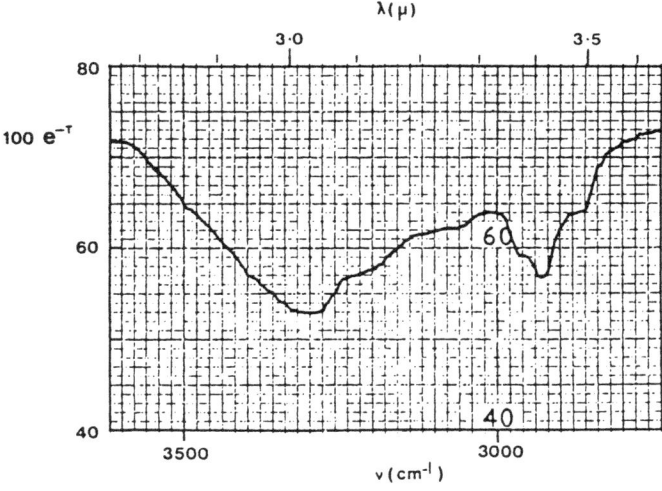

Figure 8: *Enlarged laboratory spectrum giving transmittance values over the 2.6-3.6 μm waveband for E. coli heated to 350°C*

the Planck distribution for this temperature. Multiplication by $\exp[-Q_{sca}(\lambda)]$ has the effect over a limited wavelength range of yielding an intensity distribution $I(\lambda) \exp[-Q_{sca}(\lambda)] \cong B(\lambda, T_c)$, where $B(\lambda, T_c)$ is the Planck function for a suitably chosen colour temperature T_c. We assume $T_c = 1100°K$, which is consistent with the envelope of the observations of Fig. 6. Hence (2) can be written as

$$AB(\lambda, 1100) \exp[-Q_{abs}(\lambda)] \, d\lambda, \qquad (3)$$

where A is a constant depending on the intrinsic emission and distance of the source. If $Q_{abs}(\lambda)$ arises from the absorption values given in Fig. 8, then (3) takes the form

$$AB((\lambda, 1100) \exp[-\alpha\tau(\lambda)] \, d\lambda, \qquad (4)$$

with $\exp[-\tau(\lambda)]$ given by the curve of Fig. 8 and α the factor by which the quantity of absorbing material along the astronomical line of sight exceeds the amount used in the laboratory sample for which Fig. 8 was obtained.

It is worth noting that $B(\lambda, 1100)$ is nearly flat over the wavelength range from 2.8 µm to 3.6 µm, varying by about 10 percent, so that the situation is nearly the same as if the interstellar grains were in the laboratory with a flat source function used to obtain their spectrum. This is a favourable situation for using the astronomical observations to infer $\tau(\lambda)$ - i.e. for making the present comparison.

A and α must be specified before explicit numbers can be calculated from (4). The constant A disappears when (4) is normalised with respect to the scale used for cF_λ (F_λ being the flux), with α remaining as a disposable constant. We chose $\alpha = 1.3$ so as to give the correct depth of the flux curve at the 3.4 µm band centre. All that then remains to decide is the scale factor A, which can evidently be chosen to agree with the observed flux at any one wavelength, but only at one wavelength. The sensitive region for comparing the calculation of (4) with the observed fluxes is the range 3.3 to 3.5 µm, and the comparison will be all the stronger if we avoid choosing A so as to normalise to one of the data points in this critical range. Explicitly, we have chosen to normalise with respect to the data point at $\lambda = 3.562$ µm. The result is shown in Fig. 9.

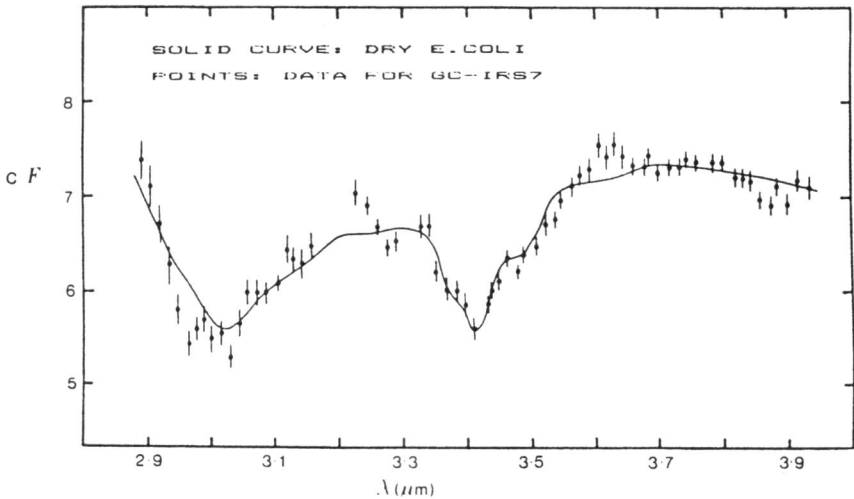

Figure 9: *The agreement between our E. coli model for the parameters given in text (solid curve) and the data for GC-IRS 7 (points with error bars) as supplied by D.T. Wickramasinghe*

The fit of our theoretical curve to the observations is seen to be exceedingly close. The requirement for any competing model is that it should have a normalised $e^{-\tau(\lambda)}$ curve that matches that of Fig. 5 to within ~ 2 percent or so at each wavelength. Fig. 10 shows the comparison between the *E. coli* transmittance curve with that for (a) a synthetic residue obtained by energetic particle irradiation of a mixture of CH_4 and HO (Strazulla, 1986), and (b) an organic residue extracted from the condensed carbonaceous matter in the Murchison meteorite (Pflug, 1982, 1983). In neither case is the agreement found to be within acceptable limits.

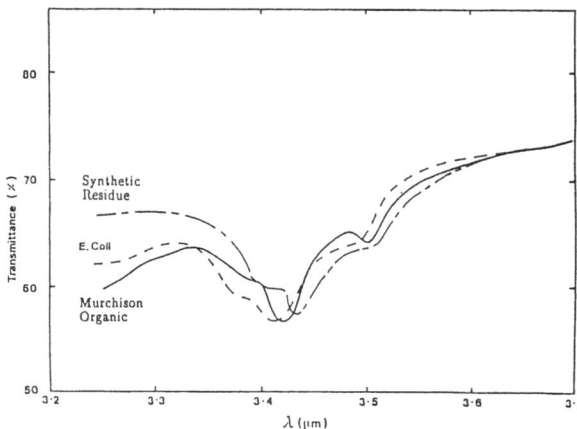

Figure 10: *Comparison between normalised transmittance curves of E coli and Strazulla's proton irradiated H_2O-CH_4 ice residues and Murchison organic matter*

314

5 THE 8-40 μm SPECTRUM OF THE TRAPEZIUM NEBULA

Our original motivation for shifting attention from inorganic grains to organic and biological dust was the defect of silicate models in explaining the 10 and 20 μm emissions from the Trapezium nebula. A combination of amorphous silica (as typified by diatom silica) and organic grains of a bacterial character, mixed in relative proportions consistent with solar abundance ratios of elements, can be shown to produce excellent agreement with the observational data (Wickramasinghe *et al.*, 1989). Fig. 11 shows the measured transmittance curves of the two components from our laboratory experiments. Fig. 12 shows a normalised flux curve calculated according to the expression

$$F_\lambda \propto \bar{\kappa}(\lambda)\, B_\lambda(T) \tag{5}$$

where $T = 170°K$, $\bar{\kappa}$ being the average opacity of the mixture and $B_\lambda(T)$ being the Planck curve. The agreement with the data points is seen to be remarkably close. This result could apply to *either* a cosmically occurring mixture of amorphous silica and bacterial grains *or* a mixture of naturally occurring carbonaceous microorganisms and diatoms as we originally found to be present in the microbial flora of local river water (Hoyle *et al.*, 1982b).

6 CLUSTERS OF AROMATIC MOLECULES

The discovery of a 3.28 μm feature in the diffuse radiation from the Galaxy (Giard *et al.*, 1988) confirms that aromatic molecules are widespread on a galactic scale. These results may be seen to complement earlier IRAS observations in the 12 μm waveband of emission from high latitude 'cirrus clouds' in the Galaxy which have also been attributed to similar molecules (Leger and Puget, 1984). Infrared emissions throughout the 3-12 μm waveband must arise from an absorption of ultraviolet starlight that is subsequently degraded into the infrared. The total fraction of carbon tied up in the form of PAH molecules may be typically in the region of a few percent (Leger and Puget, 1984).

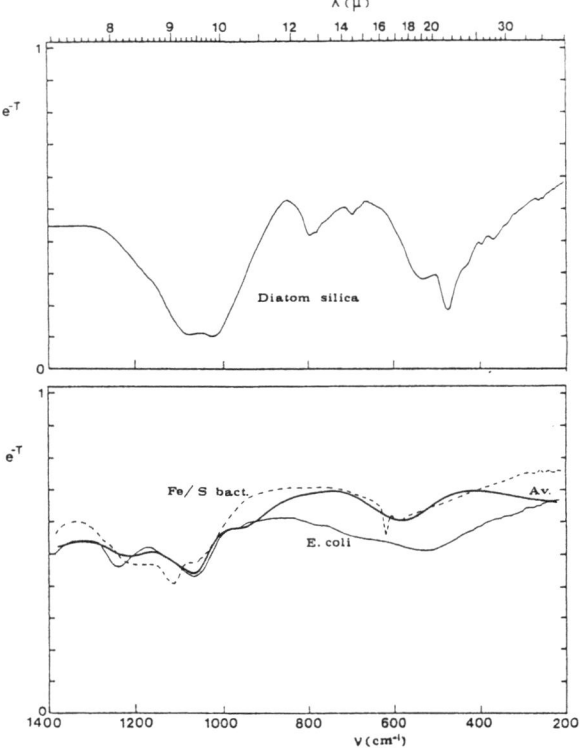

Figure 11: *Upper panel: transmittance curve for diatom silica. Lower panel: transmittance curves for E. coli (Fine solid line); for a mixed culture of iron and sulphur bacteria (dashed curve); for an 'average' interstellar organic particle (heavy solid line)*

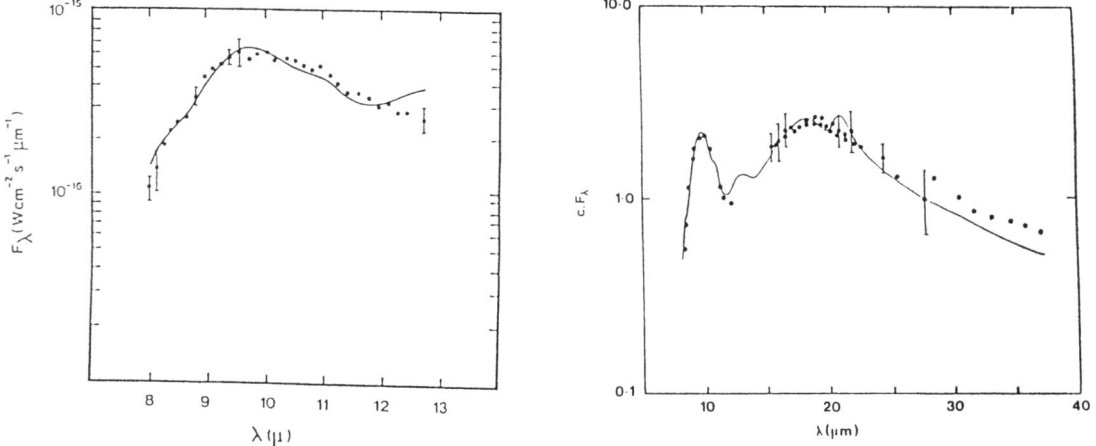

Figure 12: *Normalised flux curve for the 'average' bacterium-diatom silica mixture calculated for a temperature of 170 °K. The observational points are data for the Trapezium nebula from Forrest et al., 1975a,b; 1976*

The formation of aromatic molecules on a vast scale in the galaxy poses a major difficulty for inorganic theories of astrochemistry. Aromatic molecules, however, are commonplace in biology and such molecules are expected to arise quite naturally within fragments of biological grains as they become disrupted in conditions of intense ultraviolet or particle irradiation. It is possible that the ratio of the numbers of condensed aromatic to aliphatic molecules might increase substantially above an initial value in the parent grains under certain irradiation conditions, due to the greater stability of ring structures. In this connection we note that the insoluble organic matter in carbonaceous chondrites is comprised mainly of a polymer with an aromatic skeleton (Hayes, 1967).

Break-up of bacterial particles could lead to fragments which are widely distributed in size We are concerned here with particles that contain less than a thousand individual atoms. Such small particles would not attain a steady equilibrium temperature in most interstellar or circumstellar environments as would the larger grains. A fragment containing N atoms becomes transiently heated due to absorption of UV photons, and each absorption event leads to a thermal spike with maximum temperature T given approximately by the condition

$$E = 3NkT, \tag{6}$$

where $E = h\nu$ is the energy of the incident photon. Equation (6) can be re-written in the form

$$T = 534 \frac{E/10\,eV}{N/100} \, K, \tag{7}$$

from which a minimum value of N may be determined using the condition that the structure must survive disruption for the highest energy of incident photons. The temperature T for which disruption may be assumed to occur is controlled essentially by the average strength of the aromatic C-C bond, ~ 6 eV. This assures stability for temperatures up to $T \sim 1200$ K, and with $E \sim 10$ eV equation (7) gives a minimum N value ~ 50.

Clusters of organic molecules in the quantities involved here must contribute appreciably to the extinction of starlight at ultraviolet wavelengths. Aromatic molecules invariably possess strong absorptions in the ultraviolet with band strengths amounting to $\sim 500,000$ $cm^2 g^{-1}$ in typical cases and with the peak UV absorption wavelength varying in the range ~ 1800-2600 Å. A density of $\sim 10^{-27}$ g cm^{-3} in the form of aromatic molecules would thus produce an absorption of ~ 1.5 mag kpc^{-1} at the centre of the UV band, which is close to the value appropriate to the well-known λ2200 Å interstellar absorption feature. It would seem natural, therefore, to connect the 3.28 μm diffuse galactic emission with the λ2200 Å interstellar absorption. Any candidate molecule for the 3.28 μm emission that conspicuously fails to produce an extinction band at 2200 Å could be deemed unsatisfactory for this reason. The currently favoured explanation for the 3.28 μm galactic emission is based on the model of polycyclic aromatic hydrocarbons that are thought to be derived from interstellar graphite grains. They are essentially graphitic in their structure with each molecule containing some 50 or so atoms. Coronene $C_{24}H_{12}$ has been regarded as a useful prototype in this context because it gives a reasonable degree of correspondence in the positions and relative strengths of emissions at 3.28, 6.2, 7.7, and 8.6 μm. However, the sequence of molecules from naphtalene, tetracene, perylene, coronene and beyond is doomed to failure for reasons evident in Fig. 13. We show here the UV spectra for members of this sequence starting from naphtalene. Naphtalene has an absorption peak at ~ 2200 Å, which is indeed satisfactory, but further members have peaks that are systematically shifted to longer wavelengths, tending eventually to a peak near 2600 Å which is that appropriate to the absorption coefficient of bulk graphite. The higher-order compact PAHs, which are expected to behave like bulk graphite are, therefore, less than satisfactory models for the interstellar 3.28 μm emission band.

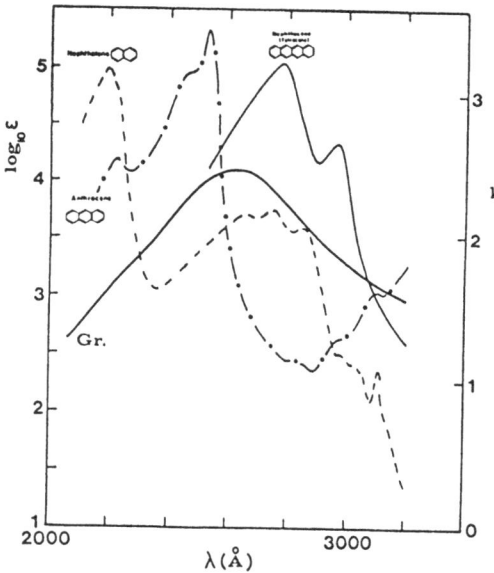

Figure 13: *The absorption spectra of naphtalene, anthracene, tetracene, and graphite. The ordinate scale for molar extinction coefficient (cm⁻¹ (g mol/litre)⁻¹) refer to the molecules; the absorption coefficient k refers to graphite (Gr.)*

We now consider a sample of naturally occurring aromatic biological molecules listed under various categories by Scott (1964). Table 1 lists the classes of compounds considered.

Table 1: *Categories of molecules considered*

α-β unsaturated acids and esters
α-β unsaturated lactones
Indole chromophores
Pyridines
Quinolines, isoquinolines, and acridines
Pyrimidines
Purines
α-oxo and γ-lactones and derivatives
Hydroxyanthraquinones

For 115 aromatic molecules in the above categories we examined infrared spectra in the Stadler Atlas (1978) and in Butterworth's Index Cards of Spectra (1966, 1970, 1972). We first constructed a histogram of the distribution of the wavelength of maximum CH absorption in the 3.2 to 3.5 μm waveband, which is shown in Fig. 14. The average band profile defined by the curve in this figure peaks at $\lambda = 3.28$ μm, in good agreement with the behaviour of an individual heterocyclic ring structure such as, for instance, quinoline.

We next proceed to estimate the average emissivity of the ensemble (including 115 molecules) listed in Table 1 over the wavelength range 2.5-12.5 μm. Each of the laboratory ε_λ curves from the standard atlases were electronically digitised and scaled so that $\varepsilon_\lambda = 1$ at the wavelength of maximum absorption in the 2.4-4 μm waveband. The set of normalised ε_λ functions were then averaged with equal weightings to yield a normalised $\overline{\varepsilon}_\lambda$ curve for the entire ensemble.

Table 2 lists the main absorption peaks in the average $\overline{\varepsilon}_\lambda$ curve compared with the astronomically occurring peaks in the spectra of NGC 7027 and NGC 2023. The general correspondences are seen to be satisfactory.

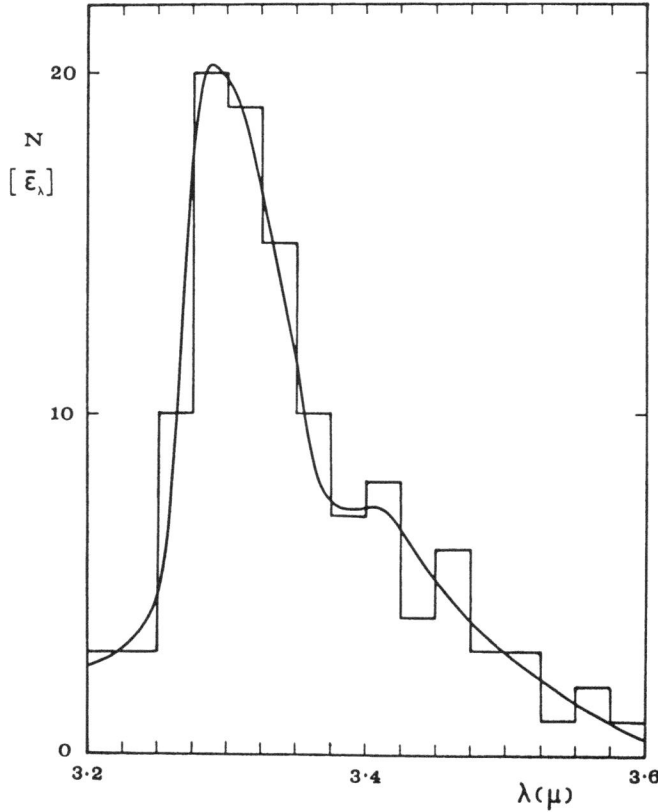

Figure 14: *The distribution of the wavelength of the maximum of the CH band in 115 aromatic compounds in categories listed in Table 1 and set out in several tables of Scott (1964). The curve represents a mean emissivity profile of the mixture in the waveband 3.2-3.6 μm*

Table 2: *Wavelengths of principal absorption peaks (μm)*

NGC 7027, NGC 2023:		3.3	3.4		6.2	7.7	8.6	11.3
Mixture of aromatics:	2.9	3.3	3.4	5.25	6.2	7.7	8.8	11.3

We next turn to the ultraviolet properties of a biologically generated aromatic ensemble. The distribution of ultraviolet absorption peaks in 115 aromatic bio-molecules of the types listed in Table 1 is plotted in the histogram of Fig. 15. The solid curve is the computed average opacity curve for this ensemble assuming that the absorption curve for each individual molecule is similar in shape to the calibrated curve for quinazoline isomers (Hoyle and Wickramasinghe, 1977).

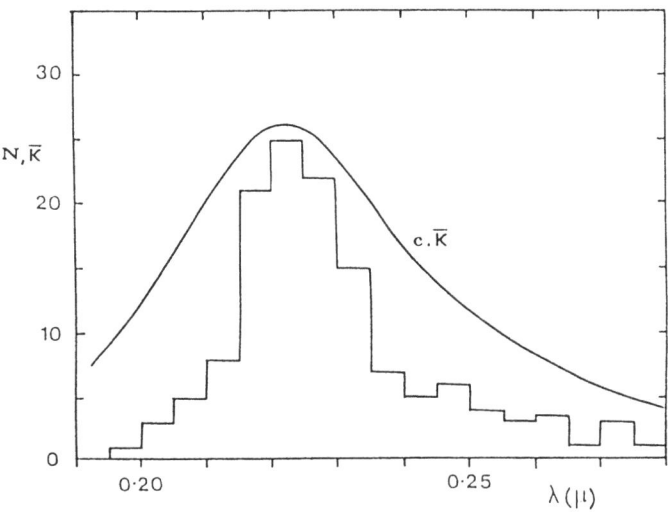

Figure 15: *Histogram shows the distribution of principal ultraviolet absorption peaks for a set of naturally occurring aromatic molecules of types listed in Table 1. The curve is the calculated mean absorption profile for the ensemble*

Combining the visual extinction arising from hollow bacterial grains as discussed earlier with small silica grains (radii 0.03 μm) to provide extinction in the far ultraviolet we are left with a total extinction curve that differs from the observations to the extent indicated by the points in Fig. 16. The solid curve here is a normalised extinction arising from aromatic clusters using the $\bar{\kappa}$ function of Fig. 16. When hollow bacterial grains, small silica grains and aromatic clusters are appropriately combined the entire extinction curve (Fig. 5 of the first contribution) could be reproduced. The inclusion of a contribution to extinction from iron whiskers of radii 0.01 μm, or iron impurities does not change this general conclusion (Wickramasinghe *et al.*, 1991; Jabir *et al.*, 1983). Indeed, variations of the relative proportions of such constituents, which are to be expected, lead to a set of extinction curves with variable characteristics in the ultraviolet waveband. Such variability is in fact fully consistent with astronomical observations (see, for example, Fig. 4 of the first contribution).

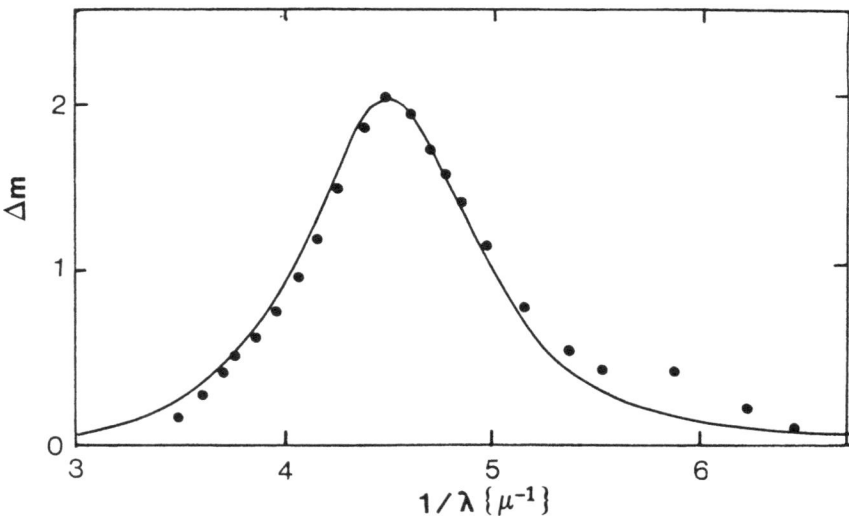

Figure 16: *The points are the excess interstellar extinction over a scattering continuum. The thin solid curve is the absorption due to aromatic molecules occurring in the form of free clusters.*

To conclude this section we note that clusters of aromatic molecules resulting from the break-up of bacterial grains are able to explain both the mid-infrared diffuse emission features and the 2175 Å interstellar extinction. In the following section we consider the role of such clusters over the visual waveband.

7 AROMATIC MOLECULES AND THE DIFFUSE OPTICAL BANDS

There are a number of diffuse lines and bands in stellar spectra which are to date unidentified and which can only be interpreted as fine structure in the wavelength dependence of extinction. They have widths ranging from ∼ 40 Å to less than 5 Å. Attempts to explain these features on the basis of inorganic models have been singularly without success. Even for the narrowest of the diffuse bands the observed widths are too large for atomic lines or lines that can be attributed to small molecules.

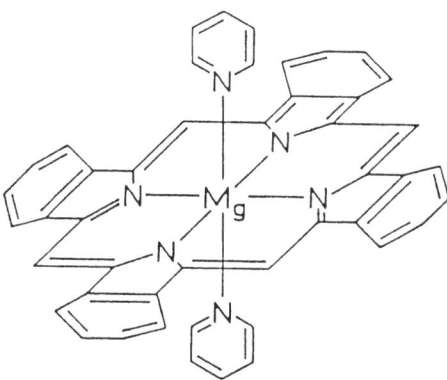

Figure 17: *Structure of dispyridyl magnesium tetrabenzoporphine*

Against such a backdrop of failure, F.M. Johnson's (1967) attempt to identify the diffuse band carrier with an aromatic macromolecule is worthy of praise (see also Johnson, 1972; Johnson *et al.*, 1973). The molecule proposed was in fact a porphyrin, dispyridyl magnesium tetrabenzoporphine ($MgC_{46}H_{30}N_6$) which belongs to the class of heterocyclic aromatic molecules. The structure of this molecule is depicted in Fig. 17. It is worth noting that Johnson's suggestion predated the radioastronomical discoveries of complex organic molecules by a few years and the now extensive discussions of polyaromatic hydrocarbon molecules (PAH's) by over two decades. It is therefore no wonder that Johnson's ingenious ideas of the 1960's scarcely received any attention and now seems to be largely forgotten. We shall point out here that the porphyrin model for the diffuse bands can now be placed in proper context and has indeed a good deal to commend it.

Firstly, let us note that porphyrins have a unique importance in living systems and therefore have a profound relevance to the biological grain model we have already discussed. Chlorophyll (a porphyrin) plays a crucial role in photosynthesis and cytochrome (also a porphyrin) is vital for respiration. Porphyrins (including non-magnesium metalloporphyrins) have predominantly planar configurations and are endowed with a high degree of thermal stability. The particular molecule considered by Johnson is reported to have a "sublimation" temperature approaching 900°K, but the dissociation temperature in vacuum could be even higher. If such heterocyclic aromatic molecules arise from the break-down of larger biological grains they are likely to have a long-term persistence under interstellar conditions.

Even a cursory glance at published spectral data of chlorophylls, metalloporphyrins and related pigments shows the possibility of the strongest visual absorptions occurring at wavelengths that overlap with λ 4430 Å and some of the other diffuse interstellar features (Hoyle and Wickramasinghe, 1979; Stern and Timmons, 1970). Although the apparent clustering of measured absorption wavelengths close to 4430 Å was an encouraging aspect at the outset, the laboratory techniques that are normally employed had introduced several limitations to modelling. In most instances spectroscopic data are secured at room temperatures for samples dispersed in a variety of solvents. Such procedures in general introduce significant wavelength shifts (up to 10 Å) as well as a merging of individual lines into relatively broad bands.

F.M. Johnson's pioneering contributions followed essentially from the development of an experimental technique that overcame some of these difficulties. The sample is suspended in an appropriately chosen paraffin matrix and cooled to 77°K. The technique effectively yields high resolution spectra which are comprised of sharp lines rather than broad bands and also permits an analysis of vibrational levels of the absorbing molecule.

Table 3 lists the principal wavelengths for the strong bands that are found in both absorption and fluorescence spectra for $MgC_{46}H_{30}N_6$. It is most remarkable that two of the strongest diffuse interstellar bands and 3-4 others are immediately matched by this model to a high degree of precision. Johnson also obtained infrared spectra of the same molecule in KBr pellets to construct a detailed energy level diagram based on experimental data of the electronic (0,0) states corresponding to wavelengths 6175 Å (state A) and 4428 Å (state B).

Table 3: *Laboratory spectra of $MgC_{46}H_{30}N_6$ compared with diffuse interstellar bands (adapted from Johnson et al., 1973)*

[f $\equiv$ strong fluorescence lines; a $\equiv$ strong absorptions]

Laboratory data for MgTBP		Astronomical data	
Central Wavelength (Å)	Range (Å)	Central Wavelength (Å)	Range (Å)
6663 (f)	1-2	6661	1-2
6614 (f)	1-2	6614	1-2
6289 (f)	1-2	6284	4
6284 (f)	1-2		
6174 (a)	14	6175	~20
4428 (a)	40	4428	~30

Fig. 18 shows Johnson's energy level configurations in which he utilises vibrational frequencies γ = 231, 280, 721, 753, 1051, 1110 cm^{-1} together with their harmonics and combinations. Johnson has shown that all the 25 or so diffuse interstellar features could arise from transitions involving various vibrationally excited levels. The question immediately arises as to why most of the diffuse interstellar bands originate from vibrationally excited levels. To resolve this matter Johnson invoked an interstellar chemistry that led to the formation of molecules preferentially in excited states. It is more likely, in our view, that spike-heating of aromatic clusters by UV photons, as discussed earlier, would lead to a distribution of internal temperatures that in turn serves to populate vibrational states. Optical photons encountering such vibrationally excited molecules could readily lead to the absorption spectra as calculated by Johnson. On the basis of such a model we expect the transitions corresponding to vibrationally excited levels to be comparatively weaker as well as narrower, as indeed they are found to be.

8 OVERVIEW AND CONCLUSION

In this final section we shall summarise the position that has been reached. Organic material spectroscopically and structurally indistinguishable from bacteria and their degradation products exists everywhere throughout the Galaxy. Under very dry conditions, at low pressure, the water which normally occupies the interior volumes of bacteria would evaporate away, leaving particles that were mostly hollow, and thus possessing low values of refractive index as are required. In addition to biological particles and aromatic molecular clusters derived from them, there is also evidence for inorganic dust particles dominated by minerals including amorphous silica.

The sites in the galaxy for active microbiology cannot be interstellar space, judging from the requirements for bacterial replication found here on the Earth. Either the presence of liquid water or an atmosphere with relative humidity above sixty percent is usually required. The temperature range for replication appears to be typically from about -20°C to +80°C. But the presence of living bacteria has been discovered at temperatures up to 300°C in deep-sea "black smoker" chimneys associated with volcanic vents, and special types of bacteria have also been recovered from deep drills into the Earth's crust down to ~ 7 km (Gold, 1992), showing clearly that the upper limit of temperature exceeds 80°C by a considerable margin under high pressure conditions.

325

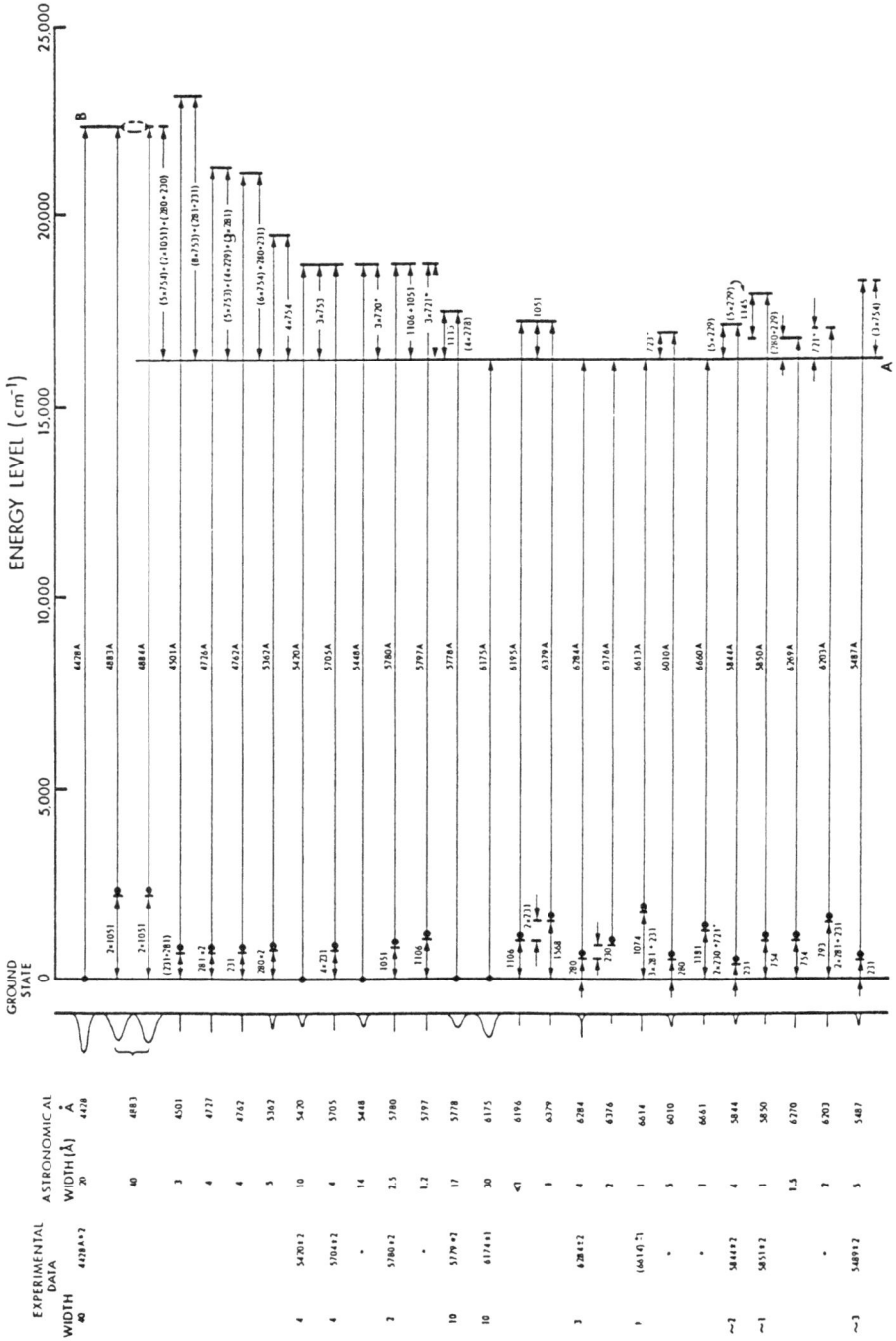

Figure 18: *Johnson's energy level diagram for disphyridylmagnesium tetrabenzoporphine compared with the wavelengths of the diffuse interstellar bands*

A supply of suitable nutrients is essential. Depending on the species of bacterium in question the nutrients can vary widely. It is the characteristic of a large class of so-called chemoautotrophic bacteria that they replicate from inorganic substances alone. Indeed, one can say that wherever inorganic substances exist under natural conditions with the possibility of energy being obtained by means of a chemical reaction, then a bacterium exists to exploit the situation, and to draw upon this store of energy. This assumes the reaction occurs only very slowly under purely inorganic conditions. Otherwise the opportunity for bacteria would not exist, because if the reaction proceeds inorganically the nutrients would be lost. The important trick deployed by chemoautotrophic bacteria is to exploit energy-producing reactions that are otherwise too slow to take place inorganically. They speed-up such reactions by enormous factors through the use of subtle catalysts in the form of proteins called enzymes.

Bacteria are exceedingly hardy in every respect one can think of. They have remarkable tolerances to temperatures, to pressure, to nutrients, and to radiation damage. The properties of bacteria in general are not of the kind that could possibly be explained by evolution in a purely terrestrial environment. In many respects their properties bear no relation whatsoever to conditions encountered naturally on the Earth, for instance in their resistance to exceedingly low temperatures and pressures, and their resistance to immense doses of destructive radiation. Consequently there has never been a terrestrial environment against which such properties could have evolved, and so according to the Darwinian mode of evolution they simply should not exist. But in fact contrary to expectations they do.

With interstellar space ruled out as replication sites of bacteria because of the hard-frozen conditions that prevail there, we are led logically to look to conditions in the vicinities of stars. We can imagine that interstellar micro-organisms, a fraction of them viable, are brought by astronomical processes to the vicinities of stars. In such locations, upon or within suitable objects, and given suitable conditions, they can replicate explosively.

Planets are much larger than comets, so that at first sight one might think planets would be more suitable places for biological replication than comets. But what a planet gains over a single comet in size and mass is lost in their relative numbers. In the solar system there are billions of times more comets than planets. Taken together, comets compare in their collective mass with large planets like Uranus and Neptune. Far from being ruled out on grounds of size, comets collectively provide places that appear far superior to planets. Comets are thought to have been assembled from much smaller cometissimals in the outer regions of the solar system some 4600 million years ago. Cometissimals inevitably combined the organic bacterial-type particles from interstellar space with icy material that condensed out of gases escaping from the primitive sun. Radioactive nuclides such as ^{26}Al would have been included in cometary bodies, and this ensured the maintenance of liquid cores for millions of years. Within such liquid cometary cores colonies of anaerobic bacteria would have found conditions favourable for rapid expansion in the early history of the solar system. Subsequently the biomaterial became hard-frozen when radioactive heat sources eventually ran out.

A fraction of comets in the solar system approach the Sun, as in the approach of Comet Halley. These special cases produce evaporation of gas and dust at a rate of about a million tonnes per day. The dust in the form of bacteria and other micro-organisms, is partly expelled from the entire solar system into interstellar space and is partly available for seeding planets like the Earth with life. Fig. 19 shows the feedback loop that controls the distribution of life in the galaxy. In all about 10^{10} circuits in this feedback loop would have been completed, one for every sun-like star.

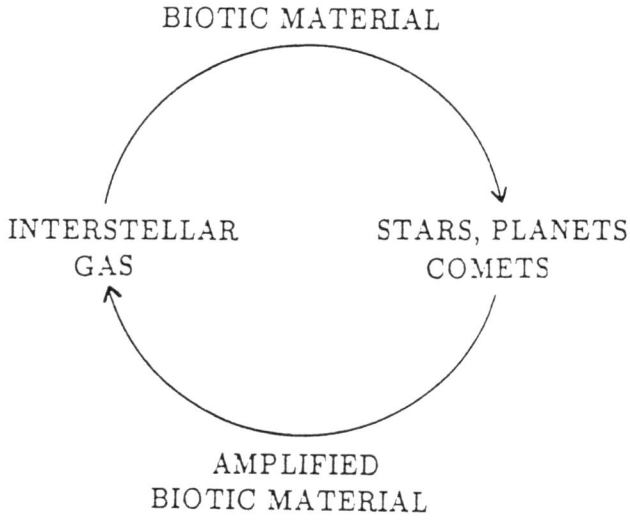

Figure 19: *Cosmic amplification cycle*

According to dogma that reigned up until the recent approach of Comet Halley, comets were supposed to be "dirty snowballs" not residues of intense biological activity. Water would of course be present in considerable quantity, whether in a snowball or in a biological comet. To decide between the two competing models it was necessary therefore to examine the composition of components of a comet other than water. The data on comets had favoured the organic hypothesis for thirty years or more, so that there was never any real support for the 'dirty snowball' dogma, if by 'dirt' one meant mineral dust. It was found long ago that present in the gases emitted by comets were all manner of carbon-based molecules containing small numbers of atoms, e.g. C_2, C_3, CN, CH, H_2CO, HCN, of a kind that would be expected to appear when complex organic molecules were broken up into fragments. The small molecules actually found in comets were not at all what would be expected from the break up of inorganic materials. For example, nothing like these molecules could be expected from the disintegration of simple inorganic ices or bits of mineral dust. The recently discovered molecule S_2 is particularly interesting in relation to disulphide linkages which play so important a role in protein structure of living material.

The recent approach of Comet Halley to the inner regions of the solar system established several important results. The nucleus of the comet was nothing like the expected bright snowfield when imaged at close quarters from the Giotto spacecraft. Rather it was described as 'black as coal', exactly as was predicted for the organic comet model (Hoyle and Wickramasinghe, 1986). Some ninety percent of the comet surface was found to be covered with an inert crust comprised of tarry organic material. The eruptions of the comet seemed mostly to be confined to a small number of craters some 100 metres deep in which the surface crust was ruptured so that small particles and gases were easily released.

A large fraction of the particles that emerged from the comet near perihelion, at a rate of about a million tonnes per day, had compositions dominated by the elements C, H, O, N. Since the particles did not evaporate away at temperatures as high as 50° C they would not be inorganic compounds of these elements like water and ammonia and so they were surely of an organic nature. Many of the particles were found to have sizes typical of bacteria. Moreover, the particles had low densities consistent with them being hollow, as bacteria would be after the water initially within them had evaporated away. *In situ* mass spectroscopy of degradation products of cometary dust, as the particles impacted instruments aboard the Giotto spacecraft, revealed beyond a shadow of doubt the presence of complex organic structures.

The types for molecules inferred from these studies, as given by Kissel and Krueger (1987), are listed in Fig. 20. These structures are of course fully consistent with the break-up of bacterial cells. A widely-stated claim that bacterial particles are excluded because the break-up fragments do not show a mass corresponding to the biologically important element phosphorus is without foundation. Mass peaks are to be interpreted in terms of plausible combinations of the atom phosphorus with other elements. The break-up of phosphate groups such as occur in biology would lead to mass peaks PO_3^+ (79), PO_2^+ (63), or PO^+ (47) rather than P^+ (31). Such evidence does indeed exist as seen in Fig. 21. The claim that mass peaks corresponding to two other biologically significant elements Na and K are too low is also open to challenge. Although it is true that freeze-dried laboratory cultures of bacteria would be a thousand times richer in Na and K, this is not true of nutrient deprived cells nor indeed of bacterial spores.

In all respects cometary particles were found to be exceedingly similar to the interstellar organic bacterial particles that we have already discussed. Particularly impressive is the infrared emission from the dust around comet Halley measured on 31 March 1986 by D. T. Wickramasinghe and D.A. Allen (1986) seen in Fig. 22. The curve is the predicted behaviour of bacterial grains. The status of the closest non-biological competitor is also shown in this figure. This comparison leaves little room for doubting the general validity of the ideas developed in the present contribution.

C—H— Compounds

(Only high-molecular probable due to volatility; hints only to unsaturated)

$HC\equiv C(CH_2)_2CH_3$ — Pentyne
$HC\equiv C (CH_2)_3CH_3$ — Hexyne
$H_2C=CH—CH=CH_2$ — Butadiene
$H_2C=CH—CH_2—CH=CH_2$ — Pentadiene

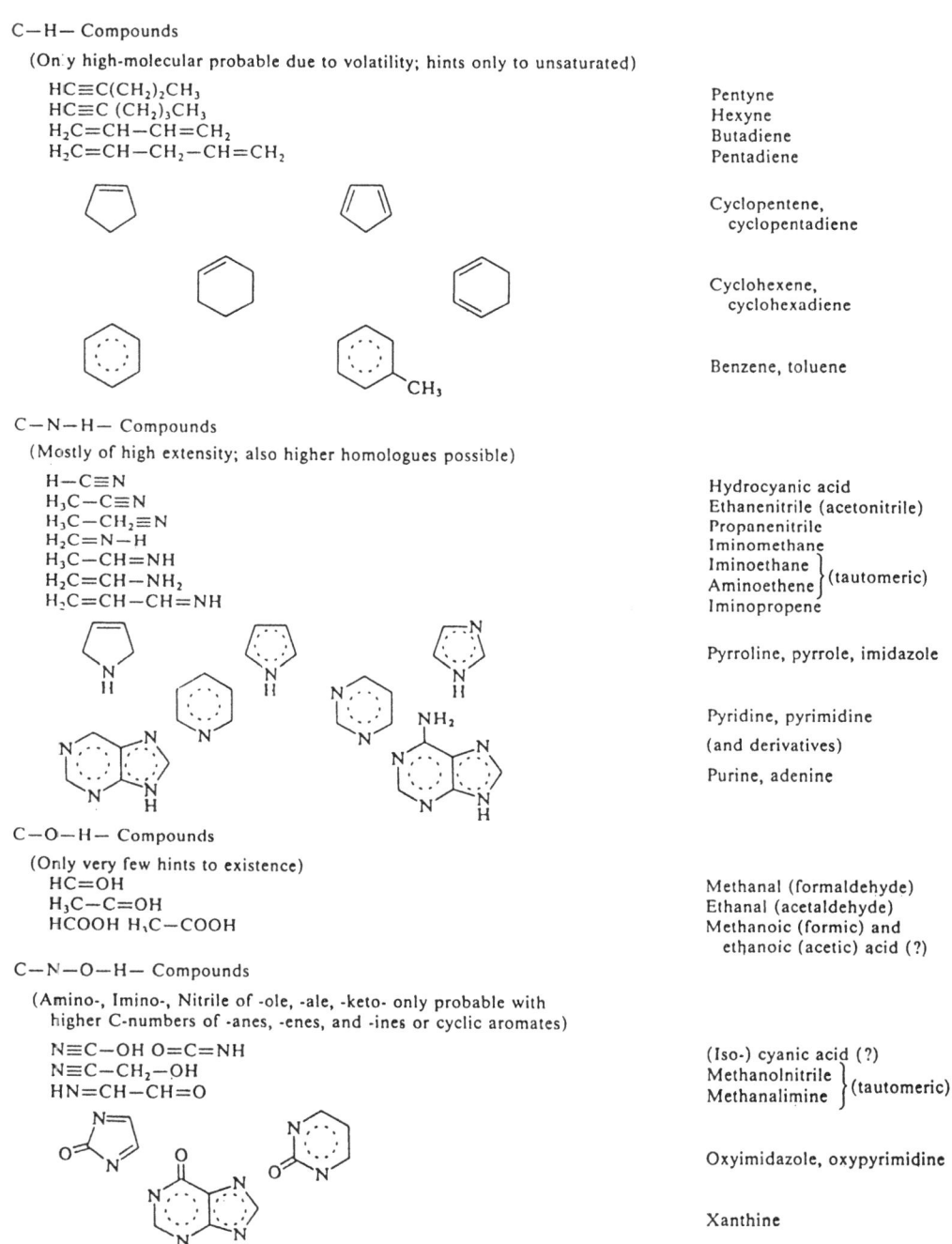

Cyclopentene, cyclopentadiene

Cyclohexene, cyclohexadiene

Benzene, toluene

C—N—H— Compounds

(Mostly of high extensity; also higher homologues possible)

$H—C\equiv N$ — Hydrocyanic acid
$H_3C—C\equiv N$ — Ethanenitrile (acetonitrile)
$H_3C—CH_2\equiv N$ — Propanenitrile
$H_2C=N—H$ — Iminomethane
$H_3C—CH=NH$ — Iminoethane
$H_2C=CH—NH_2$ — Aminoethene } (tautomeric)
$H_2C=CH—CH=NH$ — Iminopropene

Pyrroline, pyrrole, imidazole

Pyridine, pyrimidine

(and derivatives)

Purine, adenine

C—O—H— Compounds

(Only very few hints to existence)

HC=OH — Methanal (formaldehyde)
$H_3C—C=OH$ — Ethanal (acetaldehyde)
HCOOH H₃C—COOH — Methanoic (formic) and ethanoic (acetic) acid (?)

C—N—O—H— Compounds

(Amino-, Imino-, Nitrile of -ole, -ale, -keto- only probable with higher C-numbers of -anes, -enes, and -ines or cyclic aromates)

$N\equiv C—OH$ $O=C=NH$ — (Iso-) cyanic acid (?)
$N\equiv C—CH_2—OH$ — Methanolnitrile
$HN=CH—CH=O$ — Methanalimine } (tautomeric)

Oxyimidazole, oxypyrimidine

Xanthine

Figure 20: *Types of organic molecules as inferred from mass spectra (Kissel & Krueger, 1987)*

331

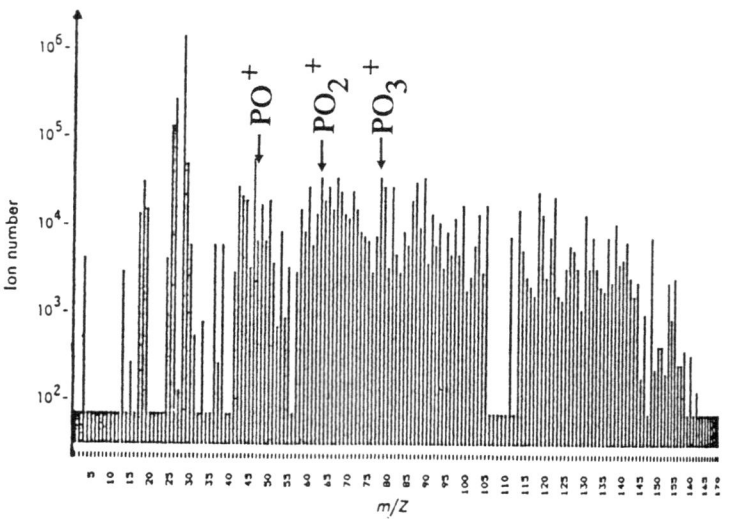

Figure 21: *Cumulative molecular ion mass spectrum on the basis of the 43 evaluated mass spectra (adapted from Kissel and Krueger, 1987)*

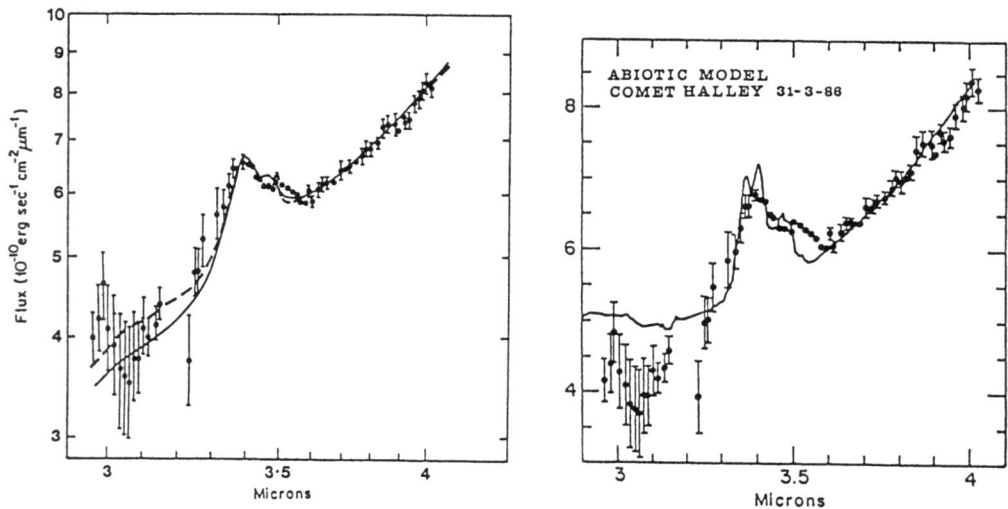

Figure 22: *Left panel: Observational data (points) for Comet Halley on 31 March 1986 compared with calculations for the biological model. The solid curve is the prediction for a single temperature model in the small particle limit; the dashed curve uses the distribution of grain sizes as detected by the Halley probes (Wallis et al., 1989). Right panel: Calculations for the Chyba-Sagan organic residue plus additions of postulated components for background thermal emission and for scattering*

REFERENCES

Allen, D.A. and Wickramasinghe, D.T.: 1981, *Nature*, **294**, 239.

Butterworths Documentation of Molecular Spectroscopy: 1966, 1970, 1972, Butterworth Scientific Publications.

Chyba, C. and Sagan, C.: 1987, *Nature*, **330**, 350.

Forrest, W.J., Gillett, F.C. and Stein, W.A.: 1975a, *Astrophys. J.,* **195**, 423.

Forrest, W.J., Gillett, F.C. and Stein, W.A.: 1975b, *Astrophys. J.,* **192**, 351.

Forrest, W.J., Houck, J.R. and Reed, R.A.: 1976, *Astrophys. J., 208*, L133.

Giard, M., Pajor, F., Lamarre, J.M., Serra, G., Caux, E., Gispert, R., Leger, A. and Rouan, D.: 1988, *Astron. and Astrophys. Let.*, **201**, L1.

Gold, T.: 1992, *Proc. Natl. Acad. Sci. USA*, **89**, 6045.

Hayes, J.M.: 1967, *Geochim. Cosmochim. Acta*, **31**, 1395.

Hoyle, F. and Wickramasinghe, N.C.: 1979, *Astrophys. Space Sci.*, **66**, 77.

Hoyle, F. and Wickramasinghe, N.C.: 1977, *Nature*, **270**, 323.

Hoyle, F., Wickramasinghe, N.C. and Al-Mufti, S.: 1982b, *Astrophys. Sp. Sci.*, **86**, 63.

Hoyle, F., Wickramasinghe, N.C., Olavesen, A.H., Al-Mufti, S. and Wickramasinghe, D.T.: 1982a, *Astrophys. Sp. Sci.*, **83**, 405.

Hoyle, F. and Wickramasinghe, N.C.: 1986, *Astrophys. Sp. Sci.,* **36**, 289.

Jabir, N.L., Hoyle, F. and Wickramasinghe, N.C.: 1983, *Astrophys. Sp. Sci.*, **91**, 327.

Johnson, F.M.: 1967, *Astron. J., 72*(3), April 1967.

Johnson, F.M.: 1972, *Ann. N. Acad. Sci.*, **187**, 186.

Johnson, F.M., Bailey, D.T. and Wegner, P.A.: 1973, in J.M. Greenberg and H.C. van de Hulst (eds) *Interstellar Dust and Related Topics*, D. Reidel.

Kissel, J. and Krueger, F.R.: 1987, *Nature*, **326**, 760.

Léger, A. and Puget, J.L.: 1984, *Astron Astrophys.,* **137**, L5.

Okuda, H., Shibai, H., Nakagawa, T., Matsuhara, H., Kobayashi, Y., Kaiful, N., Nagata, T., Gatley, I. and Geballe, T.: 1990, *Astrophys. J.*, **351**, 89.

Pflug, H.D.: 1982, Private communication.

Pflug, H.D.: 1983, in C. Wickramasinghe (ed.) *Fundamental Studies and the Future of Science*, University College Cardiff Press.

Scott, A.I.: 1964, *Interpretation of the Ultraviolet Spectra of Natural Products*, Pergamon Press, Oxford.

Stadler Handbook of Infrared Spectra, 1978 (Stadler Research Laboratories, USA).

Stern, E.S. and Timmons, C.J.: 1970, *Electronic Absorption Spectroscopy in Organic Chemistry*, Edward Arnold Ltd., London.

Strazulla, G.: 1986, *Light on Dark Matter* (ed. F.P. Israel), D. Reidel.

Wallis, M.K., Wickramasinghe, N.C., Hoyle, F. and Rabilizirov, R.: 1989, *Mon. Not. R. astr Soc.,* **238**, 1165.

Wickramasinghe, D.T. and Allen, D.A.: 1980, *Nature*, **287**, 518.

Wickramasinghe, D.T. and Allen, D.A., 1986, *Nature*, **323**, 44.

Wickramasinghe, N.C.: 1974, *Nature*, **252**, 462.

Wickramasinghe, N.C.: 1975, *Mon. Not. Roy. Astr. Soc.*, **170**, 11P.

Wickramasinghe, N.C., Hoyle, F. and Majeed, Q.: 1989, *Astrophys. Sp. Sci.*, **158**, 335.

Wickramasinghe, N.C., Jazbi, B. and Hoyle, F.: 1991, *Astrophys. Sp. Sci.*, **186**, 67.

Infrared instrumentation

Ian S. McLean

University of California, Los Angeles
405 Hilgard Ave.
Los Angeles, CA 90024, USA

1   THE INFRARED WAVEBAND

The infrared region of the electromagnetic spectrum spans a large range in wavelengths compared to that of normal visible light and it has not been easy to develop technologies which allow astronomers to study the infrared waveband. However, in the last 8 – 10 years there has been a tremendous growth in the field of infrared astronomy. This growth has been stimulated in part by the construction of infrared telescopes and by successful space missions, but the most important event has been the development of very sensitive imaging devices called *infrared arrays*. Before describing these detectors and their uses in instrumentation, it is useful to begin with a brief historical review of infrared astronomy and an explanation of the terminology of the subject.

1.1   Historical review: from Herschel to IRAS

Infrared astronomy had an early beginning when, in 1800, Sir William Herschel noted that a thermometer placed just beyond the red end of the optical spectrum of the Sun registered an increase in temperature due to the presence of invisible radiation which he called *calorific rays*. He even demonstrated that these rays were reflected and refracted like ordinary light. This discovery came 65 years before James Clerk Maxwell's theory on the existence of an entire spectrum of electromagnetic radiation.

Despite that early start, and some additional development of infrared detectors by Edison, and later by Golay, no major breakthroughs in infrared astronomy occurred until the 1950s — the era of the transistor — when simple, photoelectric detectors made from semiconductor crystals became possible. The lead sulphide cell was used by Johnson to extend the young field of photoelectric photometry, and by Neugebauer and Leighton at Caltech for a huge survey of the sky at a wavelength of $2\mu m$ with an angular resolution of about 4 arcminutes. In 1961, at the Texas Instruments Corporation, Frank Low invented the gallium-doped germanium bolometer which opened up the study of much longer infrared wavelengths, especially with the aid of rocket, balloon and airplane surveys. In the seventies, lead sulphide cells were replaced by photodiodes of indium antimonide, while improved bolometers and detectors using suitably doped silicon were introduced for longer wavelengths. By 1979, a new generation of 2 – 4 meter class telescopes dedicated to infrared astronomy were coming into operation, including the United Kingdom 3.8m Infrared Telescope (UKIRT) — the predecessor of which was the 60-inch infrared flux collector on Tenerife — and the NASA 3m Infrared Telescope Facility (IRTF), both in Hawaii.

With the launch of the Infrared Astronomical Satellite (IRAS), by the USA, Britain and the Netherlands, in 1983, the science of infrared astronomy took another leap forward. This very successful mission mapped the entire sky at wavelengths of 12, 25, 60 and 100 μm until its on-board supply of liquid helium was exhausted and the telescope and the detectors warmed-up and lost their sensitivity.

Why is the infrared waveband a key regime? I am sure you will receive detailed and convincing answers to that question at this Winter School! Briefly, to begin with, the longer wavelengths of infrared radiation penetrate through clouds of interstellar dust that are opaque at optical wavelengths because the extinction by interstellar dust grains increases steeply to shorter wavelengths. Distant objects are reddened and the dust grains warm up and re-radiate the absorbed energy at wavelengths in the far infrared. This occurs throughout the disk of our Galaxy and in regions of star formation. Furthermore, objects at lower temperatures than those of normal stars can be detected by infrared emission. There are also many spectroscopic advantages to the infrared. For example, the vibration and rotational transitions of many molecules, including the very abundant hydrogen H_2 molecule, result in low energy infrared photons. The infrared is rich in atomic and molecular transitions, as well as interesting features associated with small solids (grains). Finally, the expansion of the Universe displaces or *redshifts* key features of the spectra of the most distant galaxies into the infrared.

So what constitutes the infrared waveband? And what kind of special instrumentation is needed to perform infrared astronomy? Are there limitations to ground-based infrared astronomy? I will try to answer these questions and describe some of the practical problems facing infrared astronomers. I apologise in advance that the written version will be much more condensed and less well-illustrated that the lecture series. Some of the material presented here is based on my book *Electronic and Computer-aided Astronomy: from eyes to electronic sensors*, (McLean 1989), and a lecture course at UCLA which is currently being developed into a book.

1.2   Classical limits: near-, mid-, far-infrared

For many years the infrared part of the spectrum was considered to be the region just beyond the red limit of sensitivity of the human eye, at a wavelength of about 0.7 μm. Charge-coupled devices (CCDs) have extended optical astronomy to about 1.1 μm — a fundamental cut-off wavelength imposed by the band-gap of silicon. **Near** -infrared is generally taken to be the interval from 1 to 5 μm with

Wavelength (μm)	Symbol	Width (μm)
1.25	J	0.3
1.65	H	0.3
2.2	K	0.4
3.5	L	1.0
3.8	L'	0.6
4.8	M	0.6
11	N	2.0
20	Q	5.0

Table I: *Standard infrared windows and passbands.*

some astronomers using the term short **wavelength infrared** (SWIR) to indicate the interval from 1 to 2.5 μm. The **mid**-infrared extends from 5 to 20 μm and the **far** -infrared stretches from about 20 to 200 μm. Wavelengths longer than about 350 μm are now referred to as **sub-millimeter**, and although some of the observational techniques are similar, e.g. continuum bolometers, much of the instrumentation is really more closely related to radio astronomy.

1.3 Atmospheric extinction and "windows"

The Earth's atmosphere is not uniformly transparent at infrared wavelengths. Water vapour and carbon dioxide do an efficient job of blocking out a lot of infrared radiation. Fig. 1 shows a simplified transmission spectrum of the atmosphere from 1μm to 1mm in the upper panel and a more detailed transmission curve for the 1 – 5 μm region below. The water vapour content is particularly destructive for the longer wavelengths, but it is sensitive to the altitude of the observatory, and that is why most infrared observatories are located at high, dry sites. The absorption by these molecules occurs in certain wavelength intervals or bands, between which the atmosphere is remarkably transparent. These wavelength intervals are called atmospheric "windows" of transparency. The standard infrared windows are listed in Table I. Interference filters can be manufactured to match these atmospheric windows (e.g. by OCLI, Barr Associates) and, of course, narrower passbands can be defined within each window. Filter sets are not identical and therefore care should be taken to determine the exact effective wavelength and passband for each set. Also, some astronomers are now introducing so-called "optimised" filters, for example, a Mauna Kea K' filter (1.90 – 2.30 μm) and a K_{short} from 2.0 to 2.3 μm.

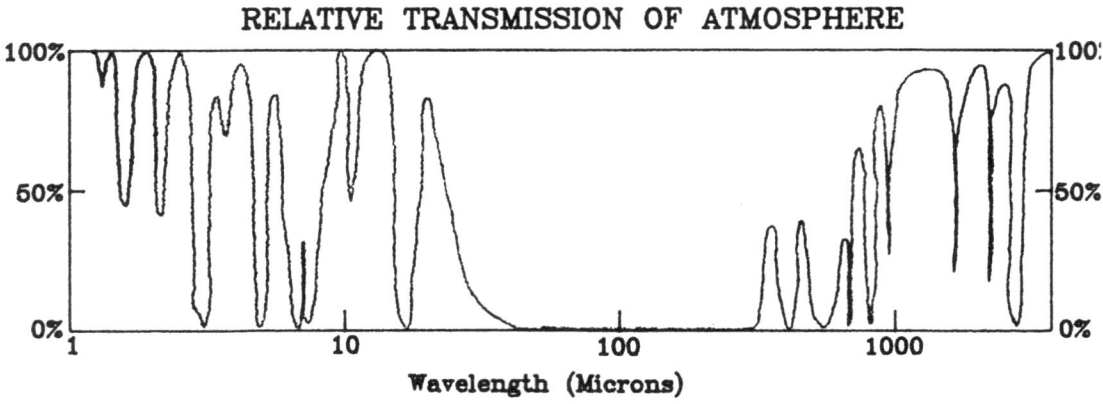

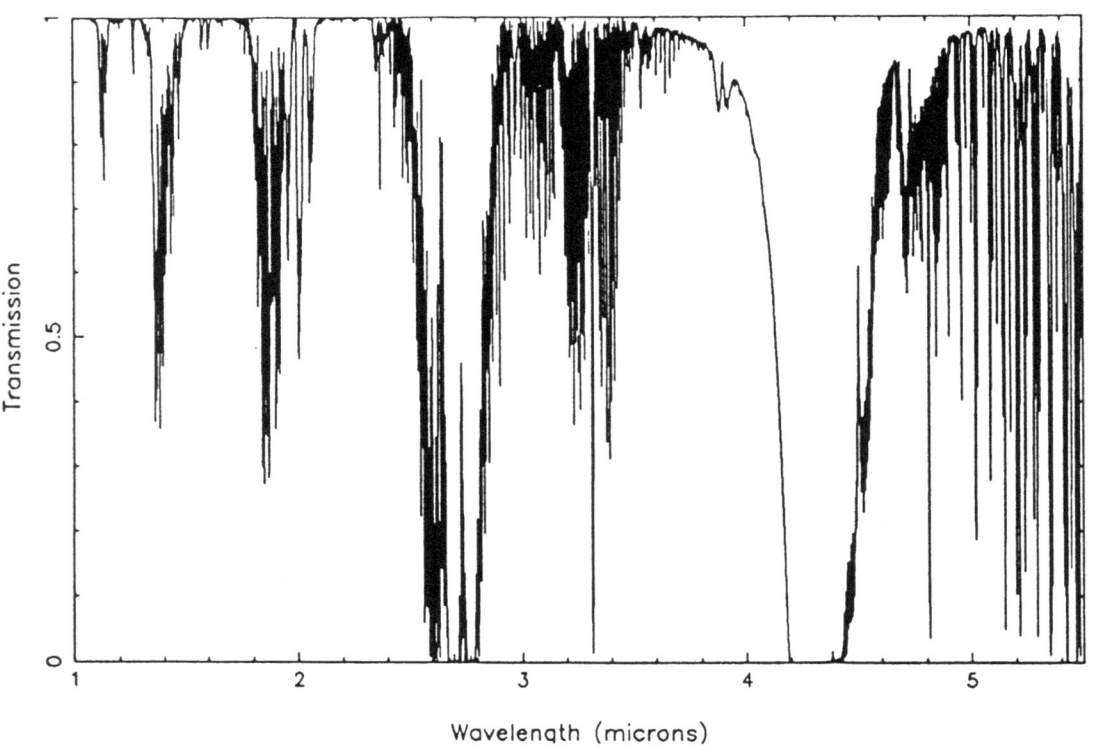

Figure 1: *Typical atmospheric transmission spectra in the infrared.*

1.4 Background radiation sources

There are other difficulties facing infrared astronomy. Everything on the Earth is warm, even the air and the telescope, and it is all a lot closer than the stars! Objects with a temperature of 300 K emit a spectrum of electromagnetic radiation which has a peak intensity at a wavelength of about 10 μm and is still very strong even at 3 μm. Between 1 and 2.5 μm this thermal radiation is replaced by emission produced by solar-induced photochemical reactions in the Earth's upper atmosphere. Together, these mechanisms produce an immensely bright "background" radiation at infrared wavelengths making it difficult to detect cosmic sources directly.

1.4.1 OH emission

There are two major sources of non-thermal radiation that dominate the near-infrared night sky from 1–2.5 μm. The first is the polar aurora, due mainly to emission from N_2, but which is negligible at mid-latitude sites such as Mauna Kea, Tenerife and Chile. The dominant problem is "airglow" which has three components:

- OH vibration-rotation bands

- O_2 IR atmospheric bands

- the near-infrared nightglow continuum

Of these, the strongest emission comes from the hydroxyl molecule which produces a dense "forest" of emission lines, especially in the 1–2.5μm region (see Fig. 2 and Fig. 3). First identified astronomically in the optical red by Meinel, these emission bands are formed as excited OH* molecules in the vibrational levels $v = 1 - 9$ relax to the ground state, OH. The excited OH is formed in a thin layer of atmosphere, called the mesopause, between the mesosphere and the lower ionosphere at an altitude of about 90 km. Since the emission occurs in a high layer, there is no decrease in the OH background by going to a high altitude site. Diurnal and seasonal variations are small, but the OH emission is time variable by a factor of 2 or more in half an hour. The source of this variability is believed to be the propagation of acoustic gravity waves in the mesopause; waves of 50 km wavelength moving across the field of view at 15 m/s could account for the observed variability on some nights. Of great importance to high resolution near-infrared spectroscopy, is the variability in the spectral structure

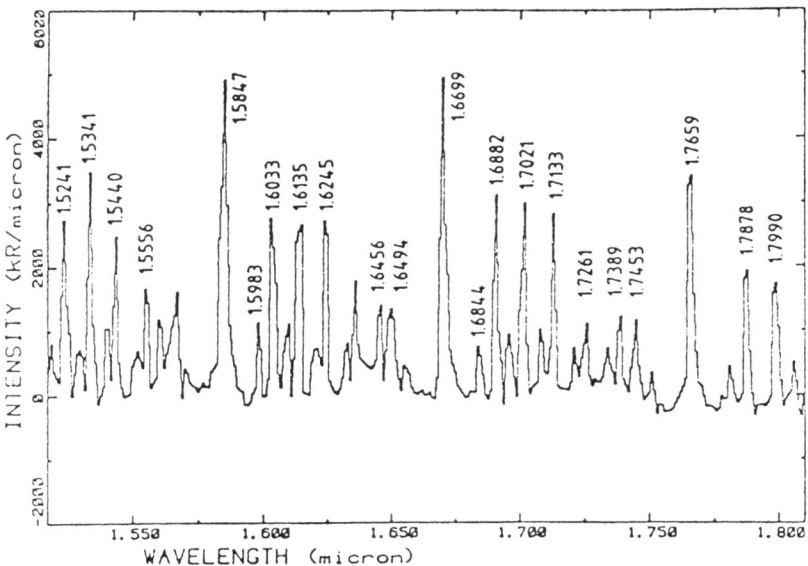

Figure 2: *The observed spectrum of OH emission in the near infrared.*

of the OH* emission. Changes in relative line and band strength occur as the rotational temperature of the OH* varies. A 50 K change in the rotational temperature is quite feasible over timescales of a few hours and this can result in large effects such as line-to-line intensity ratios changing by $\pm$ 100%.

The non-thermal emission is dominated by OH*, with the only noticeable addition being a single bright O_2 line at 1.27 μm.

The atmosphere also emits thermal (blackbody) radiation with an *emissivity* depending on the opacity in the atmosphere at that wavelength; the emissivity will be 1.0 if the atmosphere is totally opaque, but may be less than 0.1 in good windows where absorption is low. Water vapour is the main problem, as it is responsible for much of the absorption from 3–5μm, where the thermal emission of the atmosphere rises steeply. However, at wavelengths less than 13μm, the thermal background is dominated by the telescope and warm optics which are at least 20 K warmer than the effective water vapour temperature.

1.4.2 Thermal (blackbody) emission

To predict the thermal emission from the telescope optics and any other warm optics in the beam we need to know two parameters: the temperature T (K)

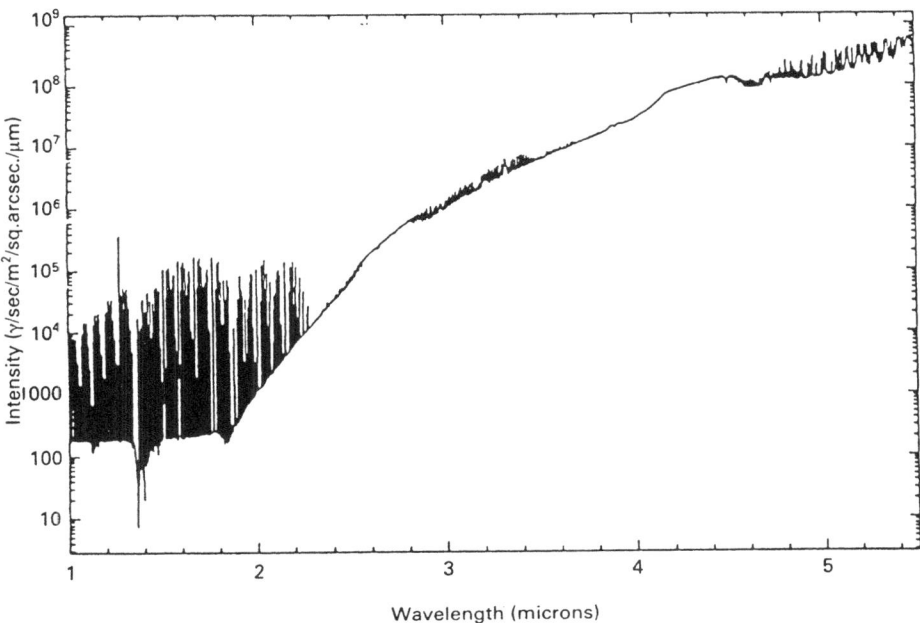

Figure 3: *Typical example of background flux as a function of wavelength in the near infrared.*

which determines the spectrum of blackbody radiation from the Planck function $B(\lambda)$, and the emissivity ϵ of each component which determines the fraction of blackbody radiation added to the beam; a truly black object has $\epsilon = 1$. The emissivity of the telescope mirrors (due to absorption by the coating or by dust on the surface) may be found from Kirchoff's law, by taking one *minus* the measured spectral reflectivity.

When the thermal and non-thermal components of background are added together we get the total background flux entering the instrument. The resulting spectrum is very steep (see Fig. 3) and has two easily recognised portions in the near infrared where the OH emission produces bright emission with a lot of spectral structure at wavelengths less than 2.5μm, whereas towards longer wavelengths it is the blackbody emission from the warm telescope and optics which dominates. Between 1 and 5 μm the background changes by many orders of magnitude.

1.5   Detector technologies

Infrared detectors are classified as *photon* detectors or *thermal* detectors.

1.5.1 Photon detectors

In photon detectors, individual incident photons interact with electrons within the detector material. For example, if a photon frees an electron from the material the process is called *photoemission*; this is how a photomultiplier tube works and it is also the underlying principle of the Schottky-Barrier diode (e.g. platinum on silicon) except that the free electron escapes from the semiconductor into a metal rather than into a vacuum. When absorption of photons increases the number of charge carriers in a material, or changes their mobility, the process is called *photoconductivity*. These devices are usually operated with an external voltage across them to separate the photogenerated electron-hole pairs before they can recombine. If absorption of a photon leads to the production of a voltage difference across a junction between differently "doped" semiconductors, the process is known as the *photovoltaic effect*. The interface is normally a simple pn junction which provides an internal electric field to separate electron-hole pairs. More often, the pn junction in this *photodiode* is operated with an externally applied voltage to cause a condition known as a "reverse bias".

There are two classes of semiconductors used in photon detection; intrinsic and extrinsic. In an intrinsic semiconductor crystal, a photon with sufficient energy creates an electron-hole pair when the electron is excited to the conduction band, leaving a hole in the valence band. Extrinsic semiconductors are doped with impurities such that a photon with insufficient energy to excite an electron-hole pair intrinsically, can still cause an excitation from an energy level associated with the impurity atom.

With the addition of trans-impedance amplifier circuits of various designs, single-element photon detectors have been, and are still being used, very successfully at many observatories. The main limitation, apart from the single element nature, is the very high impedance required in the feedback resistor. Finally, a few instruments have been developed with a small number of individual detectors arranged in an array (e.g. 1×7).

1.5.2 Thermal detectors

In thermal detectors, the absorption of photons leads to a temperature change of the detector material, which may be observed as a change in the electrical resistance of the material as in the *bolometer*; a voltage difference across a junction between two different conductors as in the *thermopile*; or a change of internal dipole moment in a temperature-sensitive ferro-electric crystal, as in the *pyroelectric* detector. Of these, only the bolometer has been widely used in infrared astronomy. The germanium bolometer is commonly used for wide band energy detection at wavelengths longer than 10μm but silicon bolometers have also been used successfully in recent years. In some instruments, many individual bolometers have been arranged in a grid pattern to produce an array of individual detectors.

1.5.3 AC detection and "chopping"

Obviously, single element detectors can only measure the infrared signal from one small patch of sky at a time. Usually, that patch will contain the source plus some unwanted background. However, the infrared background is bright and variable compared to astronomical sources and therefore some method of checking the background level very frequently is needed to ensure that the correct amount of background can be subtracted. The method used is called "chopping". The infrared beam is rapidly switched between the source position on the sky and a nearby reference position, by the use of a small plane mirror near the focal plane or a "wobbling" secondary mirror in the telescope. Chopping typically takes place at a frequency of about 10 Hz and the output signal is fed to a phase-locked amplifier with a reference signal from the chopper so that only the (AC) signal component representing the source flux is amplified.

Generally, this simple approach of chopping on and off the astronomical source is insufficient, due to asymmetries in the optics or spatial gradients in the background emission. If the source and sky beams (often called A and B) are interchanged periodically — perhaps once per minute — in a process called "nodding", then a more accurate cancellation of the background is achieved. Nodding is done by a motion of the entire telescope to make it point at a position such that the second (B position) of the chopper will now contain the object; the A position now corresponds to sky on the opposite side of the object to the original sky position. The angular displacement of the beam on the sky is called the "chop throw" and is typically less than 2–3 arcminutes. Of course, the limited chop throw reduces sensitivity to extended emission.

An infrared secondary mirror is undersized to permit chopping and therefore it is over-filled by the beam from the primary mirror. The secondary is also not baffled in the normal way which ensures that an image of the secondary is surrounded by sky, which produces a lower background that a warm, black baffle.

1.5.4 Cryogenic operation

The infrared detectors described above do not function well unless they, and their local environment, are cooled to very low temperatures. Traditionally, liquid cryogens such as liquid nitrogen (LN_2) or liquid helium (LHe) are used to obtain temperatures of 77 K and 4 K respectively. In more modern instruments, closed-cycle refrigerators (CCRs) have been used successfully. Cooling the detectors achieves two important results. Firstly, the heat or infrared emission from all the filters, lenses and supporting metal structures surrounding the infrared detector is eliminated, which greatly reduces the local background contribution. Secondly, at low temperatures the detector's own internal, thermally-generated background or *dark current* is greatly reduced which implies a large gain in sensitivity.

Of course, in order to cool the detector and all surrounding metal and glass (e.g. filters, optics) the entire unit must be placed inside a vacuum chamber with an infrared transmitting window, and the cryogenic components must be carefully isolated from the warm (radiating) walls of the enclosure and the warm (conducting) drive shafts needed to operate mechanisms such as filter wheels inside the vacuum chamber. All infrared instruments are therefore contained within a vacuum chamber which also includes a cooling system; these units are called *cryostats* or *dewars*.

1.6 The infrared "array" revolution

The kinds of infrared detector systems available to infrared astronomers up until about 1983, although sensitive, were not really ideal for making extensive infrared pictures of the sky with the fine seeing-limited detail of optical images. There were no infrared "photographic" plates or infrared "TV tubes" with sensitivities comparable to the single-element detectors used by infrared astronomers.

Any infrared images of celestial objects that had been published were really "maps" made by scanning the infrared scene point-by-point, back and forth in a sweeping pattern, called a *raster*, using a single detector with a small angular field of view on the sky determined by an aperture within the instrument; typically, the field was a circle of 2–3 arcseconds in diameter. Imagine using just *one* pixel of a CCD (charge-coupled device) which had 500×500 pixels and then moving the entire telescope (or the secondary mirror) to scan the image of the scene over that pixel; the term *pixel* is short for picture element and refers to the basic photo-sensitive "unit cell" in a 2-dimensional array detector; the term "panoramic" detector is also used. It is easy to appreciate the excitement that was felt when infrared devices similar to CCDs were introduced into astronomy.

When I carried out a survey in 1982 for the Royal Observatory Edinburgh (which runs the UKIRT) there were very few infrared array devices available for sale and virtually none of them had good performance at the (relatively) low background levels encountered in astronomy applications, as opposed to the high-background military uses for which they were developed. John Eric Arens and co-workers at the Goddard Space Flight Center had tested a 2×10 area of a 32×32 bismuth doped silicon array for 10μm as early as 1979, and Judy Pipher and Bill Forrest at the University of Rochester (USA) had been given an engineering grade 32×32 array made of indium antimonide, for the 1–5 μm region, as a "loan" from former colleague, Dr. Alan Hoffman, who was now working for the Santa Barbara Research Center (California), to see if it had potential for astronomy. I first made contact with Alan in late 1982 and by March 1984, SBRC and the Royal Observatory Edinburgh had negotiated an agreement, with input and support from Al Fowler at the US National Optical Astronomy Observatories, to develop a new and greatly improved array with 62×58 pixels specifically for infrared astronomy. About the same time, other astronomers in the USA were establishing contracts to develop different infrared array technologies for future space missions, and European sources of infrared array technology, especially for space applications, were also recognised. So excitement was high!

By late 1986 the first astronomical results from the new generation of arrays had begun to appear. An historic moment, a "turning point", occurred in March 1987 at the first conference to focus on the applications of these new infrared array detectors for ground-based astronomy. This workshop, which attracted over 200 people from all over the world was held in Hilo, Hawaii and hosted by the University of Hawaii, the UK Infrared Telescope and the Canada-France-Hawaii telescope. Although already out of date, the results presented at that meeting (Wynn-Williams & Becklin 1987) demonstrated clearly dramatically to all that were present, that a New Era in infrared astronomy had begun. For popular accounts with colour pictures see McLean (1988a, 1988b) and Gatley *et al.* (1988). By February 1990, a conference held in Tucson, Arizona (Elston 1990) demonstrated that a wide range of astrophysics was being tackled with the first generation arrays.

Over the past 6 years the pace of detector development has been rapid. Today, there are several kinds of new generation arrays in use with formats of 256×256 pixels, and even larger detectors — the *Next Generation* — are planned for the next 3 years. Experience gained in the use of silicon CCDs has been of great value in accelerating the introduction of infrared arrays. Electronic drive systems and noise reduction schemes are similar, device physics is closely related, the software and strategies for image reduction and analysis are virtually identical. The advantages of infrared array detectors (also referred to as Focal Plane Arrays or FPAs) over single element devices are:

- very time-efficient because of large numbers of elements, which also implies can reach fainter detection limits

- higher spatial resolution easily obtained; usually seeing-limited, but can be diffraction-limited in some cases

- very high sensitivity, partly due to small detector size

- high resolution spectroscopy is now viable

- "sky on the frame" introduces possibility of eliminating chopping, at least in the near infrared.

Having briefly introduced the wavebands and the basic techniques of IR astronomy in Section 1, I will describe the array detectors and IR instrumentation in more detail in the following parts.

2 INFRARED ARRAYS – BASIC PRINCIPLES

2.1 IR sensitive materials

Germanium and silicon are the two materials used as thermal detectors in astronomical bolometers, but with photon detectors there is a wider range of options. The cut-off wavelength λ_c for the absorption of infrared photons is given in terms of the band-gap energy E_g in electron volts (eV) by:

$$\lambda_c \ (\mu m) = \frac{1.24}{E_g} \tag{1}$$

Energy band gaps and long-wavelength photo-absorption limits for several semiconductors are listed in Table II. Notice that a number of infrared-sensitive semiconductors are formed from compounds containing elements in the **III** and **IV**, or the **II** and **VI** columns of the Periodic Table, *e.g.* InSb, HgCdTe respectively. In the case of HgCdTe, the ratio of mercury to cadmium determines the

Name	Symbol	Temp (K)	Band Gap (eV)	λ_c (μm)
Cadmium sulphide	CdS	295	2.4	0.5
Cadmium selenide	CdSe	295	1.8	0.7
Gallium arsenide	GaAs	295	1.35	0.92
Silicon	Si	295	1.12	1.11
Germanium	Ge	295	0.67	1.85
Lead sulphide	PbS	295	0.42	2.95
Indium antimonide	InSb	295	0.18	6.9
		77	0.23	5.4
Mercury cadmium	$Hg_x Cd_{1-x} Te$	77	0.10 ($x = 0.8$)	12.4
telluride			0.5 ($x = 0.554$)	2.5

Table II: *The band gaps and long-wavelength limits for absorption of photons for some semiconductors. Note that λ_c depends on temperature.*

effective energy band gap; with 55% Hg the longest wavelength of sensitivity is 2.5μm whereas with 80% Hg the long wavelength limit extends to 12.4μm. Each of these materials is an intrinsic semiconductor and each can be used to form a photoconductor or a photovoltaic (diode) device. Near infrared arrays are made from InSb or HgCdTe.

Currently available arrays for near IR astronomy include the 256×256 HgCdTe arrays from Rockwell International (known as the NICMOS 3 array because it was developed for the University of Arizona's Hubble Space Telescope NICMOS project) and the 256×256 InSb array from SBRC.

Extrinsic semiconductors are formed by doping intrinsic semiconductors with impurities. Table III shows the cut-off wavelengths for several extrinsic semi-conductors formed with silicon and germanium. Initially, most long wavelength arrays used extrinsic silicon photoconductors, but more recently, a technology called "impurity band conduction" (IBC) has been introduced which improves the detector performance. In these devices, the doped infrared-active layer is placed in contact with a pure epitaxial layer and the overall thickness of the device is greatly reduced. Arrays of 96x96 and 128x128 already exist in extrinsic silicon and work is under way to develop germanium IBCs for the very longest wavelengths.

Base material	:Impurity	λ_c (μm)
Germanium (Ge)	:Au	8.3
	:Hg	13.8
	:Cd	20.7
	:Cu	30.2
	:Zn	37.6
	:B	119.6
Silicon (Si)	:In	8.0
	:Ga	17.1
	:Bi	17.6
	:Al	18.1
	:As	23.1
	:P	27.6
	:B	28.2
	:Sb	28.8

Table III: *Some extrinsic semiconductors and their long-wavelength limits for absorption of photons.*

2.2 The "hybrid" structure; comparison with CCDs

Infrared array detectors are not based on the charge-coupling principle of the silicon CCD. The role of detecting infrared photons is separated from the role of multiplexing the resultant electronic signal from a pixel to the outside system. To achieve this, each device is made in two parts, an upper slab and a lower slab. The upper slab is made of the IR sensitive material and effectively subdivided into a grid of pixels by construction of a pattern of tiny photodiodes or photoconductors. In the lower slab, made of silicon, there is a matching grid but each "unit cell" contains a silicon field-effect transistor (FET) which is used as a source-follower amplifier to act as a "buffer" for the accumulated charge in the infrared pixel. Interconnecting the two slabs are tiny columns of indium, called "bumps", and the slabs are literally pressed together in a process called "bump-bonding". Finally, the upper slab is thinned to enable photons to penetrate to the pixel locations on the underside or, in the case of HgCdTe arrays, the IR detector is deposited as a thin or "epitaxial" layer on an infrared transparent substrate. The entire structure is called a "hybrid", and it is shown schematically in Fig. 4. In an infrared array detector, the absorption of a photon with a wavelength shorter than the cut-off wavelength (λ_c) generates an electron-hole pair in the semiconductor at the location of the pixel. The

350

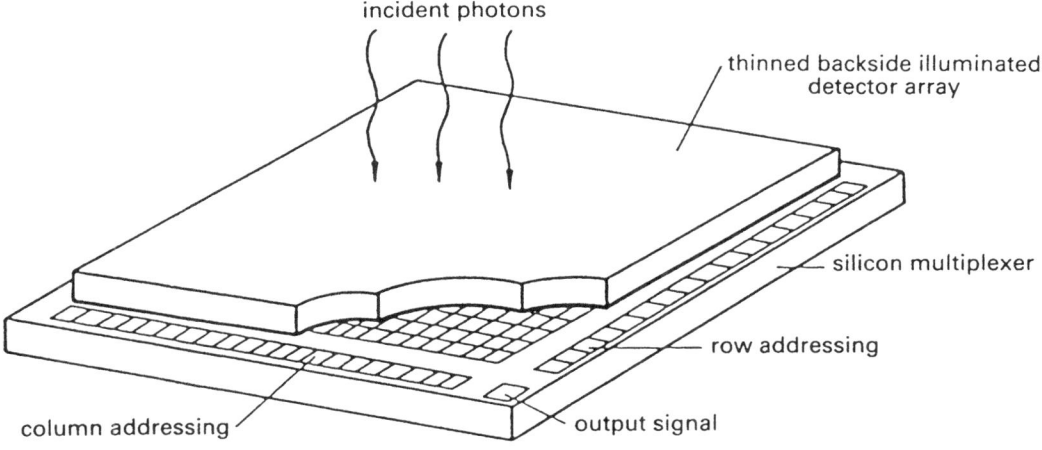

Figure 4: *Hybrid construction of infrared array detectors.*

electron-hole pair is immediately separated by an electric field which can be applied externally or internally or both. For example, in the large near infrared arrays of HgCdTe and InSb available today, the electric field is produced by a reverse-biased pn junction. The depletion region produced by the reverse-biased junction acts like a capacitor which is *discharged* from the initial reverse bias voltage by the migration of the electrons and holes; in effect, photogenerated charges are being "stored" locally at each pixel. The voltage change across the detector capacitance is applied directly to the input of the silicon FET which in turn relays the voltage change to one or more output lines when it is switched (or addressed) to do so. There is no charge-coupling between pixels since each pixel has its own FET, and there is no overflow or "charge bleeding" since the worst that can happen is that the diode becomes completely de-biased and no further integration occurs, in other words, the pixel is saturated.

Infrared arrays are backside illuminated and, with the exception of platinum silicide (PtSi) Schottky Barrier devices, all achieve excellent quantum efficiencies, typically $\geq$ 60%. The PtSi devices partially compensate for a very low quantum efficiency by having good uniformity and large formats (e.g. 640×480).

2.3 Properties of IR arrays

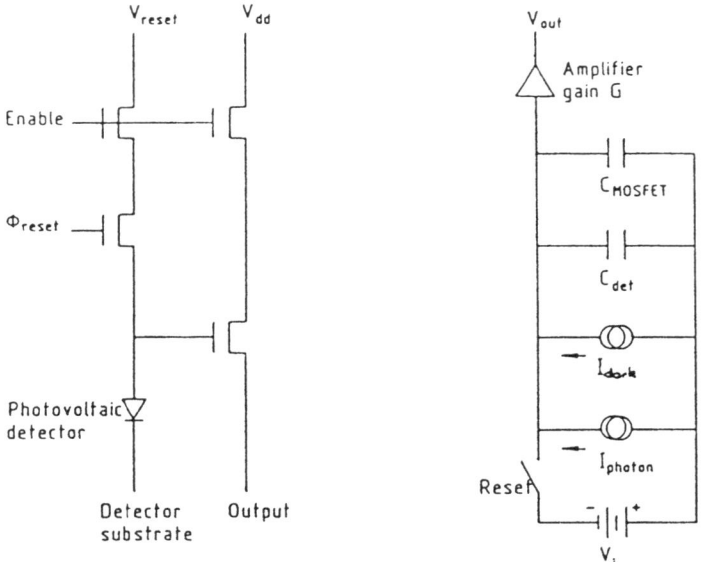

Figure 5: *Equivalent circuit for a photodiode in an IR array.*

2.3.1   Charge detection and storage

An equivalent circuit for a typical IR array with a photodiode detector is shown in Fig. 5 and the details of the process are summarised below:

- internal photoelectric effect → electron-hole pairs

- electric field separates electrons and holes

- migration of electrons across junction decreases the reverse bias, like discharging a capacitor

- amount of charge is $Q = CV/e$ electrons where $e = 1.6 \times 10^{-19}$ Coulombs per electron, V is the voltage across the detector and C is the effective capacitance

- each detector is connected to a source follower amplifier whose output voltage follows the input voltage with a small loss of gain; $V_{OUT} = A_{SF}V_{IN} \simeq 0.7V_{IN}$

- the output voltage of the source follower can be sampled or "read" without affecting the input

- after sampling, the voltage across the diode can be RESET to the full, original reverse bias in readiness for the next integration.

The reset action is accomplished with another FET acting as a simple on/off switch in response to a pulse applied to its gate. It is customary to refer to the combination of the detector node, the source follower FET and the reset FET as the "unit cell". Actually, the equivalent circuit of the unit cell reveals that there are two sources of capacitance — the pn junction and the gate of the FET. Also, there are two sources of current to drive the discharge of the reverse bias, namely photoelectrons and "dark current" electrons. There is an obvious potential for non-linearity between photon flux and output voltage in these detectors because the detector capacitance is *not* fixed, but does in fact depend on the width of the depletion region which in turn depends on the value of the reverse bias voltage. Unfortunately, this voltage changes continuously as the unit cell integrates either photogenerated charges or dark current charges. Using $Q = CV$ and taking both C and V as functions of V then

$$dQ = (C + \frac{\partial C}{\partial V}V)dV \equiv I_{det}dt \qquad (2)$$

The rate of change of voltage with time ($\frac{dV}{dt}$) is not linear with detector current I_{det} because of the term $\frac{\partial C}{\partial V}$. However, the effect is small and slowly varying ($< 10\%$ worst case) and easily calibrated to high precision.

2.3.2 Dark current

In the total absence of photons, an accumulation of electrons still occurs in each pixel and this effect is therefore known as *dark current*. There are at least three sources of dark current; diffusion, *thermal* generation-recombination (G-R) of charges within the semiconductor and leakage currents. The latter are determined mainly by manufacturing processes and applied voltages, but diffusion currents and G-R currents are both very strong (exponential) functions of temperature and can be dramatically reduced by cooling the detector to a low temperature. Dark currents as low as 1 e/s/pixel have been achieved in the most recent HgCdTe arrays at 77K and in InSb arrays at temperatures $\leq$ 35K. Note that lower band-gaps imply lower temperatures to obtain the same dark current, and that the observed residual dark current is probably due to leakage currents in most cases. Even lower temperatures are needed for extrinsic silicon arrays with long cut-off wavelengths.

2.3.3 Readout noise

Resetting the detector capacitance and sampling the output voltage cannot be done without a penalty. The term "readout noise" describes the random fluctuations in voltage (V_{noise}) which are added to the true signal and photon noise. Readout noise is always converted to an equivalent number of electrons (R) at the detector by using the effective capacitance (C), therefore

$$R = \frac{CV_{noise}}{e} \ \ electrons \qquad (3)$$

One potentially serious component of the readout noise can be eliminated easily. This is the reset or "kTC noise" associated with resetting the detector capacitance. When the reset transistor is on, the voltage across the detector increases exponentially to the reset value V_{RD} with a time constant of $R_{on}C$, where R_{on} is the "on" resistance of the transistor; this time constant is very short. Random noise fluctuations at $t \gg R_{on}C$ have a root mean square (rms) voltage noise of $\sqrt{kT/C}$ or an equivalent charge noise of $\sqrt{kTC}$, where k = Boltzmann's constant (1.38×10^{-23} joules per kelvin) and T = absolute temperature (K). After the reset switch is closed, the time constant for the decay of the (uncertain) reset voltage on the capacitor is $R_{off}C$, but $R_{off} \gg R_{on}$ and therefore the noise on the reset level will still be present long after the end of the pixel time. The unknown offset in voltage can be eliminated by taking the difference between the output voltages before and after reset. This method is called Correlated Double Sampling (CDS).

There are several possible strategies for reading out infrared array detectors. They can be categorised with reference to Fig. 6 which shows the typical voltage variation across a pixel as a function of time. Each pixel is reset to a fixed level (within the reset noise), but when the reset pulse is released there is usually a small shift to another level called the "pedestal" level before the detector begins to discharge with photocurrent. Four possible readout modes are:

- read and reset: form signal minus following reset ($S_{00} - S_{01}$)

- reset – read – read: form signal minus pedestal ($S_{10} - S_{02}$); both reads are non-destructive

- slope sampling: n reads between S_{02} to S_{10}

- multiple reads: n reads at S_{02} and n reads at S_{10}

Each method eliminates reset noise and all but the first method make use of the non-destructive readout feature of infrared arrays to further reduce noise. Typical noise values for the best arrays are $\sim$50 electrons in normal CDS mode, but $\sim$20 electrons using multiple reads.

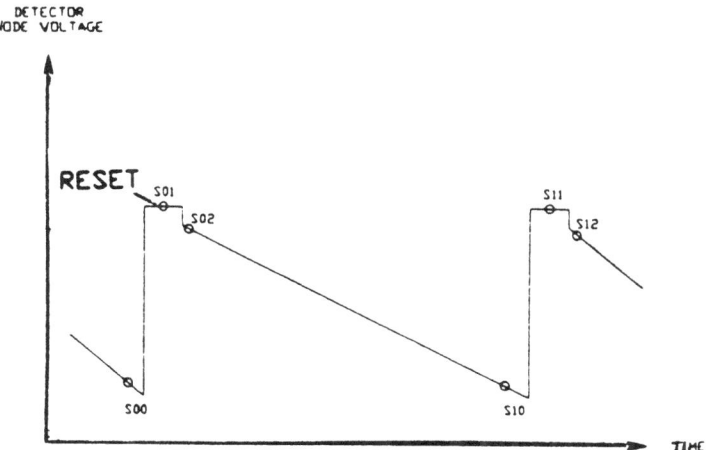

Figure 6: *Typical voltage variation as a function of time in one pixel of an IR array detector.*

2.3.4   Unusual effects

Infrared array technology is still developing and even so-called "science-grade" detectors can exhibit flaws. It is not surprising that some detectors show a variety of unusual and unwanted effects. For example, incomplete readout of an image which results in a "residual" image on subsequent readouts. Other devices show large global variations in quantum efficiency or many "hot" pixels with high dark currents or "hot" edges and corners or a rapid drop in quantum efficiency at short wavelengths. Each detector must be carefully investigated and characterized before it is used for astronomical photometry.

2.4   Calibration & characterization of arrays

2.4.1   Flat field correction

The removal of pixel-to-pixel variations in quantum efficiency to a tiny fraction of one percent of the mean sky background is a non-trivial, but necessary, exercise. In principle, the methods used at infrared wavelengths are the same as for optical CCDs. Each pixel on the array must be illuminated with the same flux level; such a uniform illumination is called a "flat field". Basically, a flat field source is obtained either by pointing the telescope at the inside of the dome, which is so far out of focus that it gives a fairly uniform illumination across the detector (in principle), or by pointing the telescope at blank sky. In practice, the latter seems to give best results for very deep imaging of faint sources whereas dome flats often work well for bright stars.

Deriving flat fields from sky frames is best done by using "median filtering" during data reduction, since no frame can be absolutely free of sources. Below is a brief summary of the steps to be followed:

- obtain numerous independent exposures each of which corresponds to a slightly different telescope pointing by a few arcseconds

- collect enough flux to be background-limited but keep exposures short enough to maximize the number of frames taken

- normalise all the frames to the same mean

- select a frame of interest and form the median frame of a set of exposures (7 – 9 is best) on either side of this frame in the sequence, but do not include the chosen frame; this is the "local" sky flat for the chosen frame

- repeat this process for all frames by dividing each individual frame by its local sky flat

- register each flat-fielded image to a common reference frame and sum all the frames together

It is important to experiment with different methods and timescales to understand the stability and repeatability of any given instrument, telescope and site.

2.4.2 Gain and the Variance versus Signal method

Usually, the numbers stored in the computer memory do not coincide with the number of photons (or even the number of photoelectrons) detected by each pixel. We need to find the conversion factor from counts (also called ADUs — Analog-Digital-Units, or DN — Data Numbers) to electrons and then to photons. The first step can be accomplished in the laboratory in two ways:
(a) by calculation
(b) by experiment (from a series of flat field exposures).
If the number of electrons per DN = g, then:

$$g = \frac{V_{f_s} C}{2^n A_G e} \qquad (4)$$

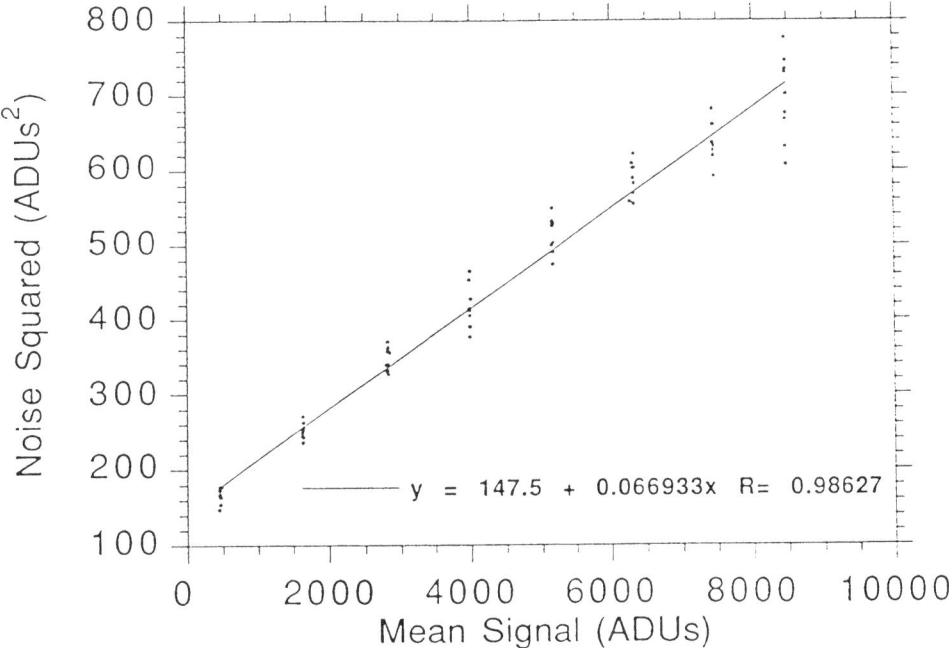

Figure 7: *A plot of Variance versus Signal for an IR array as used to derive the gain factor (g) and degree of linearity.*

where

V_{fs} = full scale voltage swing of the analogue-to-digital converter (typically 10 volts)

n = number of bits of digitisation (e.g. 14 or 16)

A_G = total gain of system (including the on-chip source follower FET and all external amplifications)

C = detector capacitance

e = charge on the electron

Often C, which is typically $\leq 0.6 \times 10^{-12}$ farads, is not known with sufficient accuracy and the second method is preferable. When the noise-squared or variance V is plotted against the signal level S, in ADUs (or DN), from a series of flat-fielded exposures of a uniform source, the data should form a curve similar to that shown in Fig. 7. It can be shown that the expected form of the linear section of the graph is given by:

$$V = \frac{1}{g}S + (\frac{R}{g})^2 \qquad (5)$$

where R = readout noise in electrons.

This equation is very important and can be used to derive other basic properties as follows:

- the inverse of the slope yields g

- the intercept yields R, given g

- departures from a straight line measure non-linearity

- turn-over point defines total saturation

The "full well" charge capacity of each pixel is given approximately by CV_{rb}/e where V_{rb} is the actual reverse bias across the junction. Setting $V = 1\mu V$, we can also derive the number of electrons per microvolt or its inverse, the number of microvolts per electron (typically $\sim 2.5\mu V/e$).

2.5   Calculating the Signal-to-Noise ratio

First we determine the signal and then identify all the noise contributions. A simple estimate of the total number of photoelectrons per second detected in the instrument from a star of magnitude **m** is:

$$S = (hc)^{-1}\ \tau\eta\lambda\Delta\lambda A F_\lambda(0) \times 10^{-0.4m}\ \ e/s \qquad (6)$$

where
h = Planck's constant and c = speed of light; $(hc)^{-1} = 5.03 \times 10^{18}$ J$^{-1}\mu$m^{-1}
τ = total transmission of optics
η = quantum efficiency of detector
λ = wavelength (μm)
$\Delta\lambda$ = passband of filter (μm)
A = collecting area of the telescope (cm^2)
$F_\lambda(0)$ = flux from a magnitude zero star (Vega) in W cm$^{-2}\mu$m^{-1}. (See Table IV).
m = magnitude in the given passband (corrected for atmospheric extinction).
    In practice, the signal (S) is spread over n$_{pix}$ pixels, each of size $\theta_{pix} \times \theta_{pix}$ arcseconds.

The "background" signal (B e/s/pixel) can be predicted from first principles as the sum of OH emission and thermal components from the sky and telescope. Alternatively, we can estimate the background using a similar expression to Eq. 6, if m is replaced with the *observed* sky magnitude per square arcsecond (m_{sky}) and the whole expression is multiplied by $(\theta_{pix})^2$ to account for the area on the sky subtended by each pixel. Dark current can be predicted from theory in some cases, but it is usually a measured quantity (D e/s/pixel). Each of

Wavelength (μm)	Symbol	$F_\lambda(0)$ ($W\ cm^{-2}\mu m^{-1}$)	$F_\nu(0)$ (Jy)
1.25	J	2.90×10^{-13}	1520
1.65	H	1.08×10^{-13}	980
2.2	K	3.80×10^{-14}	620
3.5	L	6.90×10^{-15}	280
4.8	M	2.00×10^{-15}	153
11	N	1.09×10^{-16}	37
20	Q	7.30×10^{-18}	10

Table IV: *The flux from a zeroth magnitude star in the standard infrared windows and passbands. One jansky (Jy)* $= 10^{-26}\ W\ m^{-2}Hz^{-1} = 3\times\ 10^{-16}\lambda^{-2}\ W\ cm^{-2}\mu m^{-1}$ *where* λ *is in* μm.

these sources of "signal" (S, B and D) also produce a noise term which follows Poisson statistics and goes as the square root of the total number of electrons, whether generated by dark or photon processes. Of course, the readout noise (R) is an additional noise source and we assume that all the noise sources add in quadrature.

2.5.1 The "array equation"

A full derivation of the signal-to-noise (S/N) ratio of an array measurement is straightforward but tedious (see, for example, McLean 1989). The calculation is based on the assumption that the data reduction procedure is of the form:

$$Final Frame = \frac{Source\ Frame - Dark\ Frame}{Flat\ Frame - Dark\ Frame} \qquad (7)$$

Strictly speaking, we should take into account the number of source frames (N_S), dark frames (N_D) and flat field frames (N_F) actually used in the reduction and the propagation of errors resulting from the measurment of each kind of frame. However, if we assume that flat-fielding is nearly perfect and the dark current is small and/or well-determined then the expression for S/N can be simplified. If each individual "exposure" of the source requires t seconds and N_S exposures are "co-added" to form the final frame with total integration time of $N_S t$ then,

$$\frac{S}{N} = \frac{S(N_S t)^{\frac{1}{2}}}{[S + n_{pix}\{B + D + \frac{R^2}{t}\}]^{\frac{1}{2}}} \qquad (8)$$

where the total signal S (e/s) is spread over n_{pix} pixels each of which contributes a background of B (e/s), a dark current of D (e/s) and a readout noise of R electrons. This result is known as the "array equation" and is sufficient for most purposes provided that the observer takes enough calibration data (flats and darks) to ensure that the assumptions are valid.

2.5.2   Background-limited (BLIP) case

An important simplification of the array equation occurs when

$$B \gg \frac{R^2}{t} \tag{9}$$

Notice that R^2 is the equivalent signal strength that would have a Poisson shot noise equal to the observed readout noise. This condition is known as the Background Limited Performance or BLIP condition and the time $t_{BLIP} = R^2/B$ is a useful figure of merit. When $B \gg$ S, D, and R^2/t then the array equation (Eq. 8) can be simplified and re-arranged to give

$$S = \frac{S}{N} \sqrt{\frac{n_{pix}B}{N_S t}} \quad e/s \tag{10}$$

and we can convert to measured counts (data numbers) by dividing the right hand side by g electrons/DN. Finally, this simplified expression can be used to predict the limiting magnitude of the measurement from

$$m = m_{ZP} - 2.5 \log\{\frac{1}{g}\frac{S}{N}\sqrt{\frac{n_{pix}B}{N_S t}}\} \tag{11}$$

where m_{ZP} is the "zeropoint" magnitude for the system. The zeropoint is the magnitude corresponding to 1 DN/s. By setting S = 1 in Eq. 6, an expression to predict the zeropoint can be obtained, namely,

$$m_{ZP} = 2.5 \log\{\frac{\tau\eta\lambda\Delta\lambda AF_\lambda(0)}{hcg}\} \tag{12}$$

The quantity $-2.5\log\{S(DN/s)\}$ is called the "instrumental" magnitude (IM); note that IM is a negative number. If the true magnitude of the star is known, then $m_{ZP} = m_{true}$ - IM, which is positive and numerically larger (a fainter magnitude). Having derived m_{ZP} observationally, we can invert Eq. 12 and solve for the product $\tau\eta$ which describes the "system efficiency" in this passband.

Since sources may be observed at different airmasses from standard stars, the effect of atmospheric extinction must be eliminated to obtain accurate photometric magnitudes. The true magnitude is given by

$$m = m_{ZP} - 2.5\log(S) - \alpha X + \beta(colour) \tag{13}$$

where S is the total signal in DN/s, X is the airmass (the sec of the zenith angle in a plane parallel atmosphere) and "colour" is a term like (J - K). The coefficient α is the slope of a plot of apparent magnitude versus airmass; note that the zenith corresponds to an airmass of X = 1 and above the atmosphere corresponds to X = 0.

2.5.3 Noise Equivalent Power (NEP)

Bolometers and single-element AC-coupled detectors often use a figure of merit known as the Noise Equivalent Power or NEP. A full derivation of NEP is beyond the scope of this article, but the concept is easily understood by the following discussion. If a rate of S photoelectrons per second implies S/η photons per second, then we can determine the power collected by multiplying by the energy of a single photon $E = hc/\lambda$, that is

$$P = \frac{hc}{\lambda} \frac{S}{\eta} \quad Watts \tag{14}$$

This is the signal power. When Signal = Noise we have the power "equivalent" to the noise, the NEP. In real AC-coupled systems like bolometers, the NEP is a function of the frequency bandwidth Δf and is measured in units of $W/\sqrt{Hz}$. For integrating or DC systems this terminology is not applicable although some workers force a connection by defining $\Delta f = \frac{1}{t}$ in which case the NEP may be estimated as

$$NEP = \frac{hc}{\lambda\eta}[S + n_{pix}(B + D + \frac{R^2}{t})]^{\frac{1}{2}} \quad W/\sqrt{Hz} \tag{15}$$

Again, there are two special cases.

1. BLIP $B \gg S, D, R^2/t$

2. Detector Limited $R^2/t \gg S, B, D$

As an example, if $\eta = 0.5$ at $\lambda = 2\mu m$ and the system is BLIP with $B = 10^4$ e/s/pixel then NEP $\sim 2 \times 10^{-17}$ $W/\sqrt{Hz}$.

3 DESIGNING IR INSTRUMENTS

It is impossible to provide a complete recipe for the design of an infrared instrument. My goals in this section are to alert you to some of the many differences between infrared instruments and their optical counterparts and to provide you with a basis for understanding a given design or beginning one of your own.

3.1 Systems design and constraints

Obviously, the starting point of any design is a set of specifications related to the science goals. For example, which parameters of the infrared radiation are to be measured? What signal-to-noise ratios are required? What spectral resolution, spatial resolution and time resolution are needed? In reality of course, the design will be constrained by practical considerations such as:

- size, weight, cost, and time-to-completion.

- how much automation?

- how much "on-line" data reduction and analysis?

- how many megabytes of data are to be stored?

In my view, some degree of "project planning" is absolutely essential, even for small projects involving very few people. If an instrumentation project is well-planned, you make better use of time and resources and you are forced to "think" more about the project. This results in better (and more) documentation, better trained students, and more successful observing runs because the instrument is likely to be more reliable and better understood! A method which I have found successful is to break down any project into smaller units and either assign small teams or at least identify clearly who will be responsible for each area e.g.,

- optics

- cryo-mechanics

- electronics and data acquisition

- software

- detector physics

Meet regularly as a team and be sure to record your progress in writing. A long Design Phase for everything (including — software) can be an advantage, provided it is done well enough to avoid any "do over" actions later. Debate the final design in two, well-organised review sessions, a Preliminary Design Review (PDR) and a Critical Design Review (CDR) and be absolutely honest with yourself about the progress and quality of what has been achieved.

Now it is time to look at some of the fundamental, technical issues which come up in every instrument.

3.2 Optics – basic principles

The key factor in every optical design is to "match" the spatial or spectral resolution element to the size of the detector pixel with the following constraints:

- maximum observing efficiency

- best photometry

In the absence of atmospheric "seeing", the size of an unresolved point source is determined by the Airy diffraction pattern. The Rayleigh criterion defines two objects as "resolved" when the maximum of one pattern falls on the first minimum of the second pattern; the angular resolution is $\theta = 1.22\lambda/D_{tel}$ radians. Even in the infrared, the Rayleigh criterion predicts much less than 1 arcsecond for a large telescope; for example, the result is about 0.05 *arcsec* at 2μm for the W.M. Keck 10-meter telescope. For normal imaging, or imaging of a relatively wide field of view, the angular resolution even at the best sites is usually "seeing-limited" by atmospheric turbulence to ~ 0.5–1 arcsecond, unless extremely short exposure times are used or some other image sharpening method is employed. To ensure the best photometry and astrometry, the image of a star must be well-sampled. Good "centroids" imply 5 or more pixels across image. The absolute minimum (often called the Nyquist limit) is 2 pixels across the image.

In a spectrometer, the *width* of the entrance slit is the determining factor. A narrow slit implies higher spectral resolution, but the highest efficiency is achieved when the slit is large enough to accept all of the seeing disk.

To understand the problem of optical matching to a telescope, consider first the "plate scale" of a telescope given in arcseconds per mm (arcsec/mm) by:

$$(ps)_{tel} = 206265/F_{tel} \quad (arcsec/mm) \qquad (16)$$

where F_{tel} is the focal length of the telescope and the focal ratio (or f/number) is simply F_{tel}/D_{tel}. NOTE: f/no. (or f/#) is a single symbol and not a ratio! (The numerical factor is the number of arcseconds in 1 radian.)

363

The angle on the sky subtended by the detector pixel is,

$$\theta_{pix} = (ps)_{tel} d_{pix} \tag{17}$$

Typically, $d_{pix} \sim 0.030$ mm

To determine the optical magnification factor we can follow these steps:

- choose required pixel size in arcseconds θ_{pix}

- given size of detector (d_{pix}), derive the plate scale on the detector

$$(ps)_{det} = \frac{\theta_{pix}}{d_{pix}} \tag{18}$$

- the required magnification is

$$m = \frac{(ps)_{tel}}{(ps)_{det}} \tag{19}$$

NOTE: m also defines an Effective Focal Length $(EFL = mF_{tel})$ for the new optical system. For $m < 1$, the optics are called a FOCAL REDUCER.

From optical principles, such as the Etendue (Area-Solid angle product) of the system or from the Lagrange Invariant, we can relate the pixel size in arcseconds to the f-number of the focal reducer (also known simply as the "camera").

$$\theta_{pix} = 206265 \frac{d_{pix}}{D_{tel}(f/\#)_{cam}} \tag{20}$$

where $(f/\#)_{cam} = F_{cam}/D_{cam}$.

EXAMPLE: For $d_{pix} = 30\mu m$ and $D_{tel} = 10m$: $\theta_{pix} = 0.62$ arcsec/$(f/\#)_{cam}$. Assuming seeing of 0.5 arcsec on Mauna Kea, we need pixels of less than 0.25 arcsec (2 pixel sampling) or even 0.125 arcsec (4 pixel sampling) which implies that $(f/\#)_{cam} = 1.2 - 2.4$. This is quite difficult!

To see why the above constraint is hard, consider the spherical aberration produced by a single lens. The angular blur size for a material with refractive index n = 1.5 is given by

$$\beta = \frac{0.067}{(f/\#)^3} \simeq 0.02 \; radians \tag{21}$$

which implies a linear size of $x = \beta f_{cam} = 1$ mm for $f_{cam} = 50$ mm. This is over 30 pixels! In practice, a lens train of several components can reduce this aberration to one pixel, with the consequence of greater light loss. Mirror systems with non-spherical surfaces (conic sections) and even *aspheric* surfaces are used to improve transmission.

3.3 Vacuum-cryogenics

As already noted, infrared arrays must be operated at temperatures well below room temperature to minimise intrinsic dark current. Moreover, the optical components in the beam, as well as *all* the metal parts around the optics, must also be cooled to low temperatures to eliminate thermal radiation which would otherwise saturate the infrared detector. To achieve these goals, the detector is mounted against a small heat sink of copper (or aluminium) which is connected to a low temperature source. Similarly, all optical and mechanical parts are heat sunk to a low temperature reservoir, which may be different from that of the detector. The complete assembly is placed inside a vacuum chamber with an entrance window made of an IR transmitting material.

There are two methods in use to obtain the low temperature reservoirs: liquid cryogens, such as liquid nitrogen (LN_2) which gives 77 K, and liquid helium (LHe) which gives 4 K, or closed cycle refrigerators (CCRs) which are electrically-operated heat pumps — similar to a refrigerator in the home but using the expansion of pure helium gas instead of freon — with two temperature stages, typically 77 K and 10 K. One of the smaller, least expensive units can provide 20 watts of power at 77 K and 5 W at 10 K. For liquid cryogens the cooling ability is expressed in terms of the product of the density (ρ) and the latent heat of vaporisation (L_V):

$$(\rho L_V)_{LHe} = 0.74 \ W \ hr \ l^{-1}$$

$$(\rho L_V)_{LN_2} = 44.7 \ W \ hr \ l^{-1}$$

For example, a 10W heat load boils away 1 litre (l) of LHe in 0.07 hours whereas 1l of LN_2 lasts 4.5 hr. Since liquid helium boils very rapidly and is harder to handle than liquid nitrogen, infrared instruments developed over the last three years now incorporate CCRs in preference to LHe.

Short wavelength near infrared arrays such as the NICMOS III HgCdTe arrays have a relatively large bandgap and will yield dark currents of less than 1 e/s/pixel if cooled to 77 K. This temperature is also adequate to eliminate thermal emission from the optics and mechanical structure too. For these detectors a simple, single-cryogen vessel is fine. Indium antimonide arrays are sensitive out to 5μm and the detector must be cooled to around 30 K to achieve comparable dark currents. For most applications the remainder of the system can be cooled to LN_2 temperature. Many InSb systems, and all longer wavelength instruments, achieve temperatures below 30 K with a double-cryogenic vessel which has a chamber containing LHe as well as the one containing LN_2; an example is shown in Fig. 8. This is the UKIRT IRCAM dewar. Fig. 9 shows the same camera retro-fitted with a CCR.

In a vacuum cryostat, heat may be transferred between a hot surface (T_h) and a cold surface (T_c) by conduction or radiation. For conduction through a solid of cross sectional area (A) and length (L), the heat flow $\frac{dQ}{dt}$ in watts is given by:

$$\frac{dQ}{dt} = \frac{A}{L}(I(T_h) - I(T_c)) \tag{22}$$

where I is a tabulated quantity called the thermal conductivity integral in watts/m. The important quantity for the designer is the ratio A/L.

The net radiated heat transfer in watts from a body at temperature (T_h) onto a colder body at temperature (T_c) and surface area A_c is given by:

$$\frac{dQ}{dt} = \sigma A_c F_{h-c}\{(T_h)^4 - (T_c)^4\} \tag{23}$$

where σ is the Stefan-Boltzmann constant ($5.67 \times 10^{-8} \mathrm{Wm^{-2}K^{-4}}$) and F_{h-c} is a factor which accounts for the *emissivities* of the surfaces and their geometry. Recall, an optically black surface is a good absorber and therefore a good radiator at infrared wavelengths. The following phrase is a useful reminder:

If you want it black, it must be cold

If you want it white, it should be gold!

3.4   Electronics & data acquisition

Infrared arrays, like CCDs, are digitally controlled devices. To operate each device the electronics system must supply three things : (i) a small number of very stable, low noise, direct current (dc) voltages to provide power; (ii) several input lines carrying voltages which switch between two fixed levels in a precise and repeatable pattern known as *clocks*; (iii) a low noise amplifier and analogue-to-digital conversion system to handle the stream of voltage signals occurring at the output in response to the clocking of each pixel. There are many ways in which such systems can be implemented and it is beyond the scope of this text to describe and analyze them all (see McLean 1989). The reader is referred to the publications of the SPIE which carry technical descriptions of astronomical

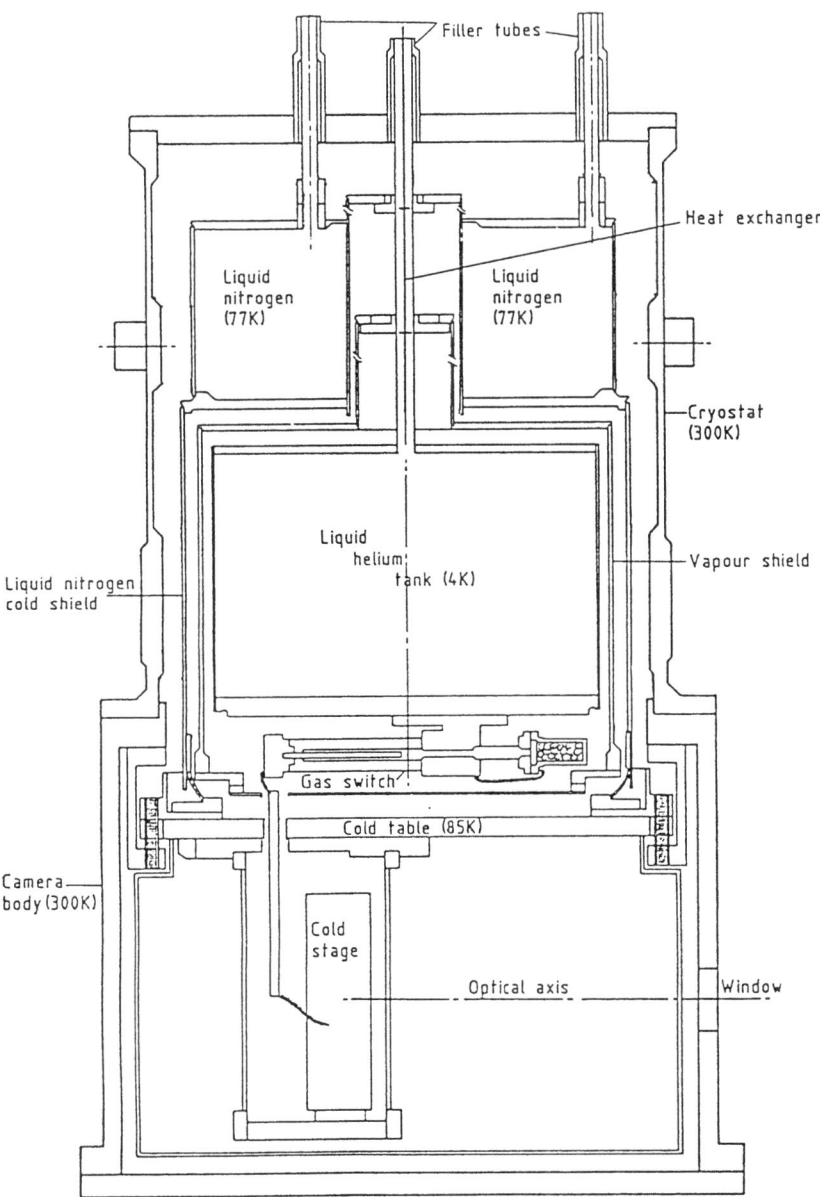

Figure 8: *The IRCAM cryostat on UKIRT: An example of a dual LN₂/LHe dewar (made by Oxford Instruments) connected to a custom vacuum chamber containing the cold optics.*

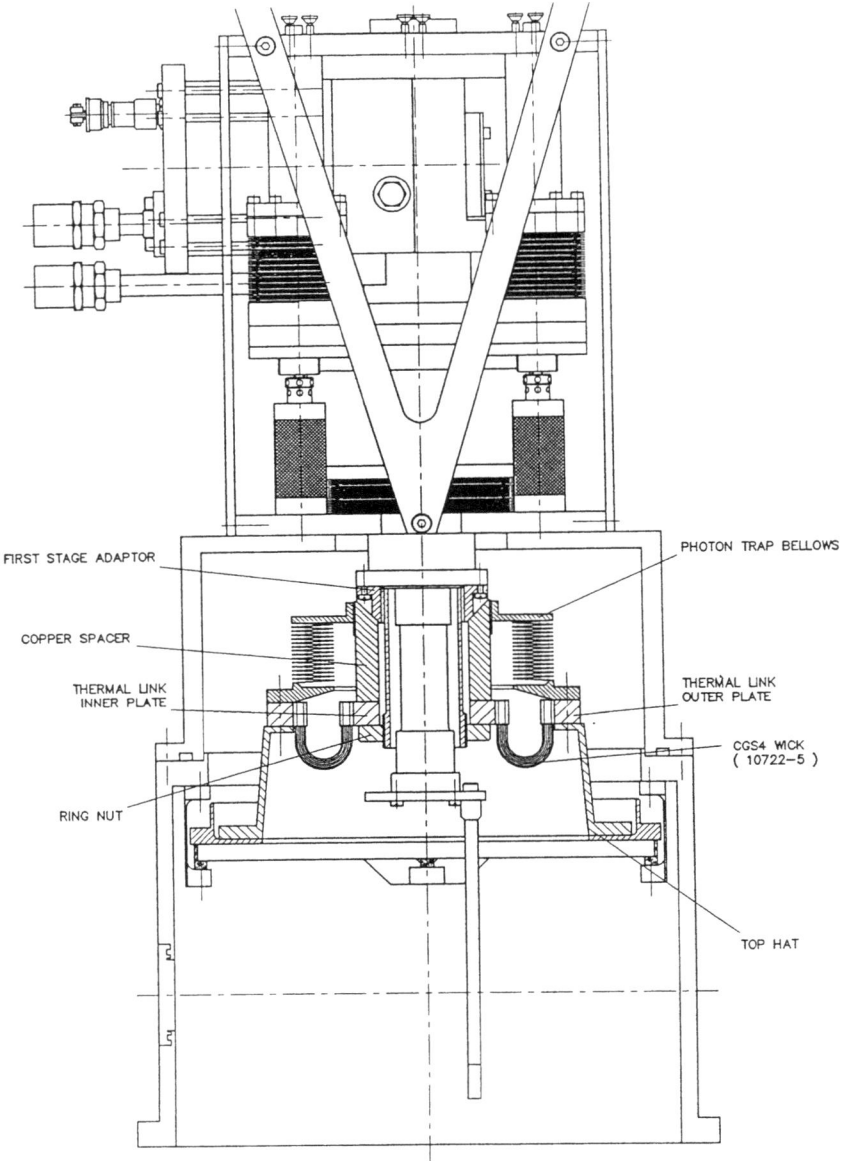

Figure 9: *The IRCAM camera on UKIRT after conversion to a closed-cycle cooler. An arrangement to damp vibrations from the piston inside the CCR has been added.*

instrumentation. Volume 1946 entitled *Infrared Detectors and Instrumentation* will be published in 1993 and will give the most recent descriptions. Whatever the scheme, the key elements are the *timing diagram* and the noise performance.

Infrared arrays have an advantage over CCDs in terms of the variety of ways in which the collected charge pattern can be read out. As with CCDs, most of the readout techniques involve a form of Correlated Double Sampling (CDS) to eliminate "kTC-noise" which is a large component of the noise unless the array is operating under very high background conditions. Double correlated sampling requires that the signal is digitized twice within one pixel time. Normally this is done using a "read-reset" method, that is, the entire array is reset pixel-by-pixel and then read out at the end of the integration time and again reset. Another way of implementing this mode is called the "reset-read-read" method. The array is reset and then immediately read out again using the *non-destructive* capability of these direct readout structures and this provides the intitial voltage level at the start of integration. Non-destructive readout occurs again at the end of the integration time and the difference between the two frames of data is formed in the host computer. This technique eliminates noise picked up unintentionally by voltage feed-through during reset.

Further use can be made of the non-destructive readout feature by taking not one, but many digital samples at the beginning and end of integration (Multiple Correlated Sampling) or by sampling each pixel periodically during the integration itself (Slope Sampling). Both methods are effective in further reducing the noise.

3.5   Software and data reduction

We have already mentioned that the analysis of infrared array data is similar to that for CCDs and involves the numerical manipulation of image frames which are merely two-dimensional arrays of numbers representing the total number of photoelectrons detected in each pixel. As with CCDs, a combined (Dark + Bias) frame must be subtracted to eliminate those additive terms — a Bias frame is one with no illumination and almost zero exposure time and a Dark frame is one with no illumination but an exposure time equal to that used in the source frame — and then the new frame must be divided by a normalised flat-field frame to correct for pixel to pixel variations. Finally, the observed signal levels in counts must be related to magnitudes or absolute flux levels. Many software packages provide tools to accomplish this reduction procedure. However, the instrument builder should ensure that the most important steps can be done on-line at the telescope to enable the astronomer to assess the quality of the data.

Photometry with an array camera is of course performed in software by creating a circular aperture around the source and summing the enclosed signal and then subtracting an equivalent area of background, either by using an annulus around the source or a blank area offset from the source. To calibrate the process the same sized aperture must be used on the standard star of known magnitude (Elias *et al.* 1982) or, alternatively, a "curve of growth" giving magnitude as a function of aperture size can be obtained using the standard star and then an "aperture correction" in magnitudes can be applied if a different aperture size is used for the unkown source. There are many internationally recognized software packages which provide facilities to perform numerical processes such as this on digitized image data. For example, IRAF (USA), STARLINK (UK) and MIDAS (Europe). It is essential to check (rather than trust) the performance of this (or any packaged) software to be sure that it handles your data correctly.

4   REVIEW OF MODERN INSTRUMENTS

4.1   Photometers and bolometers

Photometry is the measurement of the brightness of a source in a well-defined wavelength interval. Usually the source is a "point" source and the photometer has an "aperture" which encloses the point source and a minimum amount of sky background. A bolometer is an infrared instrument which relies on thermal heating of the detector by absorbed photons and which integrates the energy received over all wavelengths.

In a typical photometer, an aperture or diaphragm is placed at the focus of the telescope and the diverging beam is passed through a field lens to produce an image of the collecting aperture of the telescope on the detector, rather than an image of the star. All light rays must pass through this image. The field lens eliminates problems caused by the movement of the source within the aperture. In an infrared photometer, the aperture, the filters, the lens and the detector are all cooled to a very low temperature inside a vacuum chamber which has an infrared-transmitting entrance window. Many observatories have simple, single-element photometers and bolometers following this basic design. Many photometers contain a device called a circular variable filter (CVF) which provides a resolution of about 1% of the wavelength and is tunable in λ by rotating the long, arc-shaped filter. For the wavelength range $1 - 5\mu$m the detector is usually a single photovoltaic element of indium antimonide. At 10 and 20μm, germanium bolometers are often used. Some photometers employ an "array" of individual detectors, a good example being the bolometer array

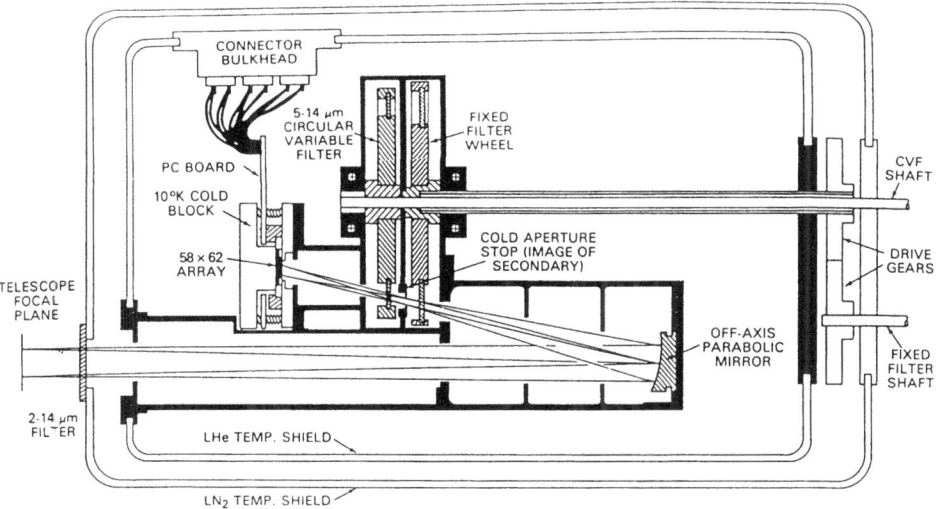

Figure 10: *The Goddard 10μm camera system developed by Dan Gezari, showing how the telescope focal plane is re-imaged onto the detector with the formation of an entrance pupil image. Note that the entire system is within the LHe temperature shield.*

developed by Dr. Telesco, but the reader should not be confused between these instruments and "cameras" using multiplexed, solid-state array devices.

4.2   Infrared Cameras

Since the goal of a camera system is to provide an image of the scene as well as a quantitative measure of brightness, the design described above must be modified to ensure that an image of the star is formed on the detector. Of course, the detector must be a two-dimensional array of pixels. The basic technique is shown in Fig. 10. A large aperture is located at the focus of the telescope to define the field of view. Rays diverging from this focal plane are "collimated" (made parallel) by a lens or mirror system which also produces an image of the entrance aperture or "pupil" of the telescope; in an optical system the entrance aperture is the primary mirror of the telescope, but in an infrared optimised telescope the secondary mirror is under-sized and therefore the primary mirror is underused and we say that the secondary mirror defines the entrance aperture. The location of the "pupil image" is approximately one focal length behind the collimator. All off-axis rays must pass through the pupil image and therefore this is a perfect location to place a circular hole called a "cold stop" or "Lyot stop". The best cold stop is one which has a central disk (of metal) to mask off

371

the image of the central hole in the primary (and the surrounding black baffle tubes).

Filters to define the wavelength interval are placed close to the pupil image in the collimated beam which provides a uniform illumination; in this system there is no effect on the image of introducing filters of different optical thicknesses into the beam. The final element is a re-imaging lens or mirror system to produce an image on the detector which has been magnified or de-magnified as required to "match" the pixel size.

More recently, we have seen the development of cameras with beam-splitters to provide two or more simultaneous observations at different wavelengths. For example, at UCLA we are building a camera with two channels (known as Gemini). Many cameras also provide "additional" options, the most common being the inclusion of a "grism" in the filter wheel. The grism provides a low-dispersion spectrum when a narrow slit is introduced into the telescope focal plane.

4.3 Cooled Grating Spectrometers

To obtain an infrared spectrum with substantially higher spectral resolution than provided by a grism, requires an optimised design. Just like at optical wavelengths, infrared spectrometers rely on diffraction gratings to provide the spectral dispersion. Unlike optical spectrographs however, the grating — which can be quite large — and all surrounding optics and metal must be cooled to cryogenic temperatures!

The basic elements in any spectrometer are well-known to every student of physics: an entrance slit, a collimating lens, a prism or diffraction grating, a camera lens and a screen or detector on which the dispersed spectrum appears.

A brief summary of some useful relations are given here for reference. The *angular dispersion* A of a grating follows from the grating equation:

$$m\lambda = d(\sin\theta + \sin i) \tag{24}$$

$$A = \frac{d\theta}{d\lambda} = \frac{m}{d\cos\theta} \tag{25}$$

therefore,

$$A = \frac{\sin\theta + \sin i}{\lambda\cos\theta} \tag{26}$$

where λ is the wavelength, θ is the angle of diffraction, i is the angle of incidence, m is the order of diffraction and d is the spacing between the grooves of the grating. Many combinations of m and d yield the same A provided the grating angles remain unchanged. Coarsely ruled reflection gratings (large d) can achieve high angular dispersion by making i and θ large, typically $\sim 60°$. Such gratings, called *echelles*, have groove densities from 30–300 lines/mm with values of m in the range 10–100. Typical first order gratings have 300–1200 lines/mm. The grooves of a diffraction grating can be thought of as a series of small wedges with a tilt angle of θ_B called the *blaze angle*. In the special case when $i = \theta = \theta_B$ the configuration is called Littrow and

$$\frac{d\theta}{d\lambda} = \frac{2\tan\theta_B}{\lambda} \tag{27}$$

How does the resolution of a spectrometer depend on its size, and on the size of the telescope? From the definition of resolving power we have:

$$R = \frac{\lambda}{\Delta\lambda} = (\lambda\frac{d\theta}{d\lambda})\frac{f_{cam}}{\Delta l} \tag{28}$$

where Δl is the projected slit width on the detector and f_{cam} is the focal length of the camera system. But $\Delta l = \theta_{res}/(ps)_{det}$ where θ_{res} is the number of arcseconds on the sky corresponding to the slit width. This angle will be matched to two or more pixels on the detector. We already know how to relate $(ps)_{det}$ to the EFL and, by definition

$$\frac{f_{cam}}{f_{coll}} = \frac{EFL}{F_{tel}} \tag{29}$$

and

$$\frac{D_{coll}}{f_{coll}} = \frac{D_{tel}}{F_{tel}} \tag{30}$$

therefore,

$$R = (\lambda\frac{d\theta}{d\lambda})\frac{206265}{D_{tel}}\frac{D_{coll}}{\theta_{res}} \tag{31}$$

For a given angular slit size, to retain a given resolving power (R) as the diameter of the telescope (D_{tel}) increases, the diameter of the collimator (D_{coll}) and hence the entire spectrometer must increase in proportion. NOTE: if W = width of grating, then $D_{coll} = W\cos i$. This equation explains why spectrometers on large telescopes are so big, and it also implies problems for infrared instrument builders who must put everything inside a vacuum chamber and cool down the entire spectrometer.

In practice, the collimator and camera (or re-imaging) optics are not simple lenses! Two designs are reproduced in Fig. 11 and Fig. 12 respectively to show the level of complexity encountered in real instruments. The first of these is the UKIRT CGS4 spectrometer and the second is the IRTF CSHELL instrument, both are spectrometers for the 1–5μm region.

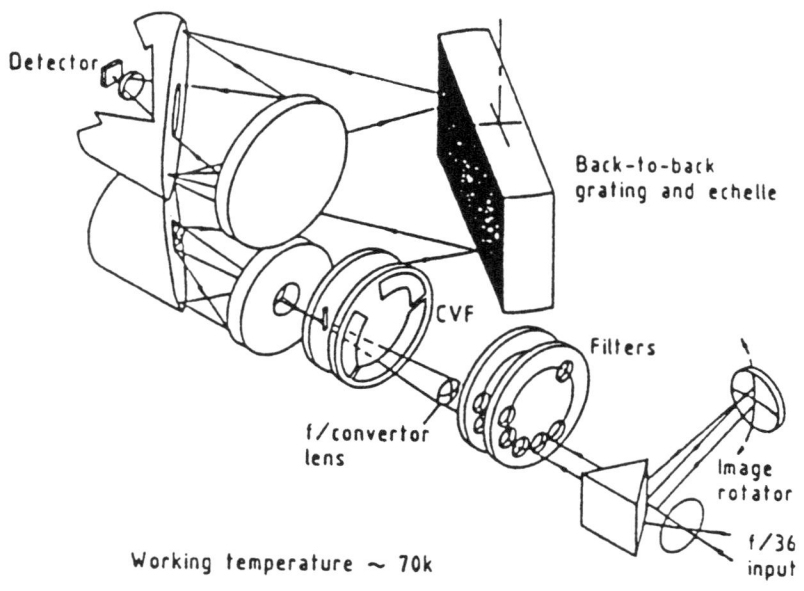

Figure 11: *The CGS4 cooled grating spectrometer on UKIRT.*

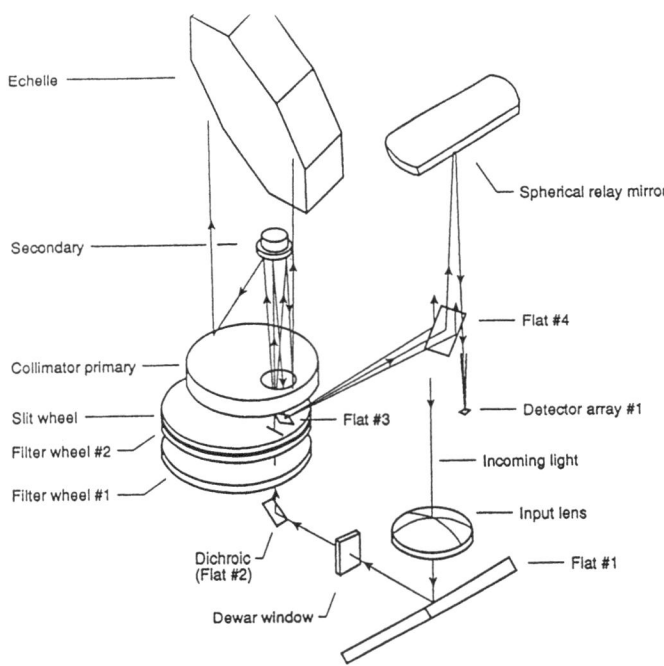

Figure 12: *The near infrared CSHELL echelle spectrometer on the IRTF.*

One of the most significant advances in infrared instrumentation in the last 2 years or so has been the development of very efficient infrared spectrometers. To obtain the required image quality and field-of-view (long slit) often requires the use of complex, aspheric, optical surfaces. Such surfaces can now be machined directly into aluminium substrates by diamond-turning techniques, and the finished surface is then gold-coated to give a very high reflectance in the infrared. The use of aluminium substrates eliminates concern about differences in the thermal contraction of glass optics and metal mountings. Also, gold-coated mirrors at 77 K are almost perfect reflectors (low emissivity) and it is therefore possible to "fold" the beam many times within such an instrument without significant loss of light.

4.4 Polarimeters

Infrared polarimetry is still a relatively unexplored subject compared to its optical counterpart. Photometers, cameras and spectrometers can all be "converted" to polarimeters (we would speak of photopolarimeters, imaging polarimeters and spectropolarimeters) by the addition of a *polarization modulator*. The basis of such a module is a device which converts the polarization information into an "intensity modulation" which can be detected. Intensity ratios yield the Stokes (or polarization) parameters of the light.

The most common approach is to install a "retardation plate" in front of the entrance window to the vacuum enclosure (dewar) *and* a fixed, perfect polarizer inside the instrument but prior to any partial polarizers such as the filters or diffraction grating. The external component is therefore not at cryogenic temperatures and will radiate infrared photons which add to the background. For measurements of linear polarization a "halfwave" retarder is used and for circular (and linear) a "quarterwave" retarder is needed. Rotation of the retarder results in rotation of the plane of polarization being fed to the fixed polarizer and hence a variation in the intensity of the beam emerging from the polarizer.

4.5 Interferometers

Two interferometry techniques are used in infrared astronomy namely, the Fabry-Perot interferometer and the Fourier Transform Spectrometer (FTS) — a scanning Michelson interferometer. The basic concept of the Fabry-Perot interferometer is a reflective cavity called an "etalon" composed of two face-to-face circular plates of high reflectivity and low absorption held parallel and flat to a tiny fraction (1/200) of the wavelength of light. Only a single wavelength or a very narrow wavelength interval is transmitted when one order of interference is selected. As the spacing or gap between the plates is changed, so the wavelength transmitted by the etalon changes. Since the etalon resembles an ordinary filter, it can be placed in the beam to enable the infrared camera to obtain an image of an entire field with the spectral purity of a very high resolution spectrometer. This instrument can be used to form data "cubes" in which wavelength (or velocity) is the third axis.

4.6 Closing remarks

Obviously it would be possible to write a book on each and every advanced instrument that can be found at a major observatory, and people do write such books — they are called Manuals! In this short written account, and in my much more *colourfully* illustrated lectures, I hope I have given you some insight into how infrared instruments work and how they are designed, and instilled in you a curiosity to find out more. Since the advent of infrared arrays has totally changed the nature of infrared astronomy, it seems appropriate to end with an infrared image. Fig. 13 is a K-band image (displayed as a negative) of the Trapezium region of the Orion nebula taken with the IRCAM instrument on UKIRT during the exciting commissioning period in 1986. Although we knew the results would be good, I don't think anyone on our team really appreciated the magnitude of the technology breakthrough until we saw this picture. I will never forget Christmas 1986! Looking at Fig. 13, the cluster in the centre is, of course, the well-known, optically-visible Trapezium stars while to the north and west lies the very bright infrared group containing the Becklin-Neugebauer (BN) object. That object, plus most of the hundreds of other stars on this 5×5 arcminute frame are not visible at optical wavelengths! This amazing picture was actually formed from a mosaic of overlapping frames, each with 62×58 pixels covering about 36 arcseconds on the sky; the complete observation took about one hour at the telescope. Already the 256×256 arrays provide over 16 times as much area in a single observation — almost the same area is replicated on the colour poster for the 1993 UCLA conference on *Infrared Astronomy with Arrays*, but that is *one* frame on a 1m telescope requiring only a 1 minute exposure! And work is under way on 1024×1024 detectors. Good observing!

I would like to thank the IAC for their hospitality, especially the conference organisers and staff, as well as all the students who made this an excellent Winter School.

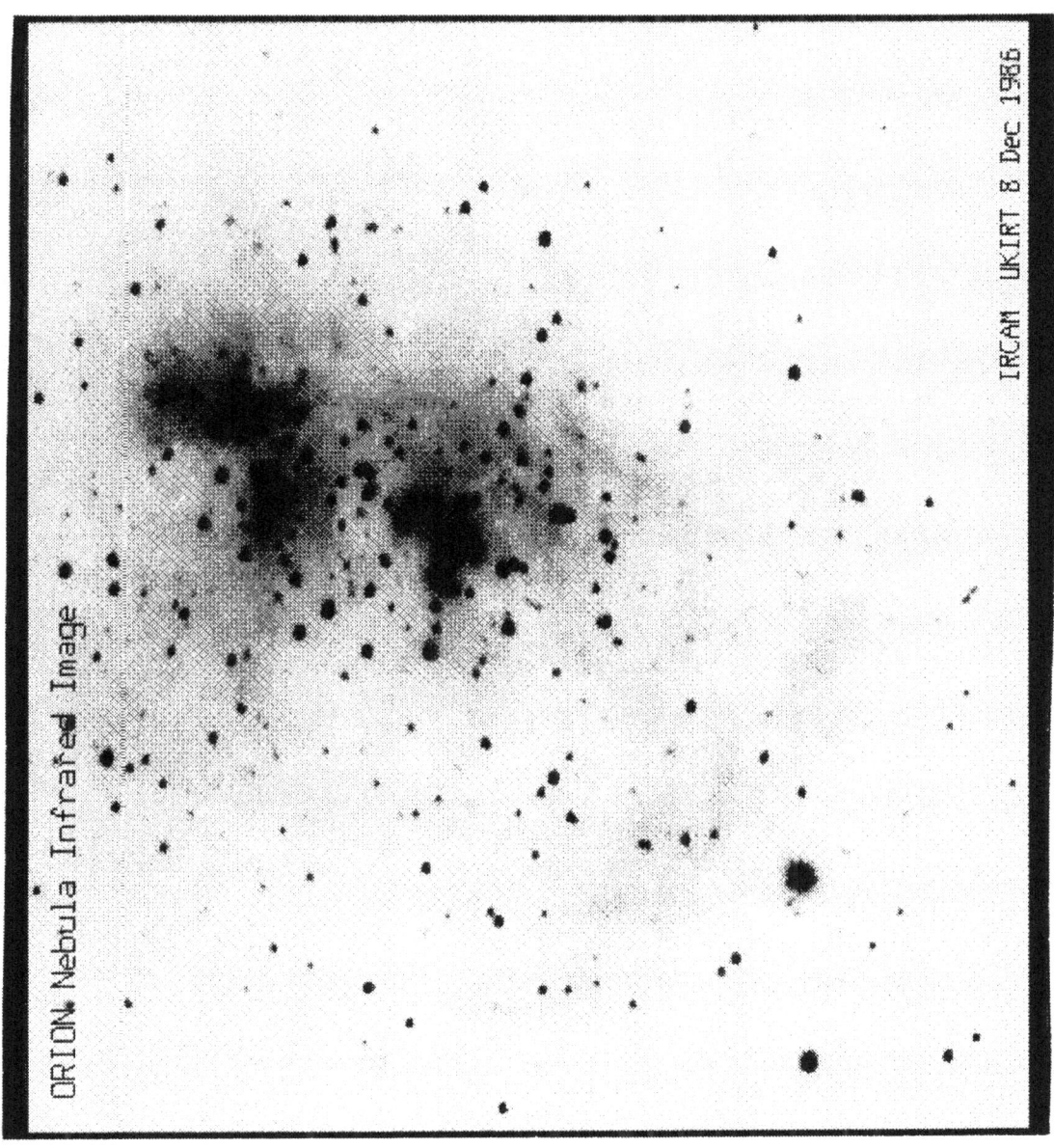

Figure 13: *The first true "image" of the Orion Nebula at 2μm with the IRCAM camera on UKIRT.*

377

5 REFERENCES

Elston. R.: 1990, ed., *Astrophysics with Infrared Arrays*, Astr. Soc. Pacific Conference Series, Volume 14.

Elias, J.H., Frogel, J.A., Matthews, K. and Neugebauer, G.: 1982, Astron. J. **87**, 1029-1034.

Gatley, I., Depoy, D.L. and Fowler, A.M.: 1988, SCIENCE, **242**, 1217-1348.

McLean, I.S.: 1988a, *Infrared Astronomy's New Image* in Sky & Telescope, **75**, No.3, 254-258.

McLean, I.S.: 1988b, *Infrared Astronomy; A New Beginning* in Astronomy Now, August issue.

McLean, I.S.: 1989, *Electronic and Computer-aided Astronomy: from eyes to electronic sensors*, Ellis Horwood Limited. Chichester.(UK).

Wynn-Williams, C.G. and Becklin, E.E.: 1987, *Infrared Astronomy with Arrays*, University of Hawaii, Institute for Astronomy, Honolulu.

Infrared Astronomy with Satellites

Thijs de Graauw

SRON Laboratory for Space Research and Kapteyn Institute
University of Groningen, P.O. Box 800
9700 AV Groningen, the Netherlands

1 ABSTRACT

This chapter gives an overview of the infrared astronomical satellites that have flown to date or are expected to be launched in the future. The IRAS, COBE, IRTS, ISO, SIRTF and FIRST missions are described together with their instrumentation. Scientific results or expected capabilities are summarized.

2 INTRODUCTION

2.1 Why Infrared Astronomy from Space?

There are two major reasons to go into space for infrared astronomical observations. First of all the earth atmosphere is not very transparent for IR radiation. Only at a few wavelength regions we can look through the atmosphere and even in those cases the transmission is far from ideal. The attenuation arises from absorption by the molecular constituents. In particular H_2O and CO_2 are strong absorbers and together with the pressure broadening effect the absorption lines are strong and wide, even for very high altitude observatories. See a.o. Morrison et al. (1973) and Cartier and Warner (1978). Figure 1 shows the global atmospheric transmission from a mountain top in the infrared.

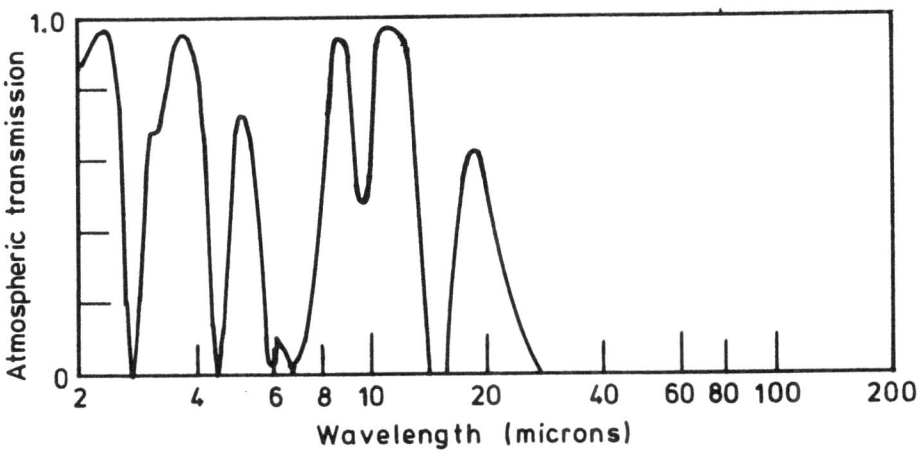

Figure 1: *Global atmospheric transmission from a high site*

The second reason to go to space is that the absorbing atmosphere, the transmission losses in the telescope optics and the cryostat window are responsible for an equivalent amount of infrared emission. This background radiation generates noise in the detectors which is proportional with $\sqrt{P_b}$, where P_b is the background radiation power. The effect on the detector Noise Equivalent Power (NEP) is given in Figure 2, where calculated NEP values are shown as function of wavelength for different temperatures of the optics. It is assumed that the emissivity is about 4% and the wavelength band ($\Delta\lambda/\lambda$) is 10%.

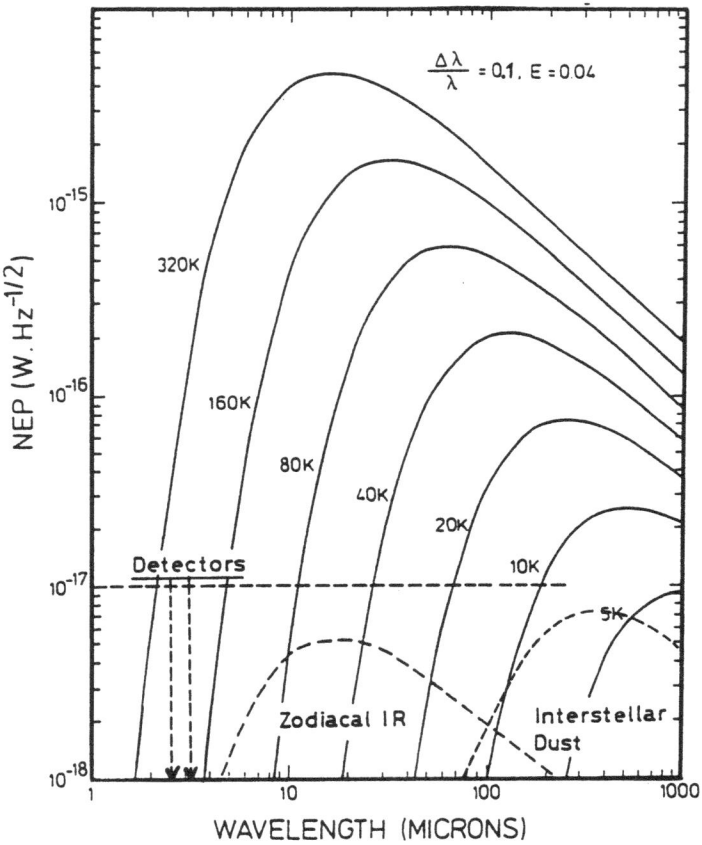

Figure 2: *Dependence of detector NEP on the temperature of the telescope assuming an optical bandwidth of 10% and an emissivity of 4%.*

In practice we usually see a total background emission (atmosphere and optics) which is about 4 times higher. This was observed a.o. at ESO's 3.6m and at the IRTF in the 10 μm atmospheric window. With detector NEP's ranging between 10^{-17} and 10^{-18} WHz$^{-1.5}$ we see that an improvement in sensitivity of a factor 100 to 1000 can be achieved. Of course this gain is much lower for high resolution spectroscopy where the optical bandwidth is a factor 100 to

1000 smaller. Another advantage of observations from space is the absence of variations in the two previously mentioned effects. This is a difficult to quantify advantage but accurate measurements as made by COBE and IRAS are totally impossible from the ground or from air and balloonborne observatories.

2.2   IR Astronomy from Air-, Balloon- and Rocketborne Platforms

IR astronomers have continuously tried to locate their telescopes at higher altitudes in order to avoid the atmospheric barrier. Observatories were or are being built at high and dry sites. In the 1970's telescopes and instrumentation were developed to operate from aircraft (learjet, KAO, Caravelle), balloonborne platforms or from rockets. The KAO has played a key role in the development of IR instrumentation and has been and still is extremely successful as an observatory for infrared and submillimeter astronomy.

With the AFGL Rocket Infrared Sky Survey in the early seventies the first sky survey was made in the thermal infrared (> 4 μm). It used a 16.5 cm liquid helium cooled telescope carried on a sounding rocket. The spinning telescope covered in nine flights about 90 percent of the sky at 11 and 20 μm , 70 percent at 4 μm and 50 percent at 27 μm. Each flight lasted about 3 minutes. A catalogue of about 2000 sources was released in 1974. AFGL continued its flights and another thousand sources were added by 1982. See Price and Murdock, 1983. With the very restricted observing time available the time was ripe to develop a satellite to do an all-sky survey. Balloonborne observations carried out also at longer wavelengths stressed the need to have long wavelength (> 50 μm) channels included. Plans in this direction were developed independently in the USA and the Netherlands. In 1975 both plans were merged which would lead to the well known IRAS mission.

2.3   IR Satellites

Figure 3 gives an overview of the IR satellites that have flown to date or are under construction, together with their wavelength coverage and year of operation. Also shown are those IR satellites that are planned to fly sometime in the next century.

Satellites to be discussed in this chapter, together with their demonstrated or expected performance are the Cosmic Background Explorer (COBE), the Infrared Astronomical Satellite (IRAS), the Infrared Telescope in Space (IRTS),

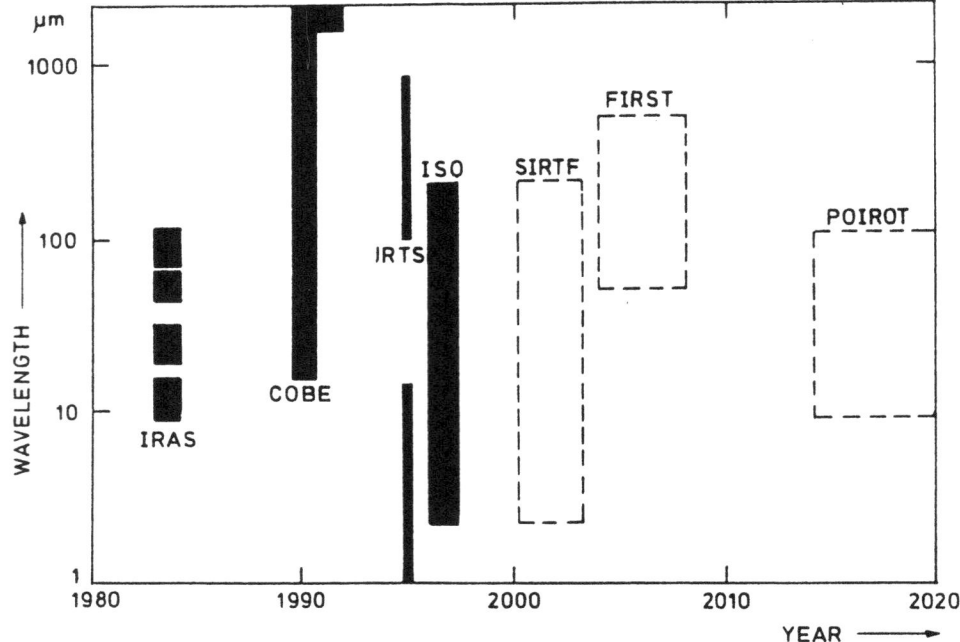

Figure 3: *Completed and planned IR satellite missions*

the Infrared Space Observatory (ISO), the Space Infrared Telescope Facility (SIRTF), the Far-InfraRed and Submm Space Telescope (FIRST) and Passively Orbiting InfraRed Telescopes (POIROT). Recommended literature and articles can be found in Davies (1988), Kondo (1990) and Bell-Burnell et al. (1992).

3   COMPLETED SATELLITE MISSIONS

3.1   IRAS

3.1.1   The IRAS satellite and mission

The primary objective of the InfraRed Astronomical Satellite (IRAS) was to conduct an unbiased all-sky survey at wavelengths centred around 12, 25, 60 and 100 μm. The project initiated in 1974 was a joint programme of the US (NASA), the Netherlands (NIVR) and the UK (SERC). It was launched in 1983 and operations lasted almost 10 months. Then it had surveyed 96% of the sky in a sunsynchronous orbit with a precession of 1° per day. The IRAS satellite consisted of two main parts. The telescope system including the survey instrument and the DAX (Dutch Additional Experiment) and the spacecraft system with the satellite control, data storage and telecommand/communications subsystems. The telescope, with a 60 cm primary mirror, was mounted in a toroidal liquid helium tank providing a 1.8K thermal environment for the detectors and optics. IRAS had two instruments on board. The survey instrument, made in the USA, with 62 detectors for the 12, 25, 60 and 100 μm survey and the DAX, built in the Netherlands, with a photometer (CPC) and low resolution spectrometer (LRS). IRAS was used in two modes: the survey mode to observe the whole sky several times and in the AO (additional observations) mode where a region of about 1°x1° was scanned several times. The IRAS survey and scan strategy was aimed for redundant coverage. The lay-out of the detectors of the survey instruments is given in Figure 4. For each spectral band there are two detector arrays so a detection in one array would have to be confirmed a few seconds later, the so-called second-confirmation. By advancing the satellite cross-scan by half the width of the focal plane on a subsequent orbit, one obtained a so-called "hours confirmation". About 70% of the sky turned out to be surveyed 3 times in that way i.e. 12 times observed. Of the Survey Instrument about 59 detectors were active. These detectors had rectangular apertures with in-scan dimensions of 0.75 to 3 arcmin and cross-scan dimensions from 4.5 to 5 arcmin. The limiting point source sensitivity in the survey mode is about 0.5 Jy at 12, 25 and 60 μm and 1.5 Jy at 100 μm.

The CPC (Chopped Photometric Channel) is a photometer operating at two wavelength bands (50 μm and 100 μm) and was designed to make high spatial (1') resolution observations in a 9'x9' field. The wavelength response of the CPC channels are shown in Figure 5 together with those of the survey instrument.

By accident the operating temperature of the CPC detectors turned out to be too low and the performance of the Germanium detectors was seriously affected (noisy, spontaneous spiking, hysteresis effects). Nevertheless a large number of high resolution maps are now available. Recently a CPC catalog of nuclei of galaxies has been finished (van Driel et al., 1993). The LRS is a slitless prism spectrometer. When an object is scanned across the aperture, the wavelength directed towards and intercepted by the detectors varies and the spectrum is recorded (see for a description Wildeman et al., 1983). The lay-out of the LRS aperture is also given in Figure 4. Its dimension is 15'x6' (cross-scan x in-scan). However, from 7.7 to 13.4 μm this aperture is divided over 3 detectors, each with an aperture of 5'x6'. From 11 to 22.6 μm the aperture is divided over 2 detectors, with apertures of 7.5'x6'. The spectral resolution ranges from 50 at 8 μm to 20 at 22 μm for point-like (<15") sources. In confused regions the LRS output is a mixture of spectral and spatial information. The LRS was in operation during the survey mode and during the AO mode for the survey instrument. So the

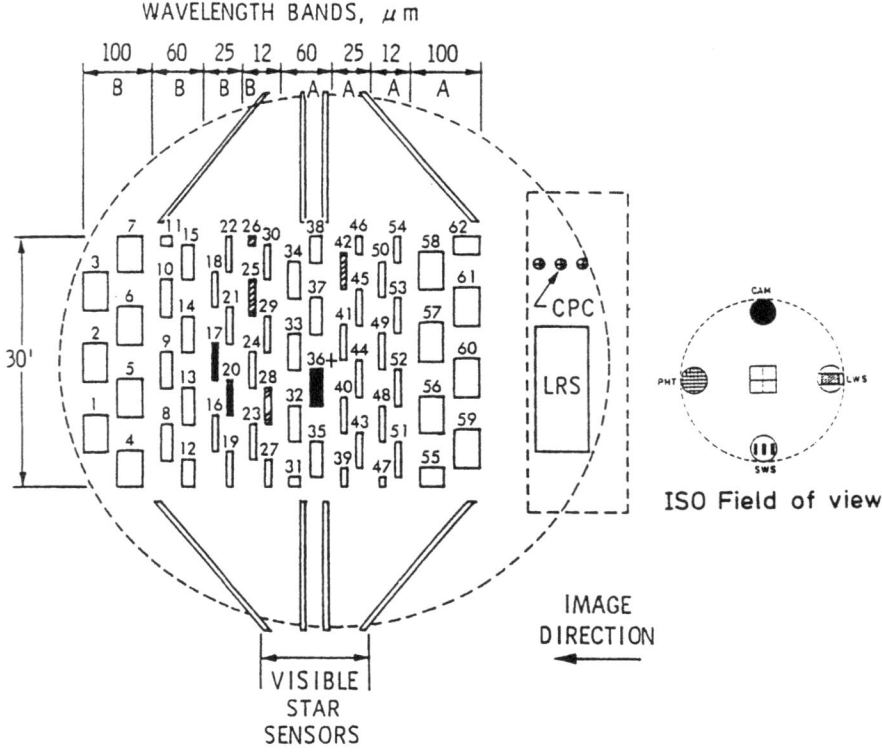

Figure 4: *Focal plane lay-out of the IRAS survey instrument*

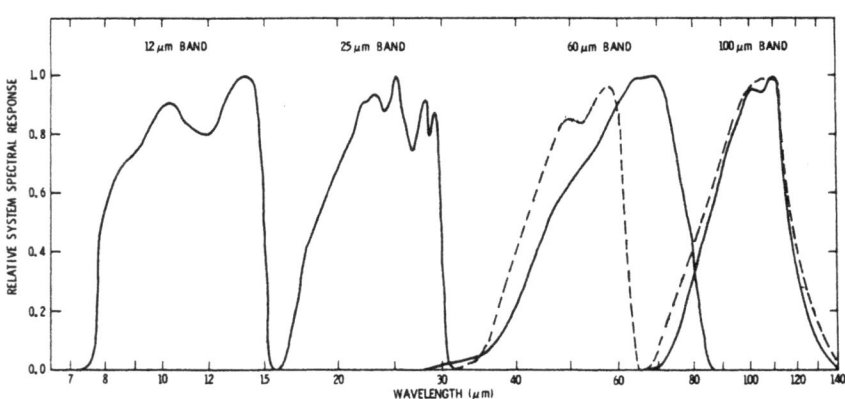

Figure 5: *Relative response of the four survey channels (drawn) and of the CPC (dotted).*

LRS observed the whole sky and, with serendipity, all AO fields.

386

3.1.2   The IRAS Data base

The IRAS mission and its data reduction system was primarily designed to produce catalogues of point sources. However with the unexpected high quality of the data other types of data processing (image processing of extended regions with low level flux, co-adding, high spatial resolution routines) appeared to be possible. See Figure 6 where a pole to pole scan recording at 12 μm is shown. Note the response of the stars on top of the zodiacal emission.

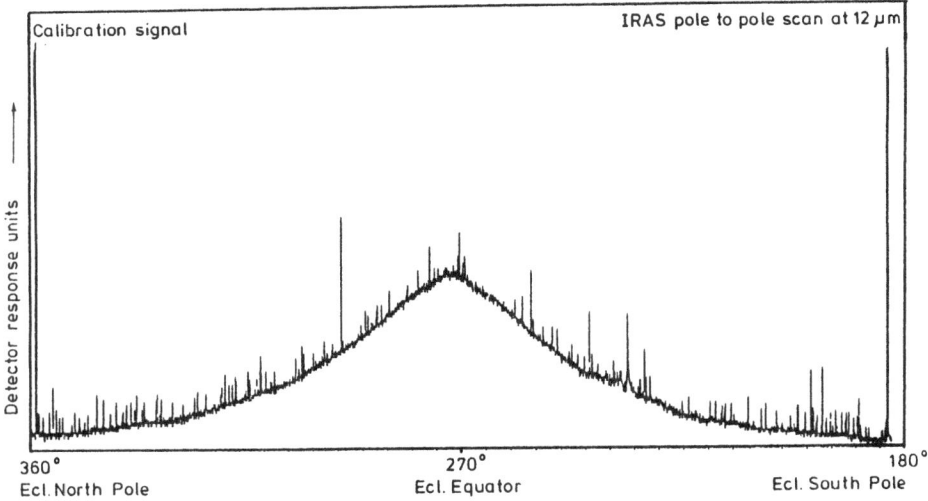

Figure 6: *An IRAS pole-to-pole scan at 12 μm*

The IRAS Processing and Analysis Centre (IPAC) at Pasadena concentrated first on the production of the point source catalogue but later more products and facilities became available. In Table I I have listed the main IRAS catalogues and products that are now available. In Groningen (see Wesselius, 1991), after having contributed to the data products from the CPC and the LRS, the instruments developed by the laboratory in Groningen, a different approach to the IRAS data processing was taken. If one is interested in the data of an object in a specific form which is not available from any of the published IRAS products, one can have access to the raw data and perform special processing. Two examples will be discussed below. To facilitate access to all IRAS data, these were ported to a rewritable optical disc jukebox. The 32 dual-sided discs with 320 Mbyte on each side contain now the raw compressed survey data (8.2 Gbyte), the raw compressed AO data (2.2 Gbyte), the raw LRS data of the survey and AO's (5 Gbyte), the splined raw survey data (1.2 Gbyte) and position and flux calibration files. Furthermore a set of software tools were developed to access, calibrate and generate images from these data. This was done under GIPSY, the Groningen

Image Processing System. This system, originally designed to process data of the Westerbork radio telescope, had a major overhaul and is now operating under UNIX. A four-dimensional internal storage format was designed for IRAS data.

Table I: *IRAS catalogues and other products*

Product	Description
IRAS Explanatory Supplement	Detailed IRAS reference book
Point Source Catalog (PSC)	245,889 point sources
Faint Source Catalog (FSC)	173,044 point sources. 12, 25 μm $\|b\| > 10°$, 60 μm $\|b\| > 20°$
Faint Source Reject File (FSR)	Point Source Rejects from FSS
Small Scale Structure Catalog (SSS)	16,740 sources with sizes point source $<$ SSS $<$ 8'
Serendipitous Survey Catalog (SSC)	43,866 point sources derived from the AOs (effective sky coverage 3%)
Catalogued Galaxies and Quasars	11,444 PSC and SSS sources associated with catalogued galaxies and quasars
IRAS Comet and Asteroid Survey	1811 asteroids and comets
Low Resolution Spectrometer (LRS) Catalog	8-22 μm spectra of 5,425 PSC sources
Zodiacal History File (ZOHF)	Time ordered data in 0°.5 x 0°.5 bins
CPC Explanatory Supplement	Detailed CPC reference book
Two arcminute In-Scan ZOHF	2' x 0°.5 bins
Bright Point Source-Removed ZOHF	0°.5 x 0°.5 bins
Additional Observations (AO's)	13,853 pointed mode observations 1° x 1° fields; 0'.25 x 1' pixels
IRAS Sky Survey Atlas (ISSA)	Co-added all-sky maps in 12°.5 x 12°.5 plates; 1'.5 pixels
Faint Source Survey (FSS) Plates	6°.6 x 6°.6 plates with 0'.25 x 0'.25 (12, 25 μm) or 0'.5 x 0'.5 (60, 100 μm) pixels

The first set of IRAS data to benefit from the renewed GIPSY was the one from the LRS. The primary LRS product is the LRS catalog (IRAS *Atlas of low-resolution spectra*, 1986) with about 5500 spectra of point sources brighter than 10 Jy at 12 or 25 μm. But many more spectra are residing in the IRAS raw data base. A preliminary overview of the results of IRAS-LRS observations and

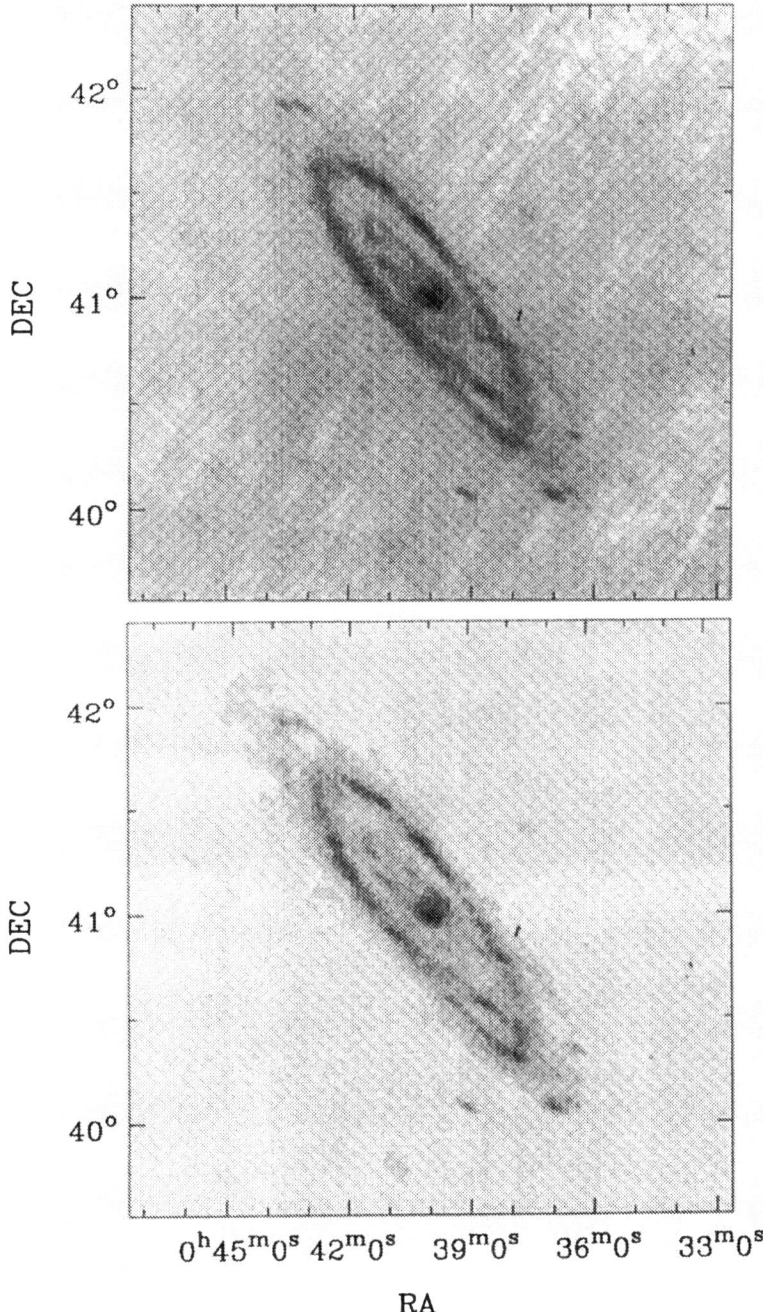

Figure 7: *IRAS 60 micron map of M31. Top figure is not yet de-striped, in the upper left one can see the remnants of the various scan directions. The bottom figure is the standard CO-ADD product.*

discoveries can be found in Jourdain de Muizon (1991). The latest extension

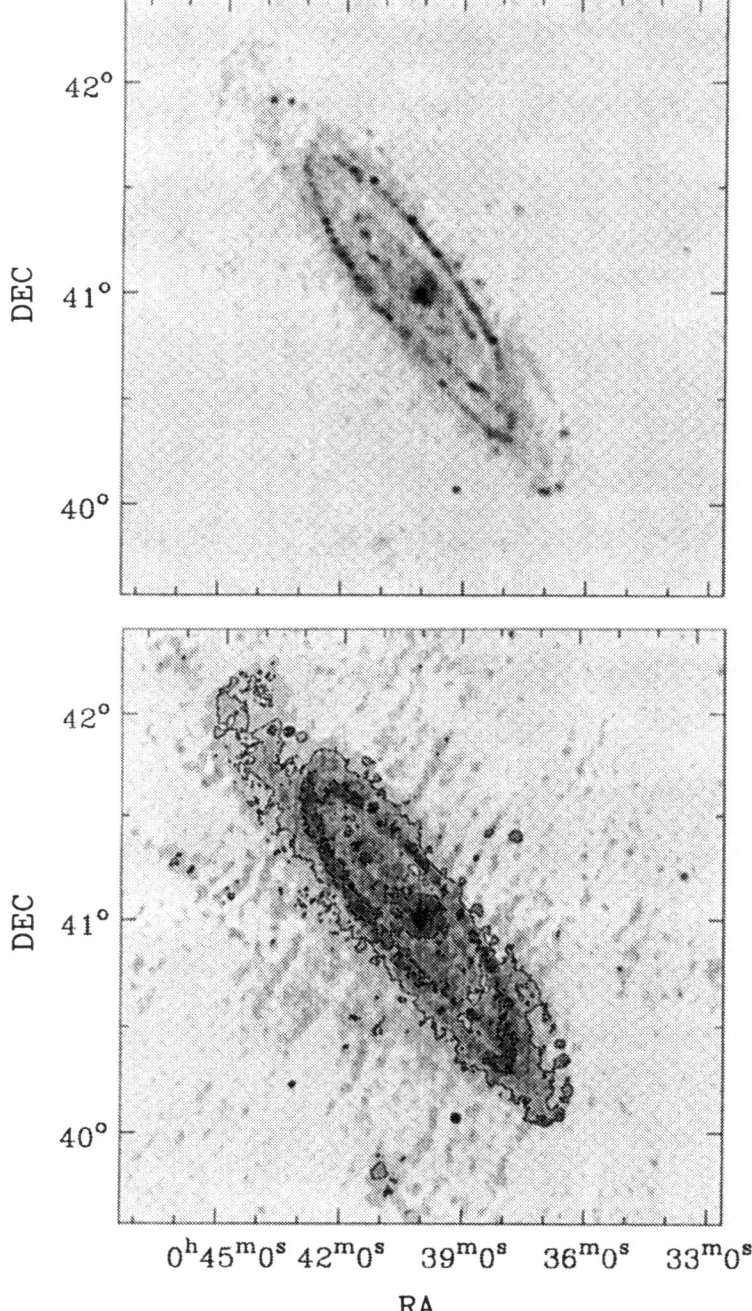

Figure 8: *IRAS 60 micron map of M31 where data from Figure 7 have been processed with a maximum entropy programme. Bottom map shows the Signal-to-Noise ratio map with 3 and 10 times r.m.s. contours.*

of the software tools in GIPSY is the implementation of a maximum entropy programme (Bontekoe et al., 1993). Figures 7 and 8 show some steps in the data processing. Figure 7 shows the 60 μm data for M31 where one can still see striping in the data due to the scanning of IRAS. In particular the upper left part shows the several scanning directions in the hours-confirmed datasets. Figure 7 shows also the so-called CO-ADD product, the standard IPAC product. Figure 8 shows the maximum entropy processed map of M31 and the Signal-to-Noise ratio map with 3 and 10 r.m.s. noise contours in the bottom map. All these GIPSY routines operate in batch mode. Even with new satellites (ISO, IRTS) to come in operation, the IRAS database will remain an important archive to continue interesting astronomical research and to prepare ISO observations.

3.2 Cosmic Background Explorer (COBE)

3.2.1 The COBE satellite and mission

The Cosmic Background Explorer (COBE) is an infrared/microwave satellite and the first of this kind that is primarily devoted to cosmology. It was initiated in 1974 and launched in 1989 by NASA (see Mather et al., 1990 and Boggess et al., 1992). It carries three scientific instruments, covering the wavelength range from 1.2 μm to 1 cm, to make accurate measurements of the spectrum of the cosmic microwave background radiation and its isotropy with an angular resolution of 0.7 degrees. The observation strategy was designed to minimize and also to determine systematic errors from spacecraft operations, the local environment of the spacecraft and the foreground astrophysical sources such as the Galaxy and the Solar System. The sensitivity of the measurements were to be limited by these astrophysical foreground sources. The selected orbit and scanning procedures provided also full sky coverage.

The satellite consists of two modules: the Instrument Module, which is the upper half of the observatory, contains three instruments which are complementary and are all needed to discriminate between cosmological emission from other astrophysical sources; the Spacecraft Module is the lower part of the satellite and includes the mechanical support structure, the power system, the solar cells, the attitude control system and the electronics of the two modules.

A summary of the instrument's characteristics is given in Table II. These instruments are: the Diffuse Infrared Background Experiment (DIRBE), the Far Infrared Absolute Spectrophotometer (FIRAS) and the Differential Microwave Radiometer (DMR). DIRBE is a cryogenically cooled photometer covering the wavelength range from 1-300 μm with 10 different wavelength channels which operate simultaneously. This design was optimized to search for the light of primordial galaxies and other objects that formed after the Big Bang. It measured, within a 0.7 degree beam, the average of millions of objects. With its line of sight 30 degrees off the satellite spin axis it made rapid comparison of the brightness of different parts of the sky at different angles away from the sun. With a number of internal reference sources on board and with a vibrating beam interrupter a high absolute photometric accuracy could be obtained.

Table II: *COBE Instrument Summary.*

Parameter	Experiments DIRBE	FIRAS	DMR
Wavelength (μm) or frequency	1.1-1.4; 14-30 2.0-2.4; 40-80 3.0-4.0; 80-120 4.5-5.1; 120-200 8-15; 200-300	1 to 20 cm^{-1} 20 to 100 cm^{-1}	90 GHz 53 GHz 31.4 GHz 23.5 GHz
Spectral Resolution	Filter bandwidth	5%	1 GHz
Field of view	0.7° diameter	7° diameter	7° diameter
Instrument Type	Filter radiometer	Fourier transform polarizing interferometer	Dicke switched differential radiometers
Flux collector	Off-axes Gregorian telescope	Flared horn	Corrugated horns
Temperature	1.7 to 2.4 K	1.7 K	300 K
Detector	Photovoltaics for λ<5.1 μm photoconductors for λ=8-120 μm bolometers for λ=120-300 μm	Bolometers (4)	Diode Mixer
Primary Science Focus	Primeval Galaxies	BB Spectrum	Isotropy

FIRAS is a cryogenically cooled Michelson interferometer covering the wavelength range from 0.1 to 10 mm which can resolve this range in 100 channels that are 5 percent wide. Its field is 7° wide and its primary objective was to determine spectra of the cosmic background radiation form the Big Bang. It could also measure the brightness and spectrum of cold interstellar dust throughout our Galaxy. The FIRAS looked out along the spin axis of the satellite and did therefore scan the sky as rapidly. A carefully designed entrance cone and an accurately calibrated black body which could be inserted on command into the entrance of the cone made absolutely calibrated observations possible with a sensitivity 100 times better than has been achieved before.

The DMR consists of three sets of microwave receivers operating at 3.3, 5.7 and 9.6 mm. The long wavelength receivers operate near 270 K. The other two sets can be cooled to about 140 K. The primary science objective is to determine whether the primeval explosion appeared to be equally bright in all directions. Each wavelength set has two receivers and two antennas, each with a 7° beam, pointed 60° apart and both 30° away from the spin axis. The antennas are switched between the two receivers and the precisely temperature stabilized instruments are frequently calibrated during flight with electronic noise sources.

The orbit of COBE is a sun synchronous orbit, as was the case for IRAS with a 900 km altitude. The satellite is spinning at a rate of 0.8 r.p.m. which was thought to be fast enough to reduce the noise and systematic errors that could arise from radiometer gain and offset variations. A sophisticated attitude control system was needed to meet the requirement of slow rotation and three axis control.

3.2.2 Preliminary Scientific results.

Preliminary analysis of the FIRAS database confirms that the cosmic microwave background (CMB) has a blackbody spectrum over the 500 μm to 1 cm wavelength range. The temperature of the CMB towards the galactic north pole is 2.735 ± 0.060 K. The 60 mK uncertainty is arising from the uncertainty in the calibration of the thermometry of the absolute calibrator. The lack of deviations from a Planck spectrum seems to rule out a hot smooth intergalactic medium that could emit more that 1% of the observed X-ray background. A dipole anisotropy of the CMB can clearly be seen in the FIRAS data and resembles the difference in spectra that is expected from two Doppler-shifted blackbody curves. This is presumably due to our peculiar motion relative to the Hubble flow. The dipole amplitude measured is 3.31 ± 0.05 mK. There are two other FIRAS results to mention. It provides the first all-sky, unbiased survey at wavelengths greater than 120 μm with a 7° resolution. It made a mean galactic spectrum between 100 μm and 1 cm with a first detection of the [NII] line at 205.3 μm. Sky maps were made in the [CII] at 158 μm and the [NII] at 205 μm. Wright et al. (1992) argue that the [CII] line is coming from photo-dissociation regions and the [NII] line is partially arising from a diffuse warm ionized medium and partially arising from dense HII regions.

The DMR results are dominated by the dipole anisotropy and Galactic plane emission. The dipole–anisotropy measured with DMR is consistent with the FIRAS results. With the dipole component and its associated kinematic quadrupole of 13 ± 4 μK, $(\Delta T/T)_Q = (4.8 \pm 1.5) \times 10^{-6}$ for galactic latitudes larger than $10°$. For the interpretation I refer to the course by dr. R. Joseph.

The preliminary analysis of DIRBE data show the infrared sky as expected. The 1.2 μm stellar emission from the galaxy and zodiacal scattered light from interplanetary dust dominate both the spectrum in the first three DIRBE bands. The infrared CIRRUS detected by IRAS is seen in all bands longward 25 μm. The DIRBE data base is a valuable new source for studies of the interplanetary medium and the Galaxy. It will allow the search for cosmic infrared background radiation which arises a.o. from pregalactic and protogalactic sources. This will enhance our understanding of the epoch between decoupling and galaxy formation. Preliminary results of DIRBE can be found in Hauser (1991). Full sky maps of all three COBE instruments are expected to be ready by mid 1994. These data will become a data base of unprecedented sensitivity for studies in cosmology, large scale galaxy and solar system studies.

4 FUTURE SATELLITE MISSIONS

4.1 The Infrared Space Observatory (ISO)

4.1.1 Introduction

The Infrared Space Observatory (ISO), a project of the European Space Agency, will operate at wavelengths from 2.4-200 μm. It will provide astronomers with a unique common user facility of unprecedented sensitivity for detailed investigations of cosmic objects ranging from solar system objects to distant galaxies. ISO was initiated in 1978, its funding approved in 1983 right after the successful start of the IRAS operations and is planned to be launched by the end of 1995. A more detailed description of ISO, its instruments and capabilities is given in *Infrared Astronomy with ISO* by Th. Encrenaz and M Kessler (1992).

The ISO satellite consists of two modules: the payload module which is essentially a large cryostat, and the service module. The service module acts as the mounting structure to the launcher, to the solar arrays fixed on the sunshield, and as a mounting plate for subsystems, such as the power supplies, data computers, the instrument warm electronics, the telecommand, attitude and orbit control units. These last units provide three-axis stabilization with an accuracy of a few arc seconds. It uses earth and sun sensors, star trackers, gyros, reaction wheels and a hydrazine gas reaction control system. The telecommand down-link bit rate is 33 kilo bits per second.

The payload module cryostat (see Figure 12 in the Appendix) encloses a toroidal tank filled with about 2300 l of superfluid helium which should provide an in-orbit operational lifetime of at least 18 months. The present tests show that the life time could be as long as 22 months. The instruments and telescope which are mounted on the so called optical support structure are cooled by the cold boil-off gas from the liquid helium which is conducted through this support structure. The anticipated temperature for the instruments is between 3 and 4 K. Some of the detectors are directly coupled to the He tank in order to be at a lowest possible temperature ($<$2K).

The telescope primary mirror with a diameter of 60 cm is made of fused silica as is also the secondary mirror. The optical quality of the telescope gives diffraction limited performance at 5 μm. By two baffles and a sunshade the straylight is kept to a minimum. Strict viewing constraints are applied during operation in order to achieve the stringent straylight conditions.

The ISO orbit has a 24 hour period, a perigee height of 1000 km and a apogee height of 72,000 km. The satellite will pass daily through the earth radiation belts decreasing the totally useful period to 16 hours per orbit. The pointing capabilities of ISO are, compared to those of the Hubble Space Telescope very limited. In optimum conditions an accuracy of two arcsec is expected to be achieved.

4.1.2 The ISO scientific Payload

The ISO will contain four instruments: two spectrometers, an imaging photo-polarimeter and a camera. These instruments are being built by international consortia and are funded by their national space agencies. An overview of ISO's four instruments is given in Table III. They are each mounted in a quadrant of a cylindrical volume behind the primary mirror. The 20' unvignetted field of view of the telescope is distributed radially to the four instruments by a pyramidal mirror whose centre has an opening to let the beam go straight into a quadrant star-sensor. This sensor is used for co–aligning the star-tracker boresight to the cryogenic telescope boresight.

Only one instrument will be operated at any time. However CAM can be used in a mode parallel to another instrument. In this so called parallel mode

Table III: *Characteristics of ISO Scientific Instruments*

Instrument Principal Investigator	λ (μm)	Spectral Resolution	Spatial Resolution	Outline Description
ISOCAM (C. Cesarsky, CEN-Saclay, F)	2.5-17	Broad-band, Narrow-band, and Circular Variable Filters	Pixel f.o.v's of 1.5, 3, 6 and 12 arcsec	Two channels each with a 32x32 element detector array
ISOPHOT (D. Lemke, MPI für Astronomie, Heidelberg, D.)	2.5-200	Broad-band and Narrow-band Filters Near IR Grating Spectrometer with R=90	Variable from diffraction - limited to wide beam	3 Sub-systems: i) Multi-band, multi-aperture photo-polarime-ter (3-110 μm) ii) FIR Camera (40-200 μm) iii) Spectrometer (2.5-12 μm)
SWS (Th. de Graauw, Lab. for Space Research, Groningen, NL)	2.5-45	1000 across wavelength range and 3×10^4 from 12-35 μm	14x20 and 20x30 arcsec	Two gratings and two Fabry-Perot Inter-ferometers
LWS (P. Clegg, Queen Mary College, London, GB)	45-180	200 and 10^4 across wave-length range	1.65 arcmin	Grating and two Fabry-Perot Inter-ferometers

it will observe an adjacent area of the sky with a reduced data transmission rate. During satellite slews PHOT will be operated at a wavelength of 200 μm in a serendipity mode, whenever possible. Tables III and IV show the overall imaging and spectroscopy capabilities of ISO's instruments.

ISO will be operated as an observatory. This requires to constrain the commanding of the satellite and its instruments and use only a limited number of observing templates. Observers are required to select from these templates their preferred observing mode and to fill in the specific information on the source and the proposed observation.

Table IV: *ISO's Imaging Capabilities*

wavelength (μm)	material array size	pixel f.o.v. (arcsec)	Instrument subunit
2.5-5	InSb 32x32	1.5, 3, 6, 12	CAM-SW
5-17	Si:Ga 32x32	1.5, 3, 6, 12	CAM-LW
50-120	Ge:Ga 3x3	43.5	PHOT-C1
120-200	Ge:Ga (stressed) 2x2	90	PHOT-C2
2.5-17	Si:Ga 1x1	10, 13.8, 18, 20x32, 23, 52, 79, 99, 180	PHOT-P1
17-24	Si:B 1x1	10, 13.8, 18, 23, 52, 79 99, 180	PHOT-P2
24-110	Ge:Ga 1x1	10, 13.8, 18, 20x32, 23, 52, 79, 99, 180	PHOT-P3

ISOCAM The camera of ISO is designed to image areas of the sky in the wavelength range from 2.5 to 7 μm with various spatial and spectral resolutions (see Cesarsky, 1991). Polarization mapping will also be possible. CAM consists of two optical sections, with one used at a time, each of which contains a 32x32 detector array. The system consists of an entrance wheel with clear apertures and a set of polarizing grids at position angles 120° apart and a selection wheel with two off-axis Fabry mirrors. Depending which mirror is in position the telescope beam is reflected in the shortwave or longwave channel. Each channel contains a lens-wheel with four lenses to match the sky (1.5", 3", 6" or 12") and a filter-wheel with fixed band-pass filters and circular variable filters (CVF). An overview of the CAM filters for the SW and LW channels is given in Figure 9. Here are also shown the PHOT filters for the same wavelength range.

The SW channel operates from 2.5 to 5.5 μm with an InSb 32x32 detector array. Very long exposures, up to 1000 s, are possible with this array. Typical integration times (between resets) for SW will be about a minute, where the readout noise is about 1200 electrons. This corresponds to NEP's at 5 μm of about 2.5×10^{-17} W/$\sqrt{Hz}$. The LW channel operates form 4 to 17 μm with a

Si:Ga 32x32 array. Typical integration times here are expected to be 10 s. The achieved NEP values are of the order of 2-3x10^{-18} W/$\sqrt{Hz}$.

CAM has three main observing modes each with different pointing strategies.

a. Photometry in the SW or LW channel with different pixel f.o.v.'s and band pass filters, carried out in one of the following 4 pointing modes: staring, beamswitching, microscanning or rasterscanning.

b. CVF spectral measurements in the SW or LW channels in one of the 4 pointing modes.

c. Polarization imaging with 3 polarizers position angles in one of the modes a or b; however microscanning and beamswitching here is not possible.

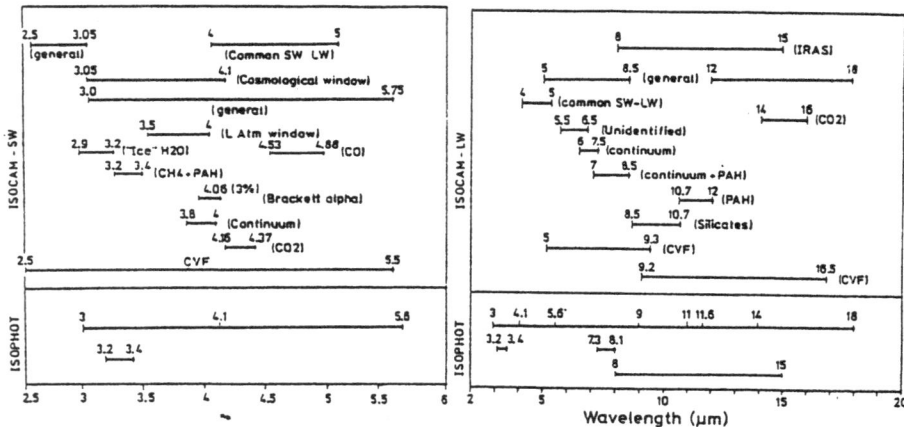

Figure 9: *Overview of the CAM bandpass filters and those of PHOT for the 2.5→20 μm range*

ISOPHOT The photometer for ISO consists of three subunits optimized for either photometry, polarimetry, imaging or spectrophotometry (Lemke, 1991). One subunit can be used at a time. These three subunits are:

ISOPHOT-P, a multiband, multi-aperture photo/polarimeter with three single detectors covering 3-120 μm (P1,P2,P3).

ISOPHOT-C, two photometric cameras with a 3x3 detector array for 50-120 μm and a 2x2 array for 120-220 μm (C1,C2).

ISOPHOT-S, a set of two grating spectrometers for the range of 2.5-5 and 6-12 μm which operate simultaneously.

The modular approach allows to minimize the number of mechanisms. ISO-PHOT has three ratchetwheels to select a certain mode and to select from a large number of filters and apertures. A focal plane chopper allows differential measurements on the sky and between calibration sources. The flexibility of the instrument requires a relatively large number of observing templates. The present ideas are that the user can select from about 18 templates to carry out the proposed investigations. ISOPHOT has the following observing modes:

Photometry:	Absolute Photometry
	Variable monitoring
	Photometry with PHT-P
	Photometry with PHT-C
Polarimetry	
Imaging:	1-D, during slews and with linear scans
	2-D, with (sparse) raster maps and special mapping procedures

As accurate photometry is one of the prime objectives for this instrument, a sophisticated calibration system is being set up to arrive at an absolute calibration sensitivity of better than 10%. For the shorter wavelengths the most suitable class of calibrators are stars that may provide known, point-source, monochromatic fluxes up to 50 μm. Known fluxes have an accuracy better than a few percent and the spectral shape is shown. Presently two ground based preparatory programmes by ISO co-Investigators are being carried out to measure near- and mid-IR fluxes, to verify the spectral shape and to support models for extrapolation to longer wavelengths. There is also a KAO programme going on to have complete calibrated spectra from 1-35 μm. At wavelengths longward of 50 μm asteroids will be used as photometric standards. Their energy distributions peak near 20 μm. They are point-like and bright. A supporting calibration activity here is undertaken by asteroid calibration measurements from the KAO. Two internal calibration sources will be used to monitor the detector responsivities in time and relate the measurements to photometric standards. These thermal calibrators will be absolutely calibrated to better than 15%, the relative error is much smaller, a few percent. The sensitivities of ISOPHOT are given in Figure 10. It shows clearly the improvement with respect to IRAS.

The ISO Spectrometers ISO has two dedicated spectroscopy instruments: the Short Wavelength Spectrometer (SWS), covering 2.4 to 45 μm, and the Long Wavelength Spectrometer (LWS), that covers 45 to 180 μm. Both spectrometers

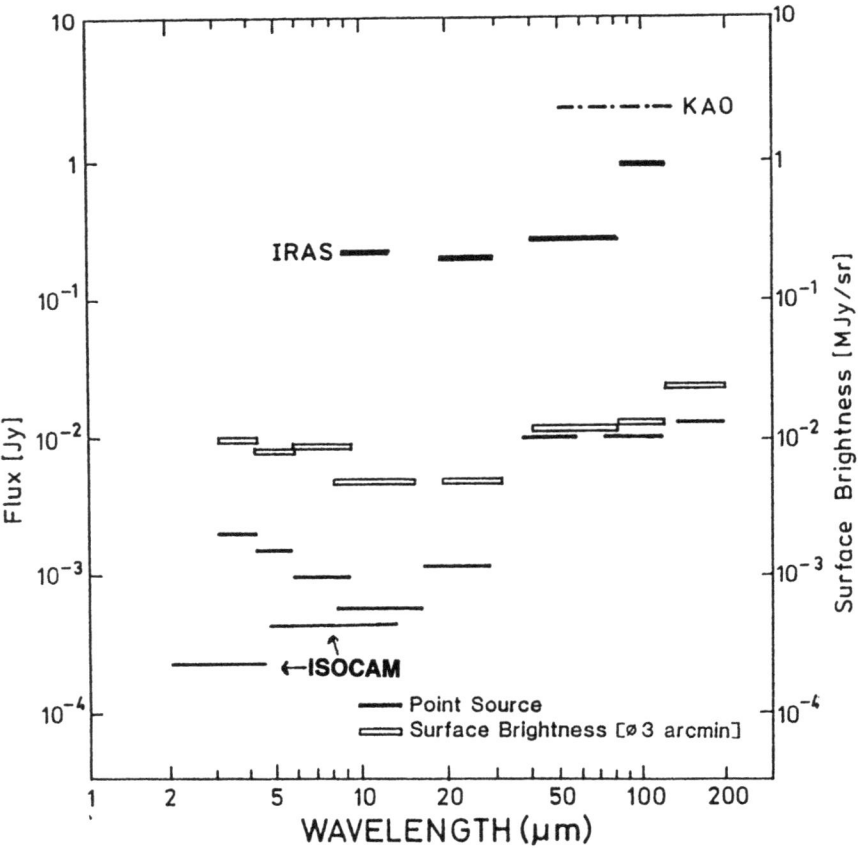

Figure 10: *ISOPHOT sensitivities for point sources and extended sources (S/N=10, T_{int}=1000 seconds). Also shown are the CAM SW and LW sensitivities for the same assumptions.*

use grating and Fabry-Perots for dispersion.

The SWS instrument (see de Graauw, 1991) consists of two nearly independent grating sections. The Short Wavelength Section (SW) covers 2.4–13 μm. The Long Wavelength section covers 12–45 μm. By inserting Fabry-Perot interferometers at the output of the LW section the resolution can be increased by about a factor 24. With the gratings a resolving power between 1000 and 2000 can be achieved. The two sections share an input unit consisting of three different apertures with shutters, three dichroic filters and two times three entrance slits. Both sections can be used simultaneously under the condition that they use the same entrance aperture. More details are given in Appendix 2.

Table V: *ISO's spectroscopic capabilities*

wavelength (μm) size	material array	entrance f.o.v. (arcsec)	$\Delta\lambda/\lambda$	Instrument
2.5-5 1x64	Si:Ga	24x24	~ 75	PHOT-S
6-12 1x64	Si:Ga	24x24	~ 75	PHOT-S
2.5-5.25 32x32	InSb	1.5, 3, 6, 12	~ 40	CAM-SW CVF
5.15-16.5 32x32	Si:Ga	1.5, 3, 6, 12	~ 40	CAM-LW CFV
2.5-4 1x12	InSb	14x20	1000-2000	SWS-G-S
4-13 1x12	Si:Ga	14x20	1000-2000	SWS-G-S
12-30 1x12	Si:As IBC	14x20	1000-2000	SWS-G-L
28-45 1x12	Ge:Be	20x32	1000-2000	SWS-G-L
12-25 1x2	Si:Sb	20x40	~ 25,000	SWS-FP-S
20-35 1x2	Ge:Be	20x40	~ 25,000	SWS-FP-L

The SWS has two main operating modes: a) Grating observations; b) Grating/Fabry Perot observations. It can also be used in a raster scan mode with fixed wavelengths. In case a) both grating sections scan spectral lines or spectral intervals with one of the two 1x12 detector arrays of each section. Such an array covers about 8 independent resolution elements. In case b) the SW section works as in a). The LW section grating scanner is set in the position to direct the spectral element of interest to the appropriate Fabry-Perot. The FP is then scanned according to its calibration table and thus scanning a spectral line. The SWS has a limited number of Astronomical Observational Templates (AOT). Two AOT's are dedicated to raster scans. One AOT is used for relatively fast scans with degraded spectral resolution for bright sources. Four AOT's are dedicated to scanning lines and spectral regions with gratings and Fabry-Perots in several combinations.

The LWS (see Clegg, 1991), as a grating spectrometer, provides a spectral resolving power of 100-200 from 45-180 μm. Two Fabry-Perot interferometers are located near the entrance optics and with one of them inserted in the collimated beam a resolving power of 10,000 can be attained. The LWS has fixed entrance aperture with a diameter of 98 arcsec. With the grating scanner spectral elements can be directed to one of the 10 detectors which together can cover the entire wavelength range. The LWS has also two prime operating modes. In the grating mode when scanning a spectral region, all detectors are used and in about 20 steps the entire 45 to 197 μm range can be covered. In the FP mode the high spectral resolution is obtained with the FP/grating combination. This mode is to be used to study individual spectral lines for wavelength regions using only one detector at a time. LWS has only 4 AOT's: Medium resolution spectrum or line scan, and the high resolution spectrum or line scan.

Both spectrometers SWS and LWS have internal illuminators that are used as a check on the responsivity of the detectors and in this way laboratory and in orbit photometric calibration measurements are connected, also with the actual observations. SWS has internal secondary calibrators for wavelength calibration which will be recalibrated in orbit against known spectral lines of celestial sources. One of the internal wavelength calibrators generates a comb of very narrow emission lines which are used for the parallelization of the two FP's. The LWS will rely completely on the use of cosmic spectral emission lines for wavelength calibration.

4.2   The Infrared Telescope in Space (IRTS)

IRTS is a cryogenically cooled small infrared telescope. It is developed in Japan and it is one of the missions to be on board of the small space platform SFU (Small Flyer Unit). The SFU is a multipurpose laboratory with more than 10 experiments, which will be launched in 1995 by a Japanese HII rocket and retrieved by the space shuttle after the mission. The SFU will be placed into a circular orbit. About three weeks will be allocated for the IRTS observations.

The IRTS is a cryogenic system cooled by super fluid helium. It has a telescope of 15 cm diameter and four instruments are installed into the focal plane. These cover a wide wavelength range from the near-IR to the submm. The characteristics of these four instruments are summarized in Table VI.

The Near Infrared Spectrometer (NIRS) and the Mid Infrared Spectrometer (MIRS) are designed to make an absolute measurement of the sky with coarse spectral resolution. Both use a one dimensional array at the focus of the spec-

Table VI: *Characteristics of IRTS Focal Plane Instruments.*

	NIRS	MIRS	FILM	FIRP
optics	grating	grating	grating	photometer
λ-range (μm)	1.4-4.0	4.5-11.7	63,158, 155,160	100-1000
$\lambda/\Delta\lambda$	12-30	20-30	400	3
beamsize	8'x8'	8'x8'	8'x20'	$\Leftrightarrow$ 0.5°
detectors	24xInSb	32xSi:Bi	1xGe:Ga 3xGe:Ga (S)	4xbolometer
science focus	X-Gal. Bgd. Stars Zodiacal	Diff. Gal. Stars IS Dust	ISM Star Form. Regions	Cosmic Bgd. X-Gal. Bgd. IS Dust

trometer with integrating amplifiers. The Far Infrared Line Mapper is a medium dispersion spectrometer to observe the CII (158 μm) and the OI (63 μm) lines from interstellar clouds together with two continuum channels next to the CII line. The sensitivity will allow to measure CII emission in cirrus at high galactic latitude. The Far Infrared Photometer is an absolute photometer with four $\lambda/\Delta\lambda \sim 3$ pass-bands, which cover 100 μm - 1 mm, and a He–3 refrigerator to cool bolometric detectors. FIRP will have higher sensitivity in the 300-500 μm window as compared to COBE with a 0.5° angular resolution.

Details of the NIRS, MIRS, FILM and FIRP are described in papers by Matsumoto (1990). The instruments have been developed by Japanese and US teams and are now completed. Figure 11 shows the sensitivity of IRTS and its instrument against the natural background for the various wavelength regions.

The IRTS is a mission optimized to observe diffuse extended sources such as: the extragalactic background, the interstellar medium (PAH's, silicates, cirrus) and interplanetary dust. However, it will also be possible to observe many stars with a wider wavelength coverage and better detection limits than the Low Resolution Spectrometer of the IRAS.

4.3   The Space Infrared Telescope Facility (SIRTF)

SIRTF is to be a cryogenically cooled one-meter class observatory for infrared astronomy. This proposed NASA mission is almost two decades under study. Recently (mid '92) the design has been changed considerably in order to reduce

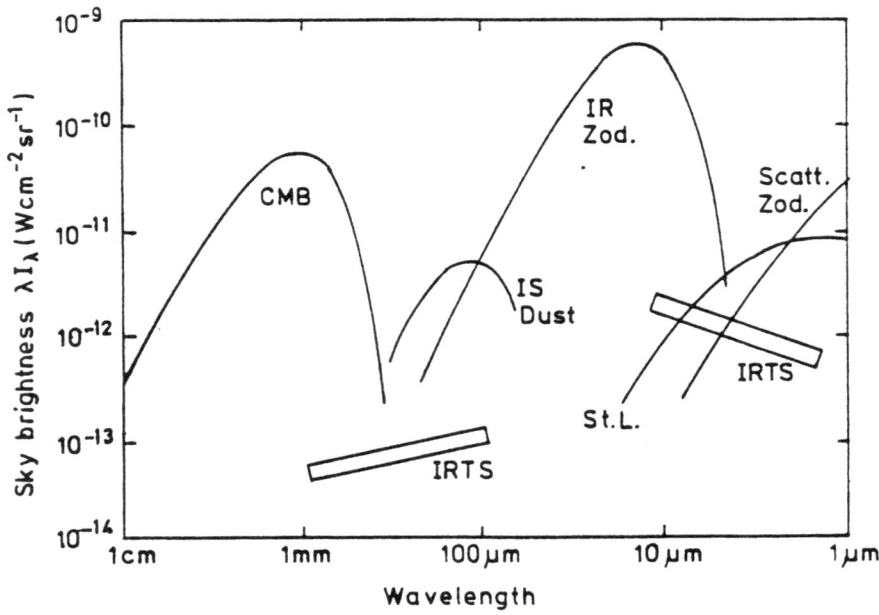

Figure 11: *Background radiation levels and IRTS sensitivities. COBE sensitivities were several orders of magnitude better, however in much wider beams (7° or 0.7°)*

cost. The present concept of SIRTF, to be launched on an Atlas rocket, will be described here. One of the larger changes in the SIRTF concept has been in the orbit. The SIRTF orbit has already been changed twice, from a low earth orbit via an ISO-type elliptical orbit into a high altitude circular orbit. Now the proposed orbit is a heliocentric, earth trailing orbit. Here the almost constant direction towards the sun and the earth with respect to the satellite allows a considerable reduction in the dimensions and complexity of the cryogenic system. See Appendix 1. The second proposed change is in the wavelength coverage. The long-wavelength limit is reduced from 700 μm to 200 μm. Also the telescope primary mirror diameter is somewhat reduced, 70 cm instead of 95 cm. Furthermore, the complexity of the instruments was considerably reduced. Less modules, less mechanisms, smaller number of detector arrays which on the other hand have more pixels.

Although SIRTF is to have similar wavelength coverage and telescope diameter as ISO, due to the availability of recently developed large 2-D detector arrays SIRTF will out-perform ISO with one or more orders of magnitude in scientific return.

Table VII: *Imaging and Spectroscopic Capabilities for SIRTF.*

wavelength range (μm)	material array size	field of view pixel size	instrument
2.5-5.3	InSb 256x256	5'x5' 1.2"	IRAC
4-15	Si:As IBC 128x128	5'x5' 2.4"	IRAC
15-36	Si:Sb IBC 128x128	5'x5' 2.4"	IRAC
40-120	Ge:Ga 32x32	5'x5' 9"	MIPS
120-200	Ge:Ga (stressed) 2x16	0'.6x5' 19"	MIPS
		RESOLVING-POWER	
2.5-5	InSb (IRAC) 256x256	150	IRAC
4-12	Si:As IBC 128x128	1000-2000	IRS
12-36	Si:Sb IBC 128x128	1000-2000	IRS
40-120	Ge:Ga 4x32	1000-2000	IRS
120-200	Ge:Ga (stressed) 2x16	500-1000	IRS

SIRTF is to have three instruments: an infrared spectrograph (IRS) covering 4 to 200 μm, an infrared array camera (IRAC) which will provide 5'x5' field imaging over the spectral range from 5–36 μm, and a multiband imaging photometer (MIPS) providing 5'x5' images over the spectral range from 40–200 μm. Table VII describes the imaging and spectroscopic instrumentation as is foreseen now for SIRTF.

4.4 FIRST

FIRST is a submillimeter spectroscopic mission and will open up unexplored wavelength bands in the region between 100 μm and 1 mm, see a.o. the conference of the 29th Liege International Astrophysical Colloquium (ESP-SP-314).

405

Although the baseline was a 8 m telescope, cost reasons have forced a descope of this cornerstone mission to a 3 m class antenna. As FIRST is to be an observatory serving an entire astronomical community, it will have a payload with photometric, imaging and spectroscopic capabilities to provide unprecedented information on physics, chemistry and dynamics of a wide variety of cosmic objects such as galaxies and quasars, birth and evolution of stars, and planetary and cometary gas and dust. The scientific payload is presently in a definition phase and these results are being used to define the requirements of the satellite system. An overview of such a payload is given in table VIII. However, this complement is going to be revised depending on the timing of this mission and on the development of the detection techniques and the satellite system.

Table VIII: *Overview of a model payload for FIRST.*

Instrument	Frequency-GHz (microns)	Resolving power	Description
Sub THz receiver	820-980 (305-365)	10^6	SIS mixers and Solid State LO
THz receiver	980-1150 (260-305)	10^6	SIS mixers and Solid State LO
Far-IR Spectrometer	1600-4300, (85-190)	1000	Dual channel double FP with 7x7 array
Far-IR Photometer	425-4300 (85-700)	1-3	Photoconductor and bolometer array

4.5 Passively Cooled Orbiting Telescopes (POIROT's)

Cryogenically cooled space telescopes, such as IRAS and ISO, are extremely efficient in reducing the atmospheric, telescopic and instrumental emission below that of the photon background of celestial sources. But in order to have a practical mission, the telescope aperture is to be very modest. On the other hand the IR wavelength range is well behind in spatial resolution when compared to neighbouring wavelength ranges. At FIR wavelengths source confusion due to distant galaxies and cirrus sets a strong limit to the broadband sensitivity (Helou and Beichman, 1990). Future IR missions should therefore have substantially larger telescope apertures and one has to adopt a different cooling technique to achieve sufficient low temperatures of these larger optical systems. Whether the temperature of the background and the optics is sufficiently low to operate in detector noise limited conditions depends of course also on the spectral resolution used in the observation. In very high resolution spectroscopy, the sensitivity depends mainly upon the light gathering power of the telescope.

The alternative technique which has been proposed for cooling infrared space telescopes is to utilize radiative cooling. Properly shielded telescopes may cool well by radiation when pointed into deep space and they may attain temperatures of the order of 30-80 K. These passively cooled systems could allow substantially larger telescopes without a lifetime limitation as is the case for cryogenically cooled telescopes.

The best detailed proposal where the effect of radiative cooling is maximized is EDISON. Extensive studies of thermal models, orbits, optical design, launch vehicles and spacecraft considerations of radiatively cooled space telescopes are described in the proceedings of the workshop on "The Next Generation Infrared Space Observatory" (Bell-Burnell et al., 1992). EDISON is proposed to feature a 2-3 m aperture with a primary mirror equilibrium temperature of about 40 K to be reached in less than 3-6 months after the observatory reaches its orbit. As is the case for SIRTF, the selection of orbit for EDISON will have an important impact. If a heliocentric, earth trailing orbit is selected, the dominant sources of heating can be restricted to one side of the spacecraft and the radiating heatshields can be made larger and more efficient. Computer models show that temperatures as low as 25 K at the primary mirror can be reached. ¿From Figure 2 one can identify where the optimum wavelength range for a mission like EDISON is located: the range from 10 to 50 μm. The proposed instruments are to cover the wavelength range of 2-100 μm or more, the spectral resolving powers are to range from 5 to 10^4 with imaging resolution of 0.2 arcsec at 6 μm. The uniqueness of EDISON depends critically on the success of its preceding missions: (ISO-FIRST-SIRTF) and on the cooling efficiency of the EDISON thermal system. With its predecessor to be launched in the period up to 2005, EDISON is truly a next generation infrared space mission to be used by the next generation infrared astronomers. Development of such a challenging project for such a distant future requires extremely enthusiastic scientists and engineers.

APPENDIX 1. IR SATELLITE TECHNOLOGY

A.1.1. Cryogenic considerations

As discussed before, one of the main reasons to use satellites for IR observations is the opportunity to use a cooled telescope, viewing directly into space without the need for a window in a cryostat. So far satellites have been used or designed using a liquid He cryostat. An illustration of a typical system is given in Figure 12. IRAS, ISO, IRTS and SIRTF all have a similar layout. The difference is mainly in the volume of the liquid He tank arising from the aimed mission lifetime.

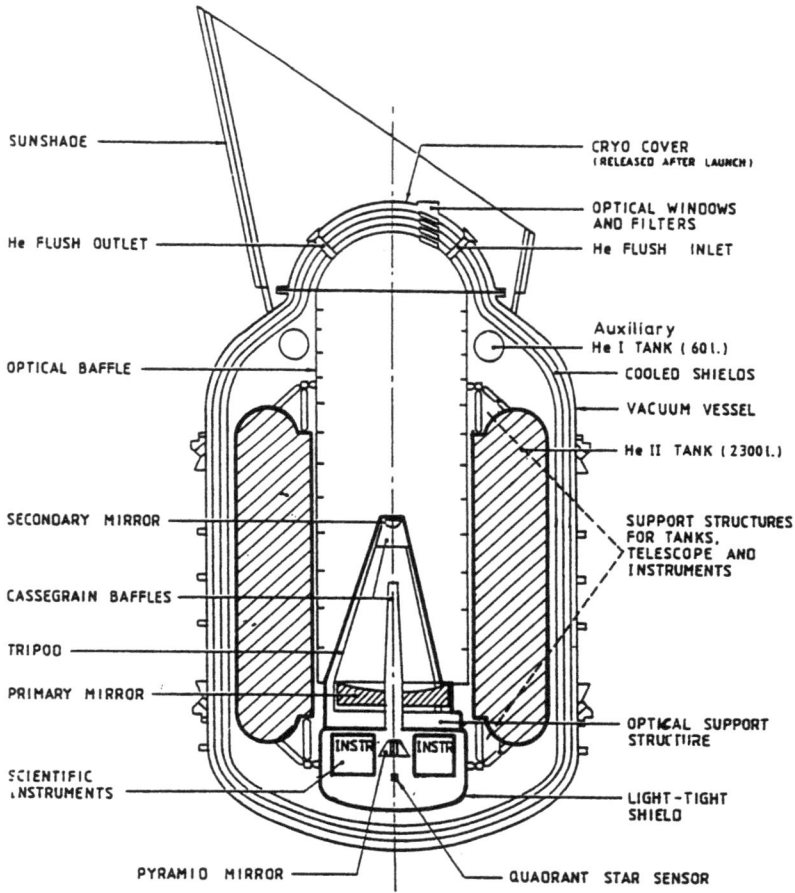

Figure 12: *Schematic drawing of the ISO Payload Module, which is characteristic for almost all IR satellite missions.*

The toroidal tank is held in place by low conducting supports and straps connected to the outside wall of the vacuum tank which is at ambient temperature. Heat conduction through these supports is to be minimized by reducing their cross sections as much as possible and this requires a minimal mass that is to be supported: i.e. mainly tank walls and structure, telescope and instruments. On the other hand, the vibration levels during launch require sufficient stiffness of the tank and telescope system, and a balance between the various requirements is to be established by computer models with a verification of the result in vibration tests. During the tests of the ISO cryostat it turned out that the tankstructure was too floppy; this was verified in a computer model and subsequently a few extra supports were designed and implemented. A second heat conduction input arises from the electrical leads from the vacuum tank wall to the instruments and other sensors, the so called harness. Going from the originally planned 500 single wires to the actual 700 wires ISO's lifetime is hardly shortened which indicates that this effect is relatively small. The third major heat source arises from the power dissipation of the instruments. ISO's power consumption budget for the instrument package is about 20 mW at the 3K level and 1 mW at the 1.8K level.

The radiation input from the vacuum tankwall onto the cryotank system is minimized by using a set of multilayers of aluminized mylar. Radiation arising from these walls has to be intercepted efficiently by these radiation shields and by a light cover around the instruments and telescope. Verification of this shielding system is one of the major test items for an IR satellite as most of the incentive to go into space might become destroyed by a failure in this system.

A.1.2. Orbit considerations

Heat input onto the vacuum tank is mainly from solar radiation. In low earth orbits the earth radiation input is of a similar level. Putting a satellite into a very high orbit has the advantage of eliminating the earth heat input completely but also eliminating its variation as a function of orbit position which is a.o. responsible for pointing errors. In the last years we have seen an increased awareness of the impact of the orbit onto an IR satellite system. ISO, planned to be in a 12 hour orbit (perigee 1000 km, apogee 36,000 km), is now to be put into a 24 hour orbit (perigee 1000 km, apogee 72,000 km). SIRTF's orbit changed from low-earth (600 km) via a 24 hour orbit, as for ISO, into an earth trailing orbit with a distance to the earth of more than 500,000 km. The impact on the SIRTF system is considerable. Figure 13 gives a comparison in size between IRAS, ISO, SIRTF in the Titan and Atlas launcher configuration.

Another major advantage of an earth trailing, high orbit is the sky visibility. ISO has a significant unaccessible area of the sky and the instantaneous access to the sky is limited as well, making observation planning difficult.

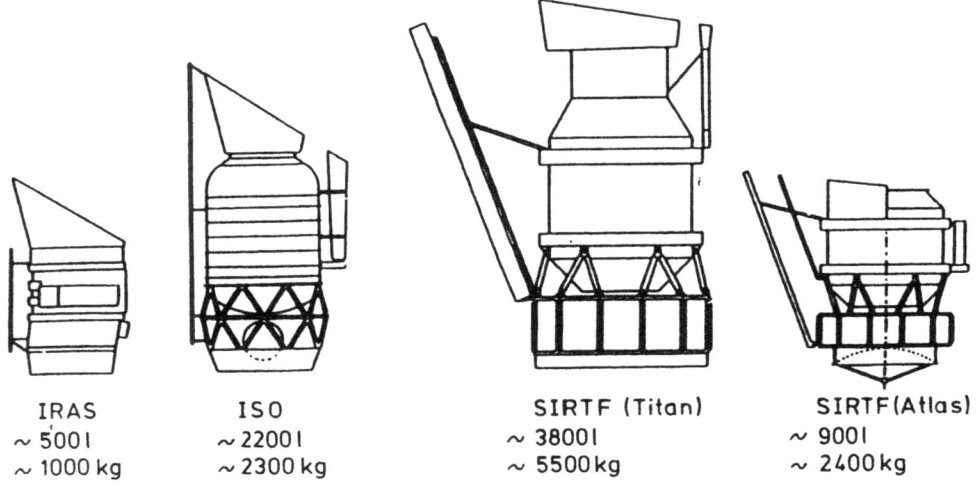

IRAS
~ 500 l
~ 1000 kg

ISO
~ 2200 l
~ 2300 kg

SIRTF (Titan)
~ 3800 l
~ 5500 kg

SIRTF (Atlas)
~ 900 l
~ 2400 kg

Figure 13: *Schematic presentation of the size of four infrared satellite designs.*

APPENDIX 2. IR SPACE INSTRUMENTATION

In general the development of scientific instruments for satellites is very challenging. This is particularly the case for IR satellites where one has some additional constraints arising from the cryogenic conditions. Below is a list of two classes of constraints for the ISO-SWS focal plane unit (FPU) and its warm electronics unit (WE). The numbers in paranthesis are the actually achieved values.

Design Constraints of the SWS:

1. Wavelength Range: 2.4 - 45 μm: (4 octaves!)
2. Resolving Power: 2.4 - 45 μm: $\geq$ 1000
   12 - 35 μm: $\geq$ 20,000
3. Sensitivity: detector NEP ($\leq 10^{-17}$ WHz$^{-1/2}$) limited
   i.e. optimum transmission

Satellite Interface Constraints:

4. Available Volume FPU (cm^3):   20x25x30
    Available Volume EU (cm^3):   20x15x15
5. Mass Limit FPU: 7.5 kg (9.3)
    Mass Limit EU: 9.3 kg (14)
6. Power Dissipation for the allowed FPU at 3 K: 10 mW (6)
                          at 1.8 K: 1 mW (1)
              allowed for the WE   14W (20)
7. Available electrical lines into cryostat: 120 (180)
8. Telemetry data rate: 32 kbits/s
9. Optical interfaces such as F/D ratio and focus location.
10. Pointing performance defining aperture diameter.
11. Vibration levels.

In retrospect the 4 octave wavelength range has driven the SWS optical design extremely close to the limits of what is feasible. This has generated during the development phase, a number of serious troubles which have been overcome thanks to a large amount of additional effort (see Wildeman et al., 1993). In order to accomodate the optical system in the given volume an anamorphic optic was used (see Figure 14) with toroidal and paraboloidal cylindrical mirrors. This reduced the diameter of the 70 mm colllimated beam in one direction down to 30 mm. A schematic of the official layout is given in Figure 14 and a photograph of the SWS-FPU Flight Module is given in Figure 15. At the right side of the Figure one can identify the gold-plated copper straps between the liquid He tank and the optical support structure. These straps run from detector blocks to the optical support structure (at 3.3K) or to the liquid He tank (1.8K).

A comparable ground-based instrument built by John Lacy (1980) but operating between 4 and 20 μm has about 5 times larger size and one can say without exaggeration, that the SWS is about an order of magnitude more compact.

Besides the compact optical design the electromagnetic conditions inside the detector mounts are extremely demanding. Inside these detector blocks the detector amplifiers have to generate clean signals for power levels of the order of 10^{-18} W, while, on the other hand, the amplifiers are operating at 55K due to a heater dissipation of 10^{-4} W. The constraints given here are similar for all four instruments together with the technical complication.

ACKNOWLEDGEMENT

The hospitability and stimulating environment provided by the IAC and the organizers and the students of the Winterschool are well appreciated.

411

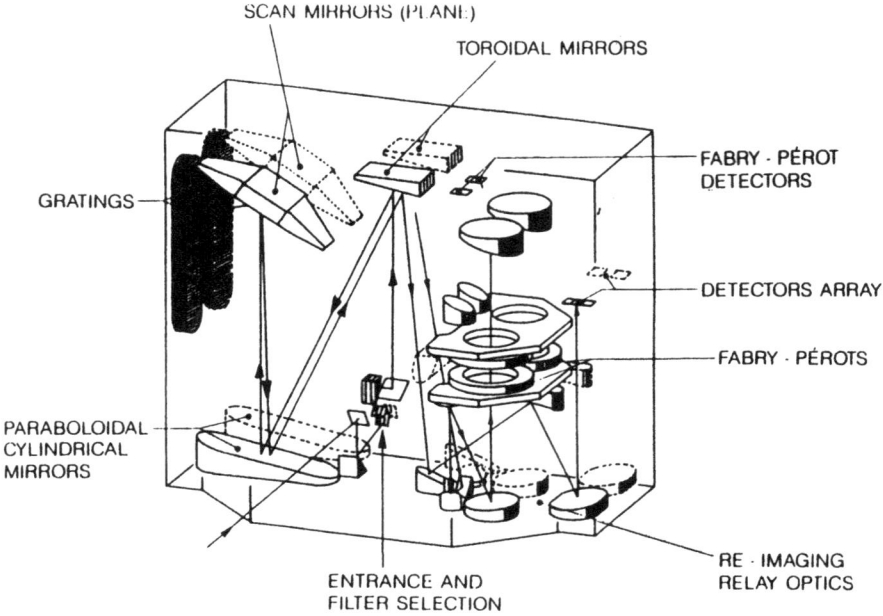

Figure 14:

References

[1] Bennett, C.L., Boggess, N.W., Cheng, E.S., Hauser, M.G., Mather, J.C., Smoot, G.F., Wright, E.L.: 1992, Third Teton Summer School: *The Evolution of Galaxies and their Environment*, H.A. Thronson and J.M. Shull, eds.

[2] Bell-Burnell, S.J., Davies, J.K., Stobie, R.S.: 1992, *Next Generation Infrared Space Observatory*, Kluwer Ac. Publ.

[3] Boggess, N.W. et al.: 1992, *Astrophys. J.* **397**, 420

[4] Bontekoe, T.R., Kester, D.J.M., Price, S.D., Jonge, A.R.W. de, Wesselius, P.R.: 1991 *Astron. Astrophys.* **248**, 328

[5] Cartier, J.T., Warner, J.W.: 1978, *Publ. Astron. Soc. Pacific* **90**, 607

[6] Cesarsky, C.J.: 1991, in *Proc. Infrared Astronomy with ISO*, Les Houches, Nova Science Publ., p. 31

[7] Clegg, P. 1991, in *Proc. Infrared Astronomy with ISO*, Les Houches, Nova Science Publ., p. 87

[8] Davies, J.K.: 1988, *Satellite Astronomy*, Ellis Horwood Ltd. Publ.

412

Figure 15: *A photograph of the SWS Flight Module FPU*

[9] De Graauw, Th.: 1991, in *Proc. Infrared Astronomy with ISO*, Les Houches, Nova Science Publ., p. 105

[10] Encrenaz, Th., Kessler, M.F.: 1991, *Proc. Infrared Astronomy with ISO*, Les Houches, Nova Science Publ.

[11] Hauser, M.G.: 1991 *Proc. Infrared Astronomy and ISO.*, Les Houches, Nova

Science Publ. (p. 479)

[12] Helou, G., Beichman, G.: 1990, Liege International Astrophysical Colloquium (ESP-SP-314), p. 117

[13] IRAS Explanatory Supplement, Beichman, C.A., Neugebauer, G., Habing, H.J., Clegg, P., Chester, T.J.: 1985

[14] IRAS Catalogues and Atlases. Atlas of low-resolution spectra, prepared by Olon, F.M. and Raimond, E.: 1986 *Astron. Astrophys. Suppl.* **65**, 607

[15] IRAS-CPC Explanatory Supplement, Wesselius, P.R., Beintema, D.A., de Jonge, A.R.W., Jurriens, T.A., Kester, D.J.M., van Woerden, J.E., de Vries, Perault, M.: 1986

[16] IPAC Newsletter, available from IPAC, Caltech, Pasadena, Cal 91125, USA

[17] Jourdain de Muizon, M. : 1991, in *Proc. Infrared Astronomy with ISO*, Les Houches, Nova Science Publ., p. 489

[18] Kondo, Y.: 1990, IAU Coll. 123, *Observatories in Earth orbit and beyond*, ed. Y. Kondo.

[19] Lacy, J.: 1989, *Publ. Astron. Soc. Pacific* **101**, 1166

[20] Lemke, D.: 1991, in *Proc. Infrared Astronomy with ISO*, Les Houches, Nova Science Publ., p. 53

[21] Mather, J.C. et al.: 1990, IAU Coll. 123, *Observatories in Earth orbit and beyond*, ed. Y. Kondo, p. 9

[22] Matsumoto, T.: 1990, IAU Coll. 123, *Observatories in Earth orbit and beyond*, ed. Y. Kondo, p. 215

[23] Morrison, D., Murphy, B.E., Creukshank, D.P., Sintom, W.M., Martin, T.Z.: 1973, *Publ. Astron. Soc. Pacific* **85**, 255

[24] Price, S.D., Murdock T.L.: 1983, The Revised AFGL Infrared Sky Survey Catalogue (AFGL-TR-83-0161)

[25] van Driel, W., Graauw, Th. de, Jong T. de, Wesselius, P.R.: *Astron. Astrophys. Suppl.* , to be published

[26] Wesselius, P.R.: 1991, in *Proc. Infrared Astronomy with ISO*, Les Houches, Nova Science Publ., p. 509.

[27] Wildeman, K.J., Luinge, W., Beintema, D.A., Graauw, Th. de, Ploeger G.R., Haser, L., Katterloher, R.L., Feuchtgruber, H.F., Melzner, F., Stöcker, J. : 1993, *Cryogenics*, **33**, 402

[28] Wildeman, K.J., Beintema, D.A., Wesselius, P.R.: 1983, *J. Br. Interplan. Soc.*, **32**, 21

[29] Wright, E.L. et al.: 1991, *Astrophys. J.* **381**, 200

For EU product safety concerns, contact us at Calle de José Abascal, 56–1°, 28003 Madrid, Spain or eugpsr@cambridge.org.

www.ingramcontent.com/pod-product-compliance
Ingram Content Group UK Ltd.
Pitfield, Milton Keynes, MK11 3LW, UK
UKHW060312090126
466816UK00021B/453